CAMBRIDGE STUD[...]
ADVANCED MATH[...]

EDITORIAL BOARD
D. J. H. GARLING, D. GORENSTEIN, T. TOM DIECK, P. WALTERS

Ergodic theory

Already published

1. W.M.L. Holcombe *Algebraic automata theory*
2. Karl Petersen *Ergodic theory*
3. Peter T. Johnstone *Stone spaces*
4. W.H. Schikhof *Ultrametric calculus*
5. J-P. Kahane *Some random series of functions*
6. H. Cohn *Introduction to the construction of class fields*
7. J. Lambek and P.J. Scott *Introduction to higher order categorical logic*
8. H. Matsumura *Commutative ring theory*
9. C.B. Thomas *Characteristic classes and the cohomology of finite groups*
10. M. Aschbacher *Finite group theory*
11. J.L. Alperin *Local representation theory*
12. Paul Koosis *The logarithmic integral: 1*
13. A. Pietsch *Eigenvalues and s-numbers*
14. S.J. Patterson *An introduction to the theory of the Riemann zeta-function*
15. H-J. Baues *Algebraic homotopy*
17. Warren Dicks and M.J. Dunwoody *Groups acting on graphs*

Ergodic theory

KARL PETERSEN
Professor of Mathematics, University of North Carolina

CAMBRIDGE UNIVERSITY PRESS
Cambridge
New York   Port Chester
Melbourne   Sydney

Published by the Press Syndicate of the Unversity of Cambridge
The Pitt Building, Trumpington Street, Cambridge CB2 1RP
40 West 20th Street, New York, NY 10011-4211 USA
10 Stamford Road, Oakleigh, Victoria 3166, Australia

© Cambridge Unveristy Press 1983

First published 1983
First paperback edition (with corrections) 1989
Reprinted 1991

Printed in Great Britain at the
Athenaeum Press Ltd, Newcastle upon Tyne

Library of Congress catalogue card number: 82-4473

British Library Cataloguing in publication data

Petersen, Karl
Ergodic theory.-(Cambridge studies in
advanced mathematics; 2)
1. Ergodic theory
I. Title
515.4′2   QA313

ISBN 0 521 23632 0 hardback
ISBN 0 521 38997 6 paperback

Contents

Preface — ix

1 Introduction and preliminaries — 1
1.1 The basic questions of ergodic theory — 1
1.2 The basic examples — 5
A. Hamiltonian dynamics. B. Stationary stochastic processes. C. Bernoulli shifts. D. Markov shifts. E. Rotations of the circle. F. Rotations of compact abelian groups. G. Automorphisms of compact groups. H. Gaussian systems. I. Geodesic flows. J. Horocycle flows. K. Flows and automorphisms on homogeneous spaces.
1.3 The basic constructions — 10
A. Factors. B. Products. C. Skew products. D. Flow under a function. E. Induced transformations. F. Inverse limits. G. Natural extensions.
1.4 Some useful facts from measure theory and functional analysis — 13
A. Change of variables. B. Proofs by approximation. C. Measure algebras and Lebesgue spaces. D. Conditional expectation. E. The Spectral Theorem. F. Topological groups, Haar measure, and character groups.

2 The fundamentals of ergodic theory — 23
2.1 The Mean Ergodic Theorem — 23
2.2 The Pointwise Ergodic Theorem — 27
2.3 Recurrence — 33

2.4	Ergodicity	41
2.5	Strong mixing	57
2.6	Weak mixing	64

3 More about almost everywhere convergence — 74

- 3.1 More about the Maximal Ergodic Theorem — 74
  A. Positive contractions. B. The maximal equality. C. Sign changes of the partial sums. D. The Dominated Ergodic Theorem and its converse.
- 3.2 More about the Pointwise Ergodic Theorem — 90
  A. Maximal inequalities and convergence theorems. B. The speed of convergence in the Ergodic Theorem.
- 3.3 Differentiation of integrals and the Local Ergodic Theorem — 100
- 3.4 The Martingale convergence theorems — 103
- 3.5 The maximal inequality for the Hilbert transform — 107
- 3.6 The ergodic Hilbert transform — 113
- 3.7 The filling scheme — 119
- 3.8 The Chacon–Ornstein Theorem — 126

4 More about recurrence — 133

- 4.1 Construction of eigenfunctions — 133
  A. Existence of rigid factors. B. Almost periodicity. C. Construction of the eigenfunction.
- 4.2 Some topological dynamics — 150
  A. Recurrence. B. Topological ergodicity and mixing. C. Equicontinuous and distal cascades. D. Uniform distribution mod 1. E. Structure of distal cascades.
- 4.3 The Szemerédi Theorem — 162
  A. Furstenberg's approach to the Szemerédi and van der Waerden Theorems. B. Topological multiple recurrence, van der Waerden's Theorem, and Hindman's Theorem. C. Weak mixing implies weak mixing of all orders along multiples. D. Outline of the proof of the Furstenberg–Katznelson Theorem.
- 4.4 The topological representation of ergodic transformations — 186
  A. Preliminaries. B. Recurrence along IP-sets. C. Perturbation to uniformity. D. Uniform polynomials.

	E. Conclusion of the argument.	
4.5	Two examples	209
	A. Metric weak mixing without topological strong mixing. B. A prime transformation.	

5	**Entropy**	**227**
5.1	Entropy in physics, information theory, and ergodic theory	227
	A. Physics. B. Information theory. C. Ergodic theory.	
5.2	Information and conditioning	234
5.3	Generators and the Kolmogorov–Sinai Theorem	243

6	**More about entropy**	**249**
6.1	More examples of the computation of entropy	249
	A. Entropy of an automorphism of the torus. B. Entropy of a skew product. C. Entropy of an induced transformation.	
6.2	The Shannon–McMillan–Breiman Theorem	259
6.3	Topological entropy	264
6.4	Introduction to Ornstein Theory	273
6.5	Finitary coding between Bernoulli shifts	281
	A. Sketch of the proof. B. Reduction to the case of a common weight. C. Framing the code. D. What to put in the blanks. E. Sociology. F. Construction of the isomorphism.	

References   302

Index   322

Jumalate juhatusel
Jooksvad elujoonekesed,
Voolavad õnnelainekesed.

For Elisabeth

Preface

Ergodic theory today is a large and rapidly developing subject. The aim of this book is to introduce the reader first to the fundamentals of the ergodic theory of point transformations and then to several advanced topics which are currently undergoing intense research. By selecting one or more of these topics to focus on, a student can quickly approach the specialized literature and indeed the frontier of the area of interest.

Of course the number of interesting topics that we have neglected is necessarily far greater than that of those we have been able to include. Thus we have to refer the reader elsewhere for discussions of, for example, operator ergodic theory, the existence of invariant measures, nonsingular transformations, orbit equivalence, differentiable dynamics, subadditive ergodic theorems, etc. Unfortunately, there do not exist coherent expositions of all of these topics; I invite those of my colleagues who are more expert than I in the areas I have omitted to do some more expository writing.

It should also be understood that, even for the advanced topics that we do discuss, their treatment here cannot be more than an entryway to the rapidly expanding specialized literature. Thus our presentations of multiple recurrence and the Ornstein theory, to mention two examples, are intended as introductions to the books of Furstenberg (1981) and Ornstein (1974), respectively.

From my point of view, ergodic theory consists of examples, convergence theorems, the study of various recurrence properties, and the theory of entropy. Each of these facets is given first a basic and later a more advanced treatment. At the introductory level, one can find the usual important basic topics: the standard examples, the mean and pointwise ergodic theorems, recurrence, ergodicity, strong mixing, weak mixing (all in Chapter 2) and the fundamentals of entropy (in Chapter 5). I have tried to make the writing as clear and complete as possible. Throughout we concentrate on single invertible measure-preserving maps on Lebesgue spaces. Some more advanced topics related to convergence theorems appear in Chapter 3, to recurrence in Chapter 4, and to entropy in Chapter 6. Chapter 3 presents a rather thorough analysis of maximal functions and their usefulness in the proof of the important convergence theorems of analysis (Lebesgue's Differentiation Theorem and the existence of the Hilbert transform), probability (the martingale convergence theorems), and ergodic theory (the Maximal Ergodic Theorem and its refinement to an equality, the Pointwise Ergodic Theorem, the Local Ergodic Theorem, the Dominated Ergodic Theorem and its converse, and the existence of the ergodic Hilbert transform). We also give Neveu's proof of the Chacon–Ornstein Theorem, as an introduction to operator ergodic theory. Chapter 4 begins with a direct construction of eigenfunctions for non-weakly-mixing transformations, which leads to a brief treatment of almost periodic functions and topological dynamics. In Section 4.3 we give an introduction to Furstenberg's approach to multiple recurrence and the Szemerédi Theorem, and Section 4.4 proves the Jewett–Krieger Theorem on the topological representation of measure-preserving transformations. The final section examines the important examples of Kakutani and Chacon, both of which are weakly but not strongly mixing and at least one of which is prime. Chapter 6 contains several nontrivial entropy computations (toral automorphisms, skew products, and induced transformations), the Shannon–McMillan–Breiman Theorem, and the fundamental facts about topological entropy. We also provide a brief introduction to Ornstein's isomorphism theory of Bernoulli shifts, proving the Ornstein Isomorphism Theorem by the Keane–Smorodinsky method, which produces a finitary (realizable by machine) map.

The background necessary to read this book is some knowledge of measure theory and functional analysis; for example, the contents of Royden's *Real Analysis* suffice. Each chapter begins with a short summary, and many sections end with exercises, which range from trivial to difficult. The list of references contains all works referred to in the text, and then some, but it is not a complete bibliography for the subject; starred entries are books and survey articles, several of which contain more extensive bibliographies. I apologize for any errors of fact, misattributions of results, misprints, passages of incompetent exposition, and all other mistakes and misjudgments; I will be happy to receive lists of errors (preferably accompanied by suggested corrections) from all who care to compile them.

Preface

This book grew out of courses given at the University of North Carolina and Yale University. During part of the writing I received financial support from the National Science Foundation, and the work was completed while I was on a leave supported by the Kenan Foundation at the Laboratoire de Calcul des Probabilités of the Université Pierre et Marie Curie (Université de Paris, VI). I thank all of those institutions for providing me with the opportunities to teach and learn this subject. I have had mathematical and editorial help especially from Brian Marcus, and also Shizuo Kakutani, Ulrich Krengel, Benjamin Weiss, Mike Keane, Shaul Foguel, Yves Derriennic, Bruce Kitchens, and Judy Halchin. My students listened patiently to my lectures, made many useful suggestions, and caught lots of errors. Janet Farrell typed the original manuscript with the assistance of Debbie Hanner Reives, Hazeline Lewis, and Doris Mahaffey. The editing and publishing were handled by David Tranah and John Samuel. My sincere thanks to all these people for their contributions.

Karl Petersen
Chapel Hill, N.C.
September 1981

1

Introduction and preliminaries

Without going into the details (to which the rest of the book is devoted), we mention some of the basic questions, examples, and constructions of ergodic theory, in order to provide an indication of the content and flavor of the subject as well as to establish reference points for terminology and notation. The final section presents a few facts from measure theory and functional analysis that will be used repeatedly.

1.1. The basic questions of ergodic theory

Ergodic theory is the mathematical study of the long-term average behavior of systems. The collection of all states of a system forms a space X. The evolution of the system is represented by a transformation $T: X \to X$, where Tx is taken as the state at time 1 of a system which at time 0 is in state x. If one prefers a continuous variable for the time, he can consider a one-parameter family $\{T_t : t \in \mathbb{R}\}$ of maps of X into itself. When the laws governing the behavior of the system do not change with time, it is natural to suppose that $T_{s+t} = T_s T_t$, so that $\{T_t : t \in \mathbb{R}\}$ is a *flow*, or group action of $\mathbb{R}$ on X. A single (invertible) transformation $T: X \to X$ also determines the action of a group, namely the integers $\mathbb{Z}$, on X. The actions of arbitrary groups, and even of semigroups in case the transformations may not all be invertible, are worthy of study, but we will be interested mainly in the action of the powers of a single transformation and, occasionally, of a flow.

In order to analyze a system mathematically, one needs to have some structure on X and restrictions on T. There are three major cases:

(1) X is a differentiable manifold and T is a diffeomorphism, the case of *differentiable dynamics*;
(2) X is a topological space and T is a homeomorphism, the case of *topological dynamics*;

(3) X is a measure space and T is a measure-preserving transformation, the case of *ergodic theory*.

Of course the three cases overlap extensively, and a single example can be viewed in different lights; in fact, some of the most interesting problems concern the relationships among the three areas.

Let us define more carefully the case (3) of most interest for us. Let $(X, \mathcal{B}, \mu)$ be a complete probability space, that is, a set X together with a σ-algebra $\mathcal{B}$ of measurable subsets of X and a countably additive nonnegative set function μ on $\mathcal{B}$ such that $\mu(X) = 1$ and such that $\mathcal{B}$ contains all subsets of sets of measure 0. Let $T: X \to X$ be a one-to-one onto map such that T and T^{-1} are both measurable: $T^{-1}\mathcal{B} = T\mathcal{B} = \mathcal{B}$. Since sets of measure 0 don't matter, we don't care if T only becomes well-defined and one-to-one onto after a set of measure 0 is discarded from X. Assume further that $\mu(T^{-1}E) = \mu(E)$ for all $E \in \mathcal{B}$. A map T satisfying these conditions is called a *measure-preserving transformation* (abbreviated m.p.t.), and the systems $(X, \mathcal{B}, \mu, T)$ will be our fundamental objects of study.

(Sometimes one wishes to consider possibly noninvertible maps $T: X \to X$ such that $T^{-1}\mathcal{B} \subset \mathcal{B}$ and $\mu T^{-1} = \mu$, and many of the results we present extend to such maps, but we will restrict ourselves for the most part to the invertible case. There is also an interesting theory of *nonsingular* maps ($T^{-1}\mathcal{B} \subset \mathcal{B}$, $\mu T^{-1} \ll \mu$) which we do not have space to discuss here.)

In the one-parameter case, we assume that T_t is a m.p.t. for all $t \in \mathbb{R}$, the map $(x, t) \to T_t x$ is jointly measurable from $X \times \mathbb{R}$ to X, T_0 is the identity, and $T_{s+t} = T_s T_t$ for all $s, t \in \mathbb{R}$.

If $T: X \to X$ is a m.p.t., the *orbit* $\{T^n x : n \in \mathbb{Z}\}$ of a point $x \in X$ represents a single complete history of the system, from the infinite past to the infinite future. The σ-algebra $\mathcal{B}$ is thought of as the family of observable events, with the T-invariant measure μ specifying the (time-independent) probabilities of their occurrences. A measurable function $f: X \to \mathbb{R}$ represents a measurement made on the system; $f(x), f(Tx), f(T^2x), \ldots$ may be thought of as the values of some physically interesting variable at successive instants of time, beginning with the world in initial state x. In statistical mechanics, information theory, and other areas of application it is interesting and sometimes necessary to consider the long-term time average

$$\frac{1}{N} \sum_{k=0}^{N-1} f(T^k x)$$

of a large number N of successive observations. (The average may not really be 'long-term', because the time unit involved may be very short:

1.1. The basic questions of ergodic theory

in statistical mechanics, for example, molecular collisions may occur so often that we can actually observe *only* 'long-term' averages of any variable f.) A basic question of ergodic theory is that of the *convergence of these averages*: when does

$$\bar{f}(x) = \lim_{N \to \infty} \frac{1}{N} \sum_{k=0}^{N-1} f(T^k x)$$

exist in some sense? If it exists, $\bar{f}(x)$ may be thought of as an equilibrium or central value of the variable f.

The convergence had been proved in special cases earlier (e.g. Borel's Strong Law of Large Numbers (1909) in case the fT^k are independent and identically distributed), but the general convergence in the mean square (L^2) sense was proved by von Neumann and the almost everywhere convergence by Birkhoff, both in 1931. Their results are known as the Mean Ergodic Theorem and the Ergodic Theorem (or Pointwise Ergodic Theorem), respectively. Generalizations and improvements of these results have been appearing continually since 1932; we will have space for only a few in this book.

The question of the existence of averages occurred to mathematicians because of the physicists' concern with the 'ergodic hypothesis', which was formulated in an (erroneous and unsuccessful) attempt to bring about the conclusion that the time mean

$$\lim_{N \to \infty} \frac{1}{N} \sum_{k=0}^{N-1} f(T^k x)$$

and the space mean

$$\int_X f \, d\mu$$

coincide almost everywhere. It is desirable that the equilibrium or central value of a physical variable coincide with its weighted average over all possible states of the system. Boltzmann felt that if orbits penetrated all corners of the space, then this useful and symmetrical conclusion would follow. The investigation of the conditions under which this equality of time and space means, as well as even stronger conclusions, holds forms a second major topic in ergodic theory, the study of general *recurrence properties*, by which we mean the qualitative behavior of orbits.

If the time mean of every measurable function coincides almost everywhere (a.e.) with its space mean, the system (or just T itself) is called *ergodic*. It turns out that a system is ergodic if and only if the orbit of almost every point visits each set of positive measure, or, equivalently, if $\mu(E) > 0$ and $\mu(F) > 0$ implies that $\mu(T^n E \cap F) > 0$ for some n. A recurrence property which implies this one is *strong mixing*: $\lim_{n \to \infty} \mu(T^n E \cap F) = \mu(E)\mu(F)$ for all $E, F \in \mathscr{B}$. Weak mixing lies between the two, and just plain *recurrence* – if $\mu(E) > 0$ then $\mu(T^n E \cap E) > 0$ for some n – always holds in a space of finite measure.

A third major question of ergodic theory is the *classification problem*. Let us say that two systems $(X, \mathscr{B}, \mu, T)$ and $(Y, \mathscr{C}, v, S)$ are *metrically isomorphic* if there are sets of measure zero $X_0 \subset X$ and $Y_0 \subset Y$ and a one-to-one onto map $\phi: X \backslash X_0 \to Y \backslash Y_0$ such that $\phi T = S \phi$ on $X \backslash X_0$ and $\mu(\phi^{-1} E) = v(E)$ for all measurable $E \subset Y \backslash Y_0$. (Such a map ϕ is sometimes called an *isomorphism mod 0*.) How can we tell whether two given systems are (metrically) isomorphic to one another? A classical way is by attaching isomorphism invariants to systems, of which ergodicity, weak mixing, and strong mixing are examples. These are, however, only *spectral invariants*. The map $T: X \to X$ determines a unitary operator $U = U_T$ on $L^2(X)$ by $U_T f(x) = f(Tx)$ (Koopman 1931). (We sometimes denote this map by U_T, sometimes by U, and sometimes even by T.) We can say that T and S are *spectrally isomorphic* if U_T and U_S are unitarily equivalent, in that $VU_T = U_S V$ for some unitary $V: L^2(X) \to L^2(Y)$. Then if S and T are spectrally isomorphic, either both or neither have any one of the recurrence properties mentioned above. An invariant which is sensitive to the nature of the action of T on individual points of X is the *entropy* $h(T)$ of T. The entropy can be used to distinguish some nonisomorphic systems (Kolmogorov and Sinai) and is in fact a complete isomorphism invariant within certain classes of systems (Ornstein).

In many parts of mathematics there is a construction problem as well as a classification problem. Such a question could be formulated in several ways in ergodic theory, one of which is the *realization problem*: which systems of type (3) above can be realized within type (2) (say with T preserving a unique Borel probability measure) or within type (1) (say with T preserving a measure determined by a smooth density)? The first of these questions is discussed below, and the second is the subject of current research. Still another question is that of *genericity*: which types of systems are 'typical', in various senses and in the several different settings?

Many of the important questions of ergodic theory receive scant or no attention here: the existence of invariant measures, the case of infinite measure spaces, operator ergodic theory, the actions of other groups, C^* dynamical systems, etc. And our discussion of the questions that we do treat is not alleged to be complete. The starred references in the bibliography are a good starting place for filling in any gaps that we leave.

1.2. The basic examples

Because the two major sources of ergodic theory are mathematical physics (especially statistical mechanics and Hamiltonian dynamics) and the theory of stationary stochastic processes, naturally these subjects provide a rich store of examples of measure-preserving transformations and flows. There are also several interesting and illustrative classes of abstract examples in an algebraic or geometric context. The following list will give some idea of the kinds of measure-preserving systems we will have in mind during the succeeding discussion. Obviously the various classes are not disjoint, and there are in fact inclusion relations among some of them.

A. Hamiltonian dynamics

The state at any time t of a physical system consisting of N particles can be specified by the three coordinates of position and the three of momentum of each particle, that is by a point in $\mathbb{R}^{6N}$, which is the *phase space* of the system. More generally (allowing for changes of variables and constraints on the system), let the state of the system be described by a pair of vectors (q, p), where $p = (p_1, \ldots, p_n)$ (the 'generalized momentum') and $q = (q_1, \ldots, q_n)$ (the 'generalized position') are in $\mathbb{R}^n$, in which case the phase space is $\mathbb{R}^{2n}$. There is given a ($\mathscr{C}^2$) *Hamiltonian function* $H(q, p)$, which we assume to be independent of time, and which is typically the sum of the kinetic energy $K(p)$ and potential energy $U(q)$ of the system. *Hamilton's equations* are

$$\frac{dq_i}{dt} = \frac{\partial H}{\partial p_i}, \quad \frac{dp_i}{dt} = -\frac{\partial H}{\partial q_i} \quad (i = 1, 2, \ldots, n).$$

These equations determine the state $T_t(q, p)$ at any time t if the system has initial state (q, p), by the theorem on the existence and uniqueness of solutions of first-order ordinary differential equations. We obtain in this way a one-parameter flow $\{T_t : -\infty < t < \infty\}$ on the phase space $\mathbb{R}^{2n}$.

Theorem 2.1 *Liouville's Theorem* The Hamiltonian flow $\{T_t\}$ preserves Lebesgue measure on $\mathbb{R}^{2n}$.

Sketch of Proof: Consider the vector field

$$V(q, p) = \left(\frac{\partial H}{\partial p_1}, \ldots, \frac{\partial H}{\partial p_n}, -\frac{\partial H}{\partial q_1}, \ldots, -\frac{\partial H}{\partial q_n} \right),$$

for which clearly div $(V) = 0$. Denoting the Jacobian at (q, p) of the map T_t by $JT_t(q, p)$, this implies through direct calculation that

$$\frac{\partial}{\partial t} JT_t(q, p) = 0.$$

If $E \subset \mathbb{R}^{2n}$ is measurable and μ denotes Lebesgue measure on $\mathbb{R}^{2n}$, then
$$\mu(T_t E) = \int_E JT_t(q, p) d\mu(q, p).$$
Thus
$$\frac{d}{dt}\mu(T_t E) = \int_E \frac{\partial}{\partial t}[JT_t(q, p)] d\mu(q, p) = 0.$$

(For the details, see Khintchine 1949 or Plante 1976.)

Hamilton's Equations yield immediately that
$$\frac{dH}{dt} = 0.$$

Thus the system is not free to wander all over the phase space but is restricted to surfaces of constant total energy E. Usually most of these surfaces are compact manifolds. The flow restricted to any such surface also has an invariant measure. For the proof of this Proposition see Khintchine (1949).

Proposition 2.2 The Hamiltonian flow restricted to a surface $S = \{(q, p): H(q, p) = E\}$ of constant energy preserves the measure $d\mu_S = dS/\|\operatorname{grad} H\|$, where dS is the element of surface volume.

Through this formulation physical dynamical systems ranging from gas in a container to a cluster of galaxies enter the purview of ergodic theory.

B. Stationary stochastic processes

Let $(\Omega, \mathscr{F}, P)$ be a probability space and $\ldots f_{-1}, f_0, f_1, f_2, \ldots$ a sequence of measurable functions on Ω. Suppose that the sequence is *stationary*, in that for any $n_1, n_2, \ldots n_r$, any Borel subsets $B_1, B_2, \ldots B_r$ of $\mathbb{R}$, and any $k \in \mathbb{Z}$,
$$P\{\omega: f_{n_1}(\omega) \in B_1, \ldots f_{n_r}(\omega) \in B_r\}$$
$$= P\{\omega: f_{n_1+k}(\omega) \in B_1, \ldots, f_{n_r+k}(\omega) \in B_r\}.$$
Such a stationary process corresponds to a measure-preserving system in a standard way.

Let $\mathbb{R}^{\mathbb{Z}} = \{(\ldots x_{-1}, x_0, x_1, \ldots): \text{ each } x_i \in \mathbb{R}\}$, define $\phi: \Omega \to \mathbb{R}^{\mathbb{Z}}$ by $(\phi\omega)_n = f_n(\omega)$ for all $n \in \mathbb{Z}$, and define μ on the Borel subsets of $\mathbb{R}^{\mathbb{Z}}$ by
$$\mu(E) = P(\phi^{-1}E).$$
Extend μ to the completion $\mathscr{B}$ of the Borel field. Let $\sigma: \mathbb{R}^{\mathbb{Z}} \to \mathbb{R}^{\mathbb{Z}}$ be the

1.2. The basic examples

shift transformation defined by

$$(\sigma x)_n = x_{n+1}.$$

Because of the stationarity of $\{f_n\}$, μ is shift-invariant on cylinder sets and hence on all of $\mathscr{B}$, so that we have constructed a measure-preserving system $(\mathbb{R}^{\mathbb{Z}}, \mathscr{B}, \mu, \sigma)$. Moreover, if $\pi_n : \mathbb{R}^{\mathbb{Z}} \to \mathbb{R}$ is the projection onto the nth coordinate $(\pi_n x = x_n)$, then $\{\pi_n\}$ has the same joint distributions on $\mathbb{R}^{\mathbb{Z}}$ as $\{f_n\}$ on Ω. Thus every stationary stochastic process 'comes from' some shift-invariant measure on $\mathbb{R}^{\mathbb{Z}}$.

A similar construction applies in the case of a continuous-parameter stochastic process $\{T_t : -\infty < t < \infty\}$ and in more general situations as well (see Doob, 1953).

C. Bernoulli shifts

Let $n = \{0, 1, \ldots n-1\}$ be an alphabet of finitely many symbols with weights $p_0, p_1, \ldots p_{n-1}$ such that all $p_i > 0$ and $\sum_{i=0}^{n-1} p_i = 1$. Form the product space $n^{\mathbb{Z}}$ of all two-sided sequences of the symbols in n, and give $n^{\mathbb{Z}}$ the product measure μ determined by the given probability measure on n. Thus for a typical cylinder set determined by a set of places $i_1, \ldots i_k \in \mathbb{Z}$ and elements $j_1, \ldots j_k \in n$,

$$\mu\{x : x_{i_1} = j_1, \ldots x_{i_k} = j_k\} = p_{j_1} p_{j_2} \cdots p_{j_k}.$$

Clearly the shift transformation $\sigma : n^{\mathbb{Z}} \to n^{\mathbb{Z}}$ preserves the measure μ. The resulting measure-preserving system is denoted by $\mathscr{B}(p_0, p_1, \ldots p_{n-1})$ and represents a finite-valued stationary stochastic process with independent identically distributed terms (i.i.d.). $\mathscr{B}(\frac{1}{2}, \frac{1}{2})$ models an experimenter tossing a fair coin from the infinite past on into eternity.

D. Markov shifts

Form the product space $n^{\mathbb{Z}}$ and shift transformation as in (C). We will define a different invariant measure on $n^{\mathbb{Z}}$, one for which the associated stochastic process is Markov rather than i.i.d.

Let $A = (a_{ij})$ be an $n \times n$ stochastic matrix, i.e. a matrix with nonnegative entries and each row sum equal to 1. Suppose also that $p = (p_0, p_1, \ldots p_{n-1})$ is a row probability vector (all $p_i \geq 0$, $\Sigma p_i = 1$) which is fixed by A:

$$pA = p.$$

By the Perron–Frobenius Theorem (see Varga 1962) such a vector can always be found, and in some cases it is unique. Define the measure of a cylinder set determined by consecutive indices by

$$\mu_A\{x : x_i = j_0, x_{i+1} = j_1, \ldots x_{i+k} = j_k\} = p_{j_0} a_{j_0 j_1} a_{j_1 j_2} \cdots a_{j_{k-1} j_k}.$$

(Thus p gives the a priori probabilities of the symbols and A the transition probabilities from one symbol to another.) It can be verified that μ_A extends in a well-defined way to a countably additive measure on the algebra generated by the cylinder sets, and hence, by the Carathéodory-Hopf Theorem, μ_A extends to the Borel field of $n^{\mathbb{Z}}$ and its completion $\mathscr{B}$. The resulting measure-preserving system $(n^{\mathbb{Z}}, \mathscr{B}, \mu_A, \sigma)$ models a finite-state Markov chain.

E. Rotations of the circle

From the point of view of ergodic theory, the unit circle $\mathbb{K} = \{z \in \mathbb{C} : |z| = 1\}$ is the same as the unit interval $[0, 1)$, and both are versions of $\mathbb{R}/\mathbb{Z}$, the reals 'mod 1'. Given an $\alpha \in \mathbb{R}$, we consider the map $T_\alpha : [0, 1) \to [0, 1)$ defined by

$$T_\alpha x = x + \alpha \pmod{1} = \langle x + \alpha \rangle = \text{fractional part of } x + \alpha$$
$$= x + \alpha - [x + \alpha].$$

Regarded as a map of $\mathbb{K}$,

$$T_\alpha e^{2\pi i \theta} = e^{2\pi i (\theta + \alpha)}.$$

It is clear that T_α preserves Lebesgue measure. If α is rational, then T_α is periodic, all orbits being finite and of the same cardinality. Thus T_α is most interesting when α is irrational.

F. Rotations of compact abelian groups

Let G be a compact abelian group and $g_0 \in G$. Define $T_{g_0} : G \to G$ by $T_{g_0} g = g + g_0$. Because Haar measure is translation-invariant, T_{g_0} is a m.p.t. Again the most interesting case is when G is monothetic and g_0 is a generator: $\{n g_0 : n \in \mathbb{Z}\}$ is dense in G.

G. Automorphisms of compact groups

Let G be a compact group and $T : G \to G$ a continuous automorphism. The uniqueness of normalized Haar measure implies that it is T-invariant.

A particular case of interest is when $G = \mathbb{K}^n = \mathbb{R}^n / \mathbb{Z}^n$ is the n-torus, when it can be shown that T is given by an $n \times n$ integer matrix with determinant ± 1.

H. Gaussian systems

Consider a stochastic process $\ldots f_{-1}, f_0, f_1, \ldots$ on a probability space $(\Omega, \mathscr{F}, P)$ which is Gaussian in that the joint distribution of any finite number of the f_i is Gaussian: given $i_1 < i_2 < \ldots < i_n$, there are $m_1, \ldots, m_n \in \mathbb{R}$ and a symmetric positive-definite $n \times n$ matrix $A = (A_{ij})$

1.2. The basic examples

such that for each Borel $E \subset \mathbb{R}^n$,

$$P\{\omega : (f_{i_1}(\omega), f_{i_2}(\omega), \ldots, f_{i_n}(\omega)) \in E\}$$
$$= \frac{1}{2\pi^{n/2}\sqrt{\det A}} \int_E \exp\left[-\tfrac{1}{2}(x-m)^{tr} A^{-1}(x-m)\right] dx_1 \ldots dx_n.$$

If $\int_\Omega f_i dp$ is a constant, m_0, and $A_{ij} = \int_\Omega (f_i - m_i)(f_j - m_j) dP$ depends only on $i - j$, then the process is stationary, and as in (B) determines a measure-preserving system (see Totoki 1970).

I. Geodesic flows

Let M be a compact Riemannian manifold and $UT(M)$ the unit tangent bundle of M, i.e. the collection of all (x, v), where $x \in M$ and v is a unit tangent vector to M at x. The geodesic flow $\{T_t\}$ on $UT(M)$ is defined as follows. Given (x, v), find the geodesic $\gamma(t)$ which at time 0 passes through x and is tangent to v. Flow along the geodesic at unit speed for a time t, and take for $T_t(x, v)$ the point and unit tangent vector that you finish with. The geodesic flow preserves the measure on the manifold that is determined by the Riemannian metric (see Gottschalk–Hedlund 1955).

J. Horocycle flows

Let M be a compact oriented surface of constant negative curvature. Then M is a quotient of the Poincaré disk by a discrete subgroup of isometries. Geodesics in the Poincaré disk are circular arcs that are perpendicular to the boundary circle:

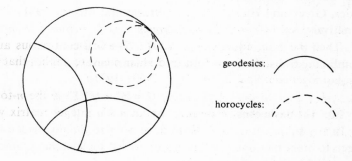

By a *horocycle* we mean a circle which is interior to the disk (except for the point of tangency) and tangent to the boundary. The *horocycle flow* on $UT(M)$ is defined as follows. Given $(x, v) \in UT(M)$, find the geodesic through x in the direction of v. Find the point $(x, v)_\infty$

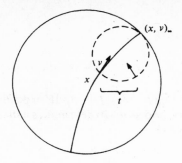

at which this geodesic intersects the boundary. Construct the horocycle tangent to this point, through x, and orthogonal to v. Flow along this horocycle (in a pre-selected sense) at unit speed for a time t, and take for $T_t(x, v)$ the equivalence class of the point and unit normal vector that you finish with. Again the natural volume is preserved.

It is important to note that the geodesic and horocycle flows on the full Poincaré disk have no recurrence whatsoever, but even strong mixing does arise when we pass to a compact quotient by a discrete group of isometries (see Gottschalk and Hedlund 1955).

K. Flows and automorphisms on homogeneous spaces

Let G be a unimodular Lie group, Γ a discrete subgroup such that G/Γ has finite volume (as determined by the Haar measure on G), and $\{g_t : t \in \mathbb{R}\}$ a one-parameter subgroup of G. Then $\{g_t\}$ determines a volume-preserving flow by left multiplication on the set of right cosets. In fact, the classical geodesic and horocycle flows arise in this way (see Auslander, Green and Hahn, 1963 and Brezin and Moore, 1981).

Alternatively, let $T: G \to G$ be a continuous automorphism such that $T\Gamma = \Gamma$. Then the map induced on G/Γ by T is a m.p.t.

By combining translations and automorphisms, one can produce affine maps on homogeneous spaces (see Parry 1969c, 1971).

1.3. The basic constructions

In any subject it is valuable to have ways to modify or combine old objects to make new ones. In this section we list some of the techniques that are available in ergodic theory.

A. Factors

Let $(X, \mathscr{B}, \mu, T)$ and $(Y, \mathscr{C}, \nu, S)$ be measure-preserving systems

1.3. The basic constructions

and $\phi: X \to Y$ a map such that

$\phi^{-1}\mathscr{C} \subset \mathscr{B}$,
$\phi T = S\phi$ a.e.,
$\mu(\phi^{-1}E) = v(E)$ for all $E \in \mathscr{C}$.

Then ϕ is called a *homomorphism* (sometimes *homomorphism mod O*) or *factor map*, $(Y, \mathscr{C}, v, S)$ (or S) is called a *factor* of $(X, \mathscr{B}, \mu, T)$ (or T), and $(X, \mathscr{B}, \mu, T)$ is called an *extension* of $(Y, \mathscr{C}, v, S)$.

B. Products

If $(X_1, \mathscr{B}_1, \mu_1, T_1)$ and $(X_2, \mathscr{B}_2, \mu_2, T_2)$ are measure-preserving systems, we define the product map $T = T_1 \times T_2$ on $X_1 \times X_2$ by $T(x_1, x_2) = (T_1 x_1, T_2 x_2)$. Then T is a m.p.t. on the (completed) product measure space $(X_1 \times X_2, \mathscr{B}_1 \otimes \mathscr{B}_2, \mu_1 \times \mu_2)$.

C. Skew products

Let $(X, \mathscr{B}, \mu, T)$ be a measure-preserving system, and suppose that $\{S_x : x \in X\}$ is a family of m.p.ts of another probability space $(Y, \mathscr{C}, v)$. Assume that $S_x y$ is a jointly measurable map $X \times Y \to Y$. Then $\tau : X \times Y \to X \times Y$ defined by

$$\tau(x, y) = (Tx, S_x y)$$

is a m.p.t. on the product space (Exercise 6).

D. Flow under a function

Let $(X, \mathscr{B}, \mu, T)$ be a measuring-preserving system and $f: X \to (0, \infty)$ a measurable function. We construct a one-parameter flow on the region

$$\Gamma = \{(x, t) : 0 \leqslant t < f(x)\}$$

under the graph of f. Each point flows vertically at unit speed, and we identify the points $(x, f(x))$ and $(Tx, 0)$.

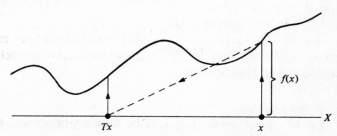

It is easy to see (Exercise 7) that this flow preserves the product of μ with Lebesgue measure. The Ambrose–Kakutani Theorem (see Jacobs 1960) says that under suitable conditions every flow can be represented as a flow under a function.

E. Induced transformations

There are two kinds of induced transformations, the derivative transformation on a subset and the primitive transformation on a superset.

1. Let $(X, \mathcal{B}, \mu, T)$ be a measure-preserving system and $A \subset X$ a measurable subset such that for a.e. $x \in A$ there is a smallest positive $n(x) \in \mathbb{N}$ such that $T^{n(x)}x \in A$. Then $T_A : A \to A$ defined by $T_A x = T^{n(x)} x$ is an invertible measurable map on A which preserves the normalized restriction of μ to A.

2. Let $X = Y_0 \supset Y_1 \supset Y_2 \supset Y_3 \supset \ldots$ be a decreasing sequence of measurable sets, for each i take a copy X_i of Y_i such that all the X_i are disjoint, and form the discrete union $\hat{X} = \bigcup_{i=0}^{\infty} X_i$. Then $\hat{X}$ inherits a σ-algebra $\hat{\mathcal{B}}$ and (possibly infinite) measure $\hat{\mu}$ from X in the obvious way. We think of $\hat{X}$ as a tower or skyscraper built over $X_0 = X$.

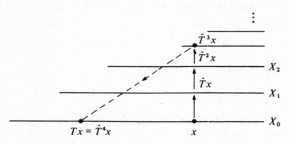

$\hat{T} : \hat{X} \to \hat{X}$ is defined by mapping each point in the tower to the one directly above it if there is one, and otherwise to the image under T in X_0 of its projection to the base. This can be made precise in the following way. We have measure-preserving injections $\phi_{j,i} : X_j \to X_i$ for $i < j$ induced by the inclusions $Y_j \subset Y_i$. We define

$$\hat{T}\hat{x} = \begin{cases} \phi_{i+1,i}^{-1}\{\hat{x}\} & \text{if } \hat{x} \in X_i \text{ and } \phi_{i+1,i}^{-1}\{\hat{x}\} \neq \phi \\ T(\phi_{i,0}\hat{x}) & \text{otherwise.} \end{cases}$$

One can show that $\hat{T}$ preserves $\hat{\mu}$.

F. Inverse limits

Let $(X_i, \mathcal{B}_i, \mu_i, T_i)$ be measure-preserving systems for $i = 0, 1, 2, \ldots$ and suppose that for each $i \geq j$ there is a homomorphism $\phi_{ij} : X_i \to X_j$ such that ϕ_{ii} = identity and $\phi_{ij} \phi_{jk} = \phi_{ik}$ if $i \geq j \geq k$. Form the subset X

of the infinite product $\prod_{i=0}^{\infty} X_i$ defined by

$$X = \{x = (x_i) : \phi_{ij} x_i = x_j \text{ for all } i \geq j\}.$$

Let $\pi_i : X \to X_i$ be the projection: $\pi_i x = x_i$. Then clearly

$$\phi_{ij} \pi_i = \pi_j \text{ on } X.$$

Let $\mathscr{B}$ be the smallest σ-algebra on X which contains all the $\pi_i^{-1} \mathscr{B}_i$. The countably additive measure μ defined on the algebra $\bigcup_{i=0}^{\infty} \pi_i^{-1} \mathscr{B}_i$ by $\mu(\pi_i^{-1} E) = \mu_i(E)$ if $E \in \mathscr{B}_i$ extends (for Lebesgue spaces – see 4.5 and 4.6) to all of $\mathscr{B}$. Complete $\mathscr{B}$ with respect to μ. Define $T: X \to X$ by

$$T(x_i) = (T_i x_i).$$

Then clearly each $\pi_j : (X, \mathscr{B}, \mu, T) \to (X_j, \mathscr{B}_j, \mu_j, T_j)$ is a homomorphism. The system $(X, \mathscr{B}, \mu, T)$ is called the *inverse limit* of the directed family $\{(X_i, \mathscr{B}_i, \mu_i, T_i) : i = 0, 1, 2, \ldots\}$. Such a construction also applies for more general partially ordered index sets.

G. Natural extensions

Let $(X, \mathscr{B}, \mu)$ be a measure space and $T: X \to X$ a possibly *noninvertible* measurable ($T^{-1} \mathscr{B} \subset \mathscr{B}$) measure-preserving ($\mu T^{-1} = \mu$) map. A measure-preserving system $(\hat{X}, \hat{\mathscr{B}}, \hat{\mu}, \hat{T})$ (with $\hat{T}$ invertible) is called a *natural extension* of $(X, \mathscr{B}, \mu, T)$ if there is a homomorphism $\phi: (\hat{X}, \hat{\mathscr{B}}, \hat{\mu}, \hat{T}) \to (X, \mathscr{B}, \mu, T)$ such that $\phi^{-1} \mathscr{B} = \hat{\mathscr{B}}$ up to sets of measure 0. A natural extension can be constructed by taking for the space the inverse limit of $(X_i, \mathscr{B}_i, \mu_i, T_i)$, where all $(X_i, \mathscr{B}_i, \mu_i) = (X, T^{-i} \mathscr{B}, \mu)$, each $T_i =$ identity, and $\phi_{ij} = T^{i-j}$, and for the transformation the (backwards unilateral) shift. The natural extension is unique up to isomorphism.

1.4. Some useful facts from measure theory and functional analysis

A. Change of variables

Let $(X, \mathscr{B}, \mu)$ be a measure space, $(Y, \mathscr{C})$ a measurable space (i.e. a set Y together with a σ-algebra $\mathscr{C}$ of subsets of Y), and $\phi: X \to Y$ a *measurable map*, in the sense that $\phi^{-1} \mathscr{C} \subset \mathscr{B}$. Then ϕ carries μ to a measure denoted by $\phi\mu$ or $\mu\phi^{-1}$ on $\mathscr{C}$, which is defined by $\mu\phi^{-1}(E) = \mu(\phi^{-1} E)$ for $E \in \mathscr{C}$.

Proposition 4.1 *If $\phi: (X, \mathscr{B}, \mu) \to (Y, \mathscr{C})$ is a measurable map and f is a real-valued measurable function on Y, then*

$$\int_Y f \, \mathrm{d}(\mu\phi^{-1}) = \int_X f \circ \phi \, \mathrm{d}\mu,$$

in the sense that if one of these integrals exists then so does the other and the two are equal.

Proof: When $f = \chi_E$ is the characteristic function of $E \in \mathscr{C}$,

$$\int_Y \chi_E \, d(\mu\phi^{-1}) = \mu\phi^{-1}(E) \text{ and } \int_X \chi_E \circ \phi \, d\mu = \int_X \chi_{\phi^{-1}E} \, d\mu$$
$$= \mu(\phi^{-1}E),$$

so that the result holds by definition of $\mu\phi^{-1}$. Hence the formula is also true for simple functions (i.e. linear combinations of characteristic functions). If f is a nonnegative measurable function, then f is the pointwise limit of an increasing sequence of simple functions, and the result follows from the Monotone Convergence Theorem. Finally, any measurable function f on Y can be written as the difference $f = f^+ - f^-$ of two nonnegative measurable functions on Y, so the formula is true in general.

B. Proofs by approximation

The argument of the preceding paragraph is of a kind frequently encountered in ergodic theory, except that, even more, it is sometimes enough to check a formula only for *certain* characteristic functions in order to be able to conclude that it holds for all $f \in L^1$ or L^2, all measurable functions, or whatever.

A *semialgebra* on X is a collection $\mathscr{S}$ of subsets of X which is closed under finite intersections and such that the complement of any $S \in \mathscr{S}$ is a finite disjoint union of members of $\mathscr{S}$. We say that $\mathscr{S}$ *generates* a σ-algebra $\mathscr{B}$ if $\mathscr{B}$ is the smallest (complete) σ-algebra which contains $\mathscr{S}$. Examples of generating semialgebras (for the usual σ-algebras) are the collection of half-open intervals $[a, b)$ in $[0, 1)$ and the family of cylinder sets

$$\{x \in n^{\mathbb{Z}} : x_{i_1} = j_1, x_{i_2} = j_2, \ldots x_{i_k} = j_k\} \text{ in } \{0, 1, \ldots n-1\}^{\mathbb{Z}}.$$

Proposition 4.2 Suppose that $\rho_n : L^2 \times L^2 \to [0, \infty)$, for $n = 0, 1, 2, \ldots$, is a function such that

(i) $\rho_n(f, g) \leq K \|f\|_2 \|g\|_2$ for all $f, g \in L^2$ and some constant K;

(ii) $\rho_n(f_1 + f_2, g) \leq \rho_n(f_1, g) + \rho_n(f_2, g)$,
$\rho_n(f, g_1 + g_2) \leq \rho_n(f, g_1) + \rho_n(f, g_2)$,
$\rho_n(af, g) \leq |a| \rho_n(f, g)$, and
$\rho_n(f, ag) \leq |a| \rho_n(f, g)$
for all $f, f_1, f_2, g, g_1, g_2 \in L^2$ and $a \in \mathbb{R}$.

If $\lim_{n \to \infty} \rho_n(\chi_E, \chi_F) = 0$ for all sets E, F in a generating semialgebra in $\mathscr{B}$, then $\lim_{n \to \infty} \rho_n(f, g) = 0$ for all $f, g \in L^2$.

1.4. Some useful facts

Proof: By (ii) it follows immediately that $\lim_{n\to\infty} \rho_n(\chi_A, \chi_B) = 0$ whenever A and B are disjoint unions of sets in $\mathscr{S}$, and also that $\lim_{n\to\infty} \rho_n(\phi, \psi) = 0$ whenever ϕ and ψ are linear combinations of characteristic functions of sets in the algebra generated by $\mathscr{S}$ (which consists of all finite disjoint unions of members of $\mathscr{S}$). The set of such simple functions ϕ is dense in L^2. Given any $f, g \in L^2$, choose such simple $\phi_k \to f$ in L^2 and $\psi_k \to g$ in L^2. Then

$$\rho_n(f, g) = \rho_n(\phi_k + f - \phi_k, g) \leq \rho_n(\phi_k, g) + \rho_n(f - \phi_k, g)$$
$$\leq \rho_n(\phi_k, \psi_k) + \rho_n(\phi_k, g - \psi_k) + \rho_n(f - \phi_k, g)$$
$$\leq \rho_n(\phi_k, \psi_k) + K\|\phi_k\|_2 \|g - \psi_k\|_2 + K\|f - \phi_k\|_2 \|g\|_2.$$

Given $\varepsilon > 0$, choose k so that each of the last two terms is less than $\varepsilon/3$, and then choose n so that the first term is less than $\varepsilon/3$.

C. Measure algebras and Lebesgue spaces

There is a coherent way to ignore the sets of measure 0 in a measure space.

A *measure algebra* is a pair $(\mathscr{B}, \mu)$, where $\mathscr{B}$ is a Boolean σ-algebra (i.e., a set with general operations $\vee, \wedge, '$ which behave the same way as union, intersection, and complementation do in a σ-algebra), and μ is a countably additive positive ($\mu(B) \geq 0$, with equality if and only if $B = 0$) extended real-valued function on $\mathscr{B}$.

If $\mathscr{B}$ is a Boolean σ-algebra and $\mathscr{I} \subset \mathscr{B}$, then we say that $\mathscr{I}$ is a σ-*ideal* in case

(i) $I \in \mathscr{I}, B \in \mathscr{B}, B < I$ (i.e., $B \wedge I = B$) implies $B \in \mathscr{I}$

(ii) $I_n \in \mathscr{I} (n = 1, 2, \ldots)$ implies $\bigvee_{n=1}^{\infty} I_n \in \mathscr{I}$.

If $\mathscr{B}$ is a Boolean σ-algebra and $\mathscr{I} \subset \mathscr{B}$ is a σ-ideal, then $\mathscr{I}$ determines an equivalence relation on $\mathscr{B}$ according to $B_1 \sim B_2$ if and only if $B_1 \Delta B_2 = (B_1 \wedge B_2') \vee (B_1' \wedge B_2) \in \mathscr{I}$. It is not difficult to check that the collection $\mathscr{B}/\mathscr{I}$ of equivalence classes forms a Boolean σ-algebra. For example, let $\mathscr{N}$ denote the collection of sets of measure zero. Then $\mathscr{N}$ is a σ-ideal in $\mathscr{B}$, μ is constant on equivalence classes of $\mathscr{B}$ modulo $\mathscr{N}$ and hence determines a countably additive positive extended real-valued function $\hat{\mu}$ on $\hat{\mathscr{B}} = \mathscr{B}/\mathscr{N}$, and $(\hat{\mathscr{B}}, \hat{\mu})$ is a measure algebra. The most important case is when $\mathscr{B}$ is the Boolean σ-algebra of measurable sets in a measure space $(X, \mathscr{B}, \mu)$; then $(\hat{\mathscr{B}}, \hat{\mu})$ is called *the measure algebra of the measure space* $(X, \mathscr{B}, \mu)$.

Let $(\mathscr{B}, \mu)$ and $(\mathscr{C}, v)$ be measure algebras. A map $T: \mathscr{B} \to \mathscr{C}$ is called a *homomorphism* (of measure algebras) in case T commutes with the Boolean σ-algebra operations (so that

$$T(B_1 \wedge B_2') = (TB_1) \wedge (TB_2)' \text{ and } T \bigvee_{n=1}^{\infty} B_n = \bigvee_{n=1}^{\infty} TB_n$$

for all $B_1, B_2, \ldots \in \mathscr{B}$)
and satisfies $\mu(B) = \nu(TB)$ for all $B \in \mathscr{B}$. If T is onto and one-to-one, then it is called an *isomorphism*.

Let $(\mathscr{B}, \mu)$ be a measure algebra. We define a metric d on $\mathscr{B}$ by $d(A, B) = \mu(A \triangle B)$ for $A, B \in \mathscr{B}$. $(\mathscr{B}, d)$ is called the *metric space of the measure algebra* $(\mathscr{B}, \mu)$.

Proposition 4.3 The metric space of a measure algebra is complete.

A measure algebra $(\mathscr{B}, \mu)$ is called *separable* in case its associated metric space is separable – that is, there is a countable dense set. If, for example, $(\mathscr{B}, \mu)$ is a finite measure algebra and is countably generated, then $(\mathscr{B}, \mu)$ is separable.

For a measure algebra $(\mathscr{B}, \mu)$, a nonzero $B \in \mathscr{B}$ is called an *atom* if whenever $A < B$ either $A = 0$ or $A = B$. A measure algebra $(\mathscr{B}, \mu)$ is called *nonatomic* if it has no atoms. The measure algebra $(\hat{\mathscr{M}}, \hat{m})$ of the unit interval is normalized ($\mu(X) = 1$), separable (since $\mathscr{M}$ is generated by a countable base for the open sets), and nonatomic. The following important theorem says that these three properties are sufficient to characterize the measure algebra of the unit interval.

Theorem 4.4 (*Carathéodory* 1939) If $(\mathscr{B}, \mu)$ is a normalized, separable, nonatomic measure algebra, then there is an isomorphism from $(\mathscr{B}, \mu)$ onto the measure algebra $(\hat{\mathscr{M}}, \hat{m})$ of the unit interval.

Thus the measure space consisting of the unit interval with Lebesgue measure is a fairly representative object, at least at the level of measure algebras. We single out the class of measure spaces that are actually isomorphic to this standard one, with the possible addition of countably many point masses, in the sense of 'point isomorphism mod 0'.

Definition 4.5 A (complete) finite measure space $(Y, \mathscr{C}, \nu)$ is called a *Lebesgue space* if it is isomorphic (mod 0) with the ordinary Lebesgue measure space $(X, \mathscr{B}, \mu)$ of a (possibly empty) interval $[a, b) \subset \mathbb{R}$ together with countably many point masses. That is, there are $y_1, y_2, \ldots \in Y$, sets of measure zero $Y_0 \subset Y$ and $X_0 \subset X$ and a one-to-one onto map ϕ:

1.4. Some useful facts

$Y\backslash(Y_0 \cup \{y_1, y_2, \ldots\}) \to X\backslash X_0$ such that ϕ and ϕ^{-1} are measurable and $\mu\phi = v|_{Y\backslash\{y_1, y_2 \ldots\}}$.

For a systematic treatment of the theory of Lebesgue spaces, see Rokhlin (1949) or Halmos and von Neumann (1942). For our purposes, it is enough to register the following two theorems of von Neumann (1932a) (see also Billingsley 1965, p. 69).

Theorem 4.6 If X is a complete separable metric space and $\mathscr{B}$ is the completion of the family of Borel sets with respect to a Borel probability measure μ on X, then $(X, \mathscr{B}, \mu)$ is a Lebesgue space.

Theorem 4.7 If $(X, \mathscr{B}, \mu)$ and $(Y, \mathscr{C}, v)$ are Lebesgue spaces and $\Phi:(\hat{\mathscr{C}}, \hat{v}) \to (\hat{\mathscr{B}}, \hat{\mu})$ is a homomorphism of their associated measure algebras, then Φ arises from a *point homomorphism mod 0*: there is a set of measure zero $X_0 \subset X$ and a measurable transformation $\phi: X\backslash X_0 \to Y$ such that ϕ^{-1} coincides with Φ as a map $(\hat{\mathscr{C}}, \hat{v}) \to (\hat{\mathscr{B}}, \hat{\mu})$.

Since for Lebesgue spaces the concepts of point homomorphism mod 0 and homomorphism of measure algebras (the measurable sets reduced modulo the σ-ideal of sets of measure 0) essentially coincide, a cavalier attitude toward sets of measure 0 can be forgiven.

In this book we will deal exclusively with Lebesgue spaces, and in fact will usually work with just the unit interval, since the action of a m.p.t. on any atomic part is relatively uninteresting.

D. Conditional expectation

Let $(X, \mathscr{B}, \mu)$ be a probability space, $f \in L^1(X, \mathscr{B}, \mu)$, and $\mathscr{F} \subset \mathscr{B}$ a sub-σ-algebra. Then

$$v(A) = \int_A f \, d\mu, \text{ for } A \in \mathscr{F},$$

defines a finite measure which is absolutely continuous with respect to the restriction of μ to $\mathscr{F}$. By the Radon–Nikodym Theorem there is a function $g \in L^1(X, \mathscr{F}, \mu)$ such that

$$v(A) = \int_A g \, d\mu \quad \text{for all } A \in \mathscr{F}.$$

Any other such function g_1 must coincide a.e. $d\mu$ with g. We use the notation $E(f|\mathscr{F})$ for g, and call $E(f|\mathscr{F})$ the *conditional expectation of f with respect to $\mathscr{F}$*.

As an element of L^1, $E(f|\mathscr{F})$ is completely characterized by the following two properties:

(i) $E(f|\mathscr{F})$ is $\mathscr{F}$-measurable;
(ii) $\int_A E(f|\mathscr{F})\,d\mu = \int_A f\,d\mu$ for all $A \in \mathscr{F}$.

This seemingly innocent application of the Radon–Nikodym Theorem in fact provides one of the most important kinds of averaging processes. The meaning of $E(f|\mathscr{F})$ can best be apprehended in case $f = \chi_A$ is a characteristic function – in which case we use the notation

$$\mu(A|\mathscr{F}) = E(\chi_A|\mathscr{F})$$

and call $\mu(A|\mathscr{F})$ the *conditional probability of A given $\mathscr{F}$* – and $\mathscr{F}$ is a finite σ-algebra $\{\phi, E, E^c, X\}$, for some $E \in \mathscr{B}$ with $\mu(E) > 0$ and $\mu(E^c) > 0$. If $E(f|\mathscr{F})$ is to be $\mathscr{F}$-measurable in this case, it must be constant on E and on E^c; then (ii) gives

$$E(\chi_A|\mathscr{F}) = \begin{cases} \dfrac{\mu(A \cap E)}{\mu(E)} = \mu(A|E) & \text{on } E \\[2ex] \dfrac{\mu(A \cap E^c)}{\mu(E^c)} = \mu(A|E^c) & \text{on } E^c. \end{cases}$$

We see that $\mu(A|\mathscr{F})(x)$ gives the probability of occurrence of A once we know to which elements of $\mathscr{F}$ the point x belongs. In the general case, too, $E(f|\mathscr{F})(x)$ represents the expected value of f if we know for each $E \in \mathscr{F}$ whether or not $x \in E$.

The following properties of conditional expectation can be proved easily by using the characterization by (i) and (ii).

Proposition 4.8
(1) If $f \geq 0$ a.e., then $E(f|\mathscr{F}) \geq 0$ a.e.
(2) If f is $\mathscr{F}$-measurable, then $E(f|\mathscr{F}) = f$ a.e.
(3) If f is $\mathscr{F}$-measurable, then $E(fg|\mathscr{F}) = f E(g|\mathscr{F})$ a.e.
(4) $(E(E(f|\mathscr{F})) = E(f)$ $(E(f) = \int_X f\,d\mu)$.
(5) If $\mathscr{F}_1 \subset \mathscr{F}$, then $E(E(f|\mathscr{F})|\mathscr{F}_1) = E(f|\mathscr{F}_1)$ a.e.
(6) If f is independent of $\mathscr{F}$, then $E(f|\mathscr{F}) = E(f)$ a.e.
(7) For each $p \geq 1$, $E(\cdot|\mathscr{F})$ is a linear operator of norm 1 on $L^p(X, \mathscr{B}, \mu)$.
(8) *Jensen's Inequality*: If $\phi: \mathbb{R} \to \mathbb{R}$ is convex $(\lambda_1, \ldots \lambda_n \geq 0$, $\sum_{i=1}^n \lambda_i = 1$, $x_1, \ldots x_n \in \mathbb{R}$ implies $\phi(\sum_{i=1}^n \lambda_i x_i) \leq \sum_{i=1}^n \lambda_i \phi(x_i))$ and $\phi \circ f$ is integrable, then $\phi(E(f|\mathscr{F})) \leq E(\phi \circ f|\mathscr{F})$ a.e.

1.4. Some useful facts

E. The Spectral Theorem

There are various ways of stating this result, of which we give three. Let $T: H \to H$ be a continuous normal ($TT^* = T^*T$) operator on a separable Hilbert space H.

(1) There is a finite measure space $(X, \mathscr{B}, \mu)$ and a function $\phi \in L^\infty(X, \mathscr{B}, \mu)$ such that T is unitarily equivalent to multiplication by ϕ on $L^2(X, \mathscr{B}, \mu)$.

(2) There are Borel measures $\sigma_1 \gg \sigma_2 \gg \ldots$ on $\mathbb{C}$ such that T is unitarily equivalent to the operator S on the direct sum $\sum_{i=1}^\infty \oplus L^2(\mathbb{C}, \sigma_i)$ defined by

$$S(f_1(z_1), f_2(z_2), \ldots) = (z_1 f_1(z_1), z_2(f_2(z_2), \ldots).$$

The σ_i are determined up to absolute-continuity equivalence. The equivalence class of σ_1 is called the *maximum spectral type* of T.

(3) There is a unique *spectral measure* E such that

$$T = \int_\mathbb{C} \lambda \, dE_\lambda.$$

That is, E is a projection-valued Borel measure on $\mathbb{C}$ with compact support such that $E(\mathbb{C}) = $ identity, and $(E \bigcup_{i=1}^\infty A_i)(f) = \sum_{i=1}^\infty (EA_i)(f)$ if the A_i are measurable and disjoint, where the series converges in the L^2 norm.

E is supported on the spectrum $\sigma(T) = \{\lambda \in \mathbb{C} : T - \lambda I \text{ has no continuous inverse}\}$ of T. If h is any bounded Borel-measurable function on $\sigma(T)$, the integral

$$h(T) = \int_{\sigma(T)} h(\lambda) \, dE(\lambda)$$

can be defined by noting that if $f, g \in H$, then $(E(\cdot)f, g)$ is a complex measure on $\sigma(T)$, and the equation

$$(Sf, g) = \int_{\sigma(T)} h(\lambda) \, d(E(\lambda)f, g) \qquad \text{for all } f, g \in L^2$$

will define $S = h(T)$ on H. In particular

$$T^k = \int_{\sigma(T)} \lambda^k \, dE(\lambda),$$

$$(T^k f, g) = \int_{\sigma(T)} \lambda^k \, d(E(\lambda)f, g).$$

There are two facts about these spectral integrals that are very useful in applications.

Proposition 4.9 If $R = \int_C f(\lambda) dE(\lambda)$ and $S = \int_C g(\lambda) dE(\lambda)$, then $RS = \int_C f(\lambda) g(\lambda) dE(\lambda)$.

Proposition 4.10 Let T have spectral measure E, let $x \in H$, and let v be a finite positive measure which is absolutely continuous with respect to the measure $(E(A)x, x)$. Then there is $y \in H$ such that $v(A) = (E(A)y, y)$ for all Borel sets A. (Actually such a y can be found in the smallest T- and T^*-invariant closed subspace containing x.)

F. Topological groups, Haar measure, and character groups

Let G be a topological group, so that G is simultaneously a group and a topological space and the map $G \times G \to G$ defined by $(x, y) \to xy^{-1}$ is continuous.

If G is locally compact, then there exists on G a *left Haar measure* m, i.e., a positive, regular, Borel measure which is finite on compact sets, positive on nonempty open sets, and *translation invariant*: $m(gA) = m(A)$ for all $g \in G$ and all measurable A. The measure m is unique up to a constant multiple. Similarly, there is a right Haar measure. Of course, on an abelian group they can be taken to be equal, and on a compact group they can be normalized to be probability measures (see Halmos 1950.).

Henceforth let G be locally compact and abelian. The *character group* of G is the set $\hat{G}$ of all continuous homomorphisms of G into the multiplicative group $\mathbb{K}$ of all complex numbers of modulus 1. $\hat{G}$ is a group under pointwise multiplication. When $\hat{G}$ is given the topology of uniform convergence on compact subsets of G, it becomes a locally compact abelian group. If G is discrete, then $\hat{G}$ is compact; if G is compact, then $\hat{G}$ is discrete. Examples: $\hat{\mathbb{K}} = \mathbb{Z}$, $(\mathbb{K}^n)\hat{} = \mathbb{Z}^n$, $\hat{\mathbb{R}} = \mathbb{R}$, $(\mathbb{R}^n)\hat{} = \mathbb{R}^n$.

The *Fourier transform* $\hat{f}$ of a function $f \in L^1(G, m)$ is defined on $\hat{G}$ by

$$\hat{f}(\chi) = \int_G f(g) \overline{\chi(g)} dm(g).$$

According to the *Plancherel theorem*, the Fourier transform $f \to \hat{f}$ is an L^2-norm-preserving map from $L^1(G) \cap L^2(G)$ onto a dense subspace of $L^2(\hat{G})$, which, therefore, extends to an isometry from $L^2(G)$ onto $L^2(\hat{G})$. Frequently, f can be recovered from $\hat{f}$ via an *inversion formula*. If G is compact, then $\hat{G}$ forms a complete orthonormal set in $L^2(G)$ and one has the Fourier series decomposition, in $L^2(G)$,

$$f(g) = \sum_{\chi \in \hat{G}} \hat{f}(\chi) \chi(g) \quad \text{for any } f \in L^2(G).$$

The *Pontryagin Duality Theorem* says that $(\hat{G})\hat{}$ can be identified in a

1.4. Some useful facts

natural way with G. Thus, $\hat{\mathbb{Z}} = \mathbb{K}$, etc. See Rudin (1967) for more details and references.

Exercises

1. Show that if T is a m.p.t., then $U = U_T$, defined on $L^2(X, \mathcal{B}, \mu)$ by $U_T f(x) = f(Tx)$, is unitary. What if T is noninvertible?

2. Show that:
   (a) The one-sided Bernoulli shift $(\sigma x)_n = x_{n+1}$ on $\prod_{i=0}^{\infty} \{0, 1\}$, where 0 and 1 both have weights $\frac{1}{2}$, is isomorphic to the map $t \to 2t \bmod 1$ on $[0, 1)$.
   (b) $\mathcal{B}(\frac{1}{2}, \frac{1}{2})$ is isomorphic to the 'baker's transformation'
   $$T(x, y) = \begin{cases} (2x \bmod 1, \frac{1}{2}y) & \text{if } 0 \leq x \leq \frac{1}{2} \\ (2x \bmod 1, \frac{1}{2}(y + 1)) & \text{if } \frac{1}{2} \leq x \leq 1 \end{cases}$$
   on $[0, 1) \times [0, 1)$.

3. Show that an automorphism of a compact group preserves Haar measure.

4. Show that a homomorphism of measure-preserving systems is onto, up to a set of measure 0.

5. Show that there is a correspondence between (isomorphism classes of) factors of a Lebesgue space and (appropriate equivalence classes of) invariant sub-σ-algebras. (You may use the fact that a factor of a Lebesgue space is a Lebesgue space.)

6. Verify that the skew product defined in 1.3.C. is a m.p.t. (Note: To get both τ and τ^{-1} measurable, extra hypotheses may be needed.)

7. Verify that the flow built under a function in 1.3.D. is measure-preserving.

8. Give a categorical definition of the inverse limit (1.3.F) of a countable directed family of measure-preserving systems and show that an object satisfying this definition is unique up to isomorphism.

9. (a) Construct the natural extension (1.3.G) as the shift on an inverse limit of measure spaces.
   (b) Prove that the natural extension is unique up to isomorphism.
   (c) Show that the natural extension of a unilateral Bernoulli shift is the corresponding bilateral Bernoulli shift.

10. Let $(X, \mathcal{B}, \mu)$ and $(Y, \mathcal{C}, \nu)$ be Lebesgue spaces and $T: L^2(Y, \mathcal{C}, \nu) \to L^2(X, \mathcal{B}, \mu)$ an isometry which is *multiplicative*: $T(fg) = Tf \cdot Tg$

whenever f, g, and $fg \in L^2(Y, \mathscr{C}, v)$. Show that there is a homomorphism $\phi: X \to Y$ such that $Tf(x) = f(\phi x)$ a.e.

11. Fill in the details in the proof of Liouville's Theorem (2.1).
12. Use the Carathéodory–Hopf Extension Theorem to prove *Kolmogorov's Consistency Theorem*: Let A be an index set, and for each $n = 1, 2, \ldots$ and each n-tuple $(\alpha_1, \ldots \alpha_n)$ of elements of A, let $\mu_{(\alpha_1,\ldots,\alpha_n)}$ be a Borel measure on $\mathbb{R}^n$. Assume that:
  (i) If τ is any permutation of the set with n elements and T_τ is the corresponding transformation of $\mathbb{R}^n$, then
  $$\mu_{\tau(\alpha_1,\ldots,\alpha_n)}(E) = \mu_{(\alpha_1,\ldots,\alpha_n)}(T_\tau^{-1}E) \quad \text{for all Borel } E \subset \mathbb{R}^n.$$
  (ii) If $\Pi_{n+k,n}: \mathbb{R}^{n+k} \to \mathbb{R}^n$ is the projection map defined by $\Pi_{n+k,n}(x_1, \ldots, x_{n+k}) = (x_1, \ldots, x_n)$, then
  $$\mu_{(\alpha_1,\ldots,\alpha_n)}(E) = \mu_{(\alpha_1,\ldots,\alpha_{n+k})}(\Pi_{n+k,n}^{-1}E) \quad \text{for all } n, k = 1, 2, 3, \ldots$$
  and all Borel $E \subset \mathbb{R}^n$.
  Then there is a probability space $(\Omega, \mathscr{F}, P)$ and a family $\{f_\alpha : \alpha \in A\}$ of measurable functions on Ω such that always
  $$\mu_{(\alpha_1,\ldots\alpha_n)}(E) = P\{\omega : (f_{\alpha_1}(\omega), \ldots f_{\alpha_n}(\omega)) \in E\}.$$
13. Prove Proposition 4.8. (Hint: For Jensen's Inequality, first consider the case when f is a simple function.)
14. Prove Proposition 4.3.
15. (Lind) Consider the two-to-one map $Tx = \frac{1}{2}(x - 1/x)$ on $\mathbb{R}$.
  (a) Show that T preserves the measure $dx/(1+x^2)$. (Hint: It is enough to show by calculus that $\int_\mathbb{R} f(\frac{1}{2}(x-1/x))(dx/(1+x^2)) = \int_\mathbb{R} f(y)(dy/(1+y^2))$ for enough functions f.)
  (b) Show that the change of variables $x = \tan t$ carries T to a Lebesgue-measure-preserving map S of $(-\pi/2, \pi/2)$.
  (c) Show that S is isomorphic to the one-sided Bernoulli shift $\mathscr{B}(\frac{1}{2}, \frac{1}{2})$.
  (d) Find a map that carries T to $Vx = 4x(1-x)$ on $[0,1]$ (which preserves $dx/\pi\sqrt{x(1-x)}$).
  (e) For each $a \in \mathbb{R}$, discuss the action of $T_a z = \frac{1}{2}(z + (a/z))$ on $\mathbb{C}$, represented as a sphere with poles at the two square roots of a. (Iteration of the map T_a is an ancient algorithm for computing square roots.)

2

The fundamentals of ergodic theory

In this chapter we prove the basic convergence theorems of ergodic theory, namely von Neumann's Mean Ergodic Theorem and Birkhoff's Pointwise Ergodic Theorem. We also discuss recurrence, ergodicity, strong mixing, and weak mixing. These properties help to understand the manner in which a m.p.t. moves points and sets through the space on which it acts, and their presence or absence provides the first test for deciding whether two transformations are isomorphic.

2.1. The Mean Ergodic Theorem

Although the Mean Ergodic Theorem and the Pointwise Ergodic Theorem bear the publication dates 1932 and 1931, respectively, von Neumann's theorem came first.

Theorem 1.1 Mean Ergodic Theorem (von Neumann 1932b) Let $(X, \mathscr{B}, \mu)$ be a σ-finite measure space, $T: X \to X$ a measure-preserving transformation, and $f \in L^2(X, \mathscr{B}, \mu)$. Then there is a function $\bar{f} \in L^2(X, \mathscr{B}, \mu)$ for which

$$\lim_{n \to \infty} \left\| \frac{1}{n} \sum_{k=0}^{n-1} f \circ T^k - \bar{f} \right\|_2 = 0.$$

(The requirement that $(X, \mathscr{B}, \mu)$ be σ-finite is not really necessary. For if $f \in L^2(X, \mathscr{B}, \mu)$, then f vanishes outside a set E of σ-finite measure. If we let

$$X' = \bigcup_{k=-\infty}^{\infty} T^k E,$$

then each $f \circ T^k$ vanishes outside X', and X' is invariant and has σ-finite measure. If the Mean Ergodic Theorem holds in $L^2(X')$, then we can find $P'f \in L^2(X')$ such that

$$\frac{1}{n} \sum_{k=0}^{n-1} f \circ T^k \to P'f$$

in $L^2(X')$. Then $P'f$ extends to $Pf \in L^2(X)$ (by putting $Pf(x) = 0$ in case $x \in X - X'$) and

$$\frac{1}{n}\sum_{k=0}^{n-1} f \circ T^k \to Pf \quad \text{in} \quad L^2(X).)$$

We will prove a more general theorem than this. Note that the m.p.t. $T: X \to X$ determines a linear operator $U: L^2 \to L^2$ according to $(Uf)(x) = f(Tx)$. U is an isometry, since (using 1.4.1)

$$\|Uf\|_2^2 = \int |Uf(x)|^2 d\mu = \int |f(Tx)|^2 d\mu = \|f\|_2^2.$$

U is also invertible ($U^{-1}f(x) = f(T^{-1}x)$), and hence U is a unitary operator (its adjoint equals its inverse). $U = U_T$ is called the *unitary operator induced by* T. The Mean Ergodic Theorem, then, asserts that

$$\frac{1}{n}\sum_{k=0}^{n-1} U^k f$$

converges in L^2; this is a consequence of the following theorem.

Theorem 1.2 *Mean Ergodic Theorem in Hilbert Space* Let U be a (linear) contraction on a Hilbert space $\mathcal{H}$ (so that $\|Uf\| \leq \|f\|$ for all $f \in \mathcal{H}$), let $\mathcal{M} = \{f \in \mathcal{H} : Uf = f\}$ (so $\mathcal{M}$ is a closed linear subspace of $\mathcal{H}$), and let $P: \mathcal{H} \to \mathcal{M}$ be the projection of $\mathcal{H}$ onto $\mathcal{M}$. Then for each $f \in \mathcal{H}$,

$$\frac{1}{n}\sum_{k=0}^{n-1} U^k f$$

converges in $\mathcal{H}$ to Pf.

Proof: Let $\mathcal{N}$ denote the closed linear span of $\{g - Ug : g \in \mathcal{H}\}$; we claim that $\mathcal{N}$ and $\mathcal{M}$ are orthogonal complementary subspaces of $\mathcal{H}$. It suffices to show that $\mathcal{N}^\perp = \mathcal{M}$, where $\mathcal{N}^\perp = \{h \in \mathcal{H} : (h, n) = 0 \text{ for all } n \in \mathcal{N}\}$ and $(\cdot, \cdot)$ denotes the inner product on $\mathcal{H}$.

Let $h \in \mathcal{N}^\perp$; then $(h, g - Ug) = 0$ for all $g \in \mathcal{H}$, so

$$0 = (h, g) - (h, Ug) = (h, g) - (U^*h, g) = (h - U^*h, g)$$

for all $g \in \mathcal{H}$, and hence $h = U^*h$. We wish to show that $Uh = h$. Now

$$\begin{aligned}
\|Uh - h\|^2 &= (Uh - h, Uh - h) \\
&= \|Uh\|^2 - (h, Uh) - (Uh, h) + \|h\|^2 \\
&= \|Uh\|^2 - (U^*h, h) - (h, U^*h) + \|h\|^2 \\
&\leq \|h\|^2 - (h, h) - (h, h) + \|h\|^2 = 0.
\end{aligned}$$

Therefore $Uh = h$, so $h \in \mathcal{M}$. We conclude that $\mathcal{N}^\perp \subset \mathcal{M}$.

2.1. The mean ergodic theorem

Conversely, if $h \in \mathcal{M}$, then $Uh = h$ and hence $U^*h = h$ by the same argument applied to U^* (which is also a contraction since

$$\|U^*f\|^2 = |(U^*f, U^*f)| = |(UU^*f, f)|$$
$$\leq \|UU^*f\| \|f\| \leq \|U^*f\| \|f\|).$$

Therefore, for any $g \in \mathcal{H}$,

$$(h, g - Ug) = (h, g) - (h, Ug) = (h - U^*h, g) = 0,$$

so $h \in \mathcal{N}^\perp$. Consequently $\mathcal{N}^\perp = \mathcal{M}$.

We will show next that if $f \in \mathcal{N}$ then

$$\frac{1}{n}\sum_{k=0}^{n-1} U^k f$$

converges to zero. First, if $f = g - Ug$ for some $g \in \mathcal{H}$, then

$$\frac{1}{n}\sum_{k=0}^{n-1} U^k f = \frac{1}{n}(g - U^n g),$$

so

$$\left\|\frac{1}{n}\sum_{k=0}^{n-1} U^k f\right\| \leq \frac{2}{n}\|g\| \to 0.$$

If $f \in \mathcal{N}$, then there is a sequence $\{g_i\}$ in $\mathcal{H}$ such that $f_i = g_i - Ug_i$ converges to f. Then

$$\left\|\frac{1}{n}\sum_{k=0}^{n-1} U^k f\right\| \leq \left\|\frac{1}{n}\sum_{k=0}^{n-1} U^k(f - f_i)\right\| + \left\|\frac{1}{n}\sum_{k=0}^{n-1} U^k f_i\right\|$$

for each i. Given $\varepsilon > 0$, choose i so that $\|f - f_i\| < \frac{1}{2}\varepsilon$, and choose n so that

$$\left\|\frac{1}{n}\sum_{k=0}^{n-1} U^k f_i\right\| < \frac{1}{2}\varepsilon;$$

then

$$\left\|\frac{1}{n}\sum_{k=0}^{n-1} U^k f\right\| < \frac{1}{2}\varepsilon + \frac{1}{2}\varepsilon = \varepsilon.$$

Now if $f \in \mathcal{H}$ is arbitrary, there is a unique $f_0 \in \mathcal{N}$ such that $f = f_0 + Pf$. Then

$$\left\|\frac{1}{n}\sum_{k=0}^{n-1} U^k f - Pf\right\| = \left\|\frac{1}{n}\sum_{k=0}^{n-1} U^k(f_0 + Pf) - Pf\right\|$$
$$= \left\|\frac{1}{n}\sum_{k=0}^{n-1} U^k f_0\right\| \to 0.$$

The proof is complete.

There are many generalizations and modifications of this theorem, of which we will consider one example. The following theorem, essentially due to K. Yosida (1938) (see also Kakutani 1938b and Yosida and Kakutani 1941) contains several basic ideas common to many of these extensions.

Theorem 1.3 *Mean Ergodic Theorem in Banach Space* Let $\mathcal{Y}$ be a Banach space and $T:\mathcal{Y} \to \mathcal{Y}$ a continuous linear map satisfying $\|T^k\| \leq M$ for all $k = 1, 2, \ldots$ for some constant M. Let $y \in \mathcal{Y}$. Then the following statements are equivalent:

(i) $\left\{\dfrac{1}{n}\sum_{k=0}^{n-1} T^k y : n = 1, 2, \ldots\right\}$ converges with respect to the norm on $\mathcal{Y}$.

(ii) $\left\{\dfrac{1}{n}\sum_{k=0}^{n-1} T^k y : n = 1, 2, \ldots\right\}$ has a weak cluster point.

(iii) T has a fixed point in the weakly closed convex hull of $\{T^k y : k = 0, 1, 2, \ldots\}$ (i.e. the intersection of all the weakly closed convex sets containing $\{T^k y : k = 0, 1, 2, \ldots\}$).

Proof. (i) $\Rightarrow$ (ii): Clear.

(ii) $\Rightarrow$ (iii): Abbreviating $A_n y = (1/n)\sum_{k=0}^{n-1} T^k y$, suppose that a subnet $A_{n_\alpha} y$ converges weakly to $\bar{y}$. Then for any continuous linear functional $\phi \in \mathcal{Y}^*$,

$$\langle T\bar{y}, \phi \rangle = \langle \bar{y}, T^*\phi \rangle = \lim \langle A_{n_\alpha} y, T^*\phi \rangle = \lim \langle T A_{n_\alpha} y, \phi \rangle$$
$$= \lim \langle A_{n_\alpha} y, \phi \rangle = \langle \bar{y}, \phi \rangle,$$

because

$$TA_n y - A_n y = \frac{T^{n+1} y - Ty}{n} \to 0.$$

Thus $T\bar{y} = \bar{y}$. Of course $\bar{y}$, being the weak limit of convex combinations of $\{T^k y\}$, is in the weakly closed convex hull of $\{T^k y\}$.

(iii) $\Rightarrow$ (i): Let $\bar{y}$ be the fixed point. Since $T^k y = T^k \bar{y} + T^k(y - \bar{y}) = \bar{y} + T^k(y - \bar{y})$, we have $A_n y = \bar{y} + A_n(y - \bar{y})$, and it is enough to prove that $A_n(y - \bar{y}) \to 0$ in $\mathcal{Y}$. Now $\bar{y}$ is a weak limit of convex combinations

$$S_\alpha y = \sum_{i=1}^{m_\alpha} \lambda_{i\alpha} T^i y, \quad \lambda_{i\alpha} \geq 0, \sum_i \lambda_{i\alpha} = 1.$$

This implies that

$$y - S_\alpha y = \sum_i \lambda_{i\alpha}(y - T^i y) = \sum_i \lambda_{i\alpha}(I - T)(I + T + T^2 + \ldots T^{i-1})y$$
$$\in \text{range}\,(I - T),$$

2.2. The pointwise ergodic theorem

so that $y - \bar{y}$ is in the weak closure of range $(I - T)$, hence also in $\overline{\text{range}\,(I - T)}$ (weak and strong closure coincide for subspaces – see Royden 1968, p. 201). Now it is easy to show that $A_n(y - \bar{y}) \to 0$. For, given $\varepsilon > 0$, choose $x \in \text{range}\,(I - T)$ with $\|x - (y - \bar{y})\| < \varepsilon$. Clearly $A_n x \to 0$, since $A_n(I - T) = (T - T^{n+1})/n$. Thus for large enough n,

$$\|A_n(y - \bar{y})\| \leq \|A_n x\| + \|A_n(x - (y - \bar{y}))\| < \varepsilon + M\varepsilon.$$

Remark 1.4 'Weakly closed' can be replaced by 'norm closed' in (iii) above, since in a locally convex topological vector space a convex set is strongly closed if and only if it is weakly closed (Royden 1968, p. 206).

2.2. The Pointwise Ergodic Theorem

We turn our attention now to Birkhoff's Ergodic Theorem, also known as the Individual (or Pointwise) Ergodic Theorem. The basic estimate needed to prove the a.e. convergence of the ergodic averages

$$\frac{1}{n}\sum_{k=0}^{n-1} f(T^k x)$$

is the content of the preliminary result known as the Maximal Ergodic Theorem, which concerns the maximum

$$f^*(x) = \sup_{n \geq 1} \frac{1}{n}\sum_{k=0}^{n-1} f(T^k x)$$

of these averages. Actually, an estimate on the lim sup rather than the sup is sufficient to prove the Pointwise Ergodic Theorem, and this was Birkhoff's method in his original proof. Later we will study the role of maximal inequalities in some detail. For now we prove the estimate by means of an idea that goes back to Kolmogorov (1937) and Yosida and Kakutani (1939) and most recently has been presented in an especially clear form by Katznelson and Ornstein.

Theorem 2.1 *Maximal Ergodic Theorem (Wiener 1939, Yosida and Kakutani 1939)* If $f \in L^1(X, \mathcal{B}, \mu)$, then

$$\int_{\{f^* > 0\}} f \,\mathrm{d}\mu \geq 0.$$

Remark: It follows then that also

$$\int_{\{f^* \geq 0\}} f \,\mathrm{d}\mu \geq 0,$$

since for each $\varepsilon > 0$ we have
$$0 \leq \int_{\{(f+\varepsilon)^* > 0\}} (f+\varepsilon) d\mu \leq \int_{\{f^* > -\varepsilon\}} f d\mu + \varepsilon,$$
and we can let ε decrease to 0 and apply a convergence theorem.
Proof: The set where $f^* > 0$ is the disjoint union of the sets
$$B_1 = \{x : f(x) > 0\}$$
$$B_2 = \{x : f(x) \leq 0, f(x) + f(Tx) > 0\}$$
$$\vdots$$
$$B_n = \{x : f(x) \leq 0, \ldots, f(x) + \ldots + f(T^{n-2}x) \leq 0,$$
$$f(x) + \ldots + f(T^{n-1}x) > 0\}$$
$$\vdots$$

If we can show that
$$\int_{B_1 \cup \ldots \cup B_n} f d\mu \geq 0 \text{ for all } n = 1, 2, \ldots,$$
then it will follow (by the Dominated Convergence Theorem applied to $\chi_{B_1 \cup \ldots \cup B_n} f$) that
$$\int_{\{f^* > 0\}} f d\mu \geq 0.$$

Let us fix an $n = 1, 2, \ldots$. The idea is to break $B_1 \cup \ldots \cup B_n$ into a union of *different* disjoint pieces, over each of which the integral of f will clearly be nonnegative. These pieces will be pictured as towers $B'_k \cup TB'_k \cup \ldots \cup T^{k-1} B'_k$.

In order to accomplish this, we make three observations:

(1) $T^k B_n \subset B_1 \cup \ldots \cup B_{n-k}$ for $k = 1, 2, \ldots, n-1$.

This is so because if $x \in B_n$, then
$$f(x) + f(Tx) + \ldots + f(T^{k-1}x) \leq 0,$$
while
$$f(x) + f(Tx) + \ldots + f(T^{k-1}x) + f(T^k x) + \ldots + f(T^{n-1}x) > 0,$$
so that we must have
$$f(T^k x) + \ldots + f(T^{n-1}x) > 0,$$
i.e.,
$$f(T^k x) + f(T(T^k x)) + \ldots + f(T^{n-k-1}(T^k x)) > 0.$$

(2) The sets $B_n, TB_n, \ldots, T^{n-1} B_n$ are pairwise disjoint. This is so because if $T^i B_n \cap T^j B_n \neq \emptyset$ for some $i < j$, then $B_n \cap T^{j-i} B_n \neq \emptyset$, contradicting (1).

2.2. The pointwise ergodic theorem

(3) If we let

$$B'_n = B_n, C_n = B_n \cup TB_n \cup \ldots \cup T^{n-1}B_n,$$
$$B'_{n-1} = B_{n-1} \setminus C_n, C_{n-1} = B'_{n-1} \cup TB'_{n-1} \cup \ldots \cup T^{n-2}B'_{n-1},$$
$$\vdots$$
$$B'_1 = B_1 \setminus (C_2 \cup \ldots \cup C_n), C_1 = B'_1,$$

then the 'columns' $C_1, C_2, \ldots, C_n$ are pairwise disjoint, and the levels $B'_k, TB'_k, \ldots, T^{k-1}B'_k$ within each column C_k are also pairwise disjoint. Thus the following picture is a correct representation of $B_1 \cup B_2 \cup \ldots \cup B_n$:

$B_1 \cup B_2 \cup \ldots \cup B_n$

$\quad\quad T^{n-1}B'_n$			
$\vdots$	$\quad\quad T^{n-2}B'_{n-1}$		
$\vdots$	$\vdots$	$\quad\quad T^{n-3}B'_{n-2}$	
$\vdots$	$\vdots$	$\vdots$	
$\quad\quad TB'_n$	$\quad\quad TB'_{n-1}$	$\quad\quad TB'_{n-2}$	
$\quad\quad B'_n = B_n$	$\quad\quad B'_{n-1} = B_{n-1}\setminus C_n$	$\quad\quad B'_{n-2} = B_{n-2}\setminus(C_n \cup C_{n-1})$ $\ldots$	$\quad\quad B'_1$
C_n	C_{n-1}	C_{n-2}	C_1

This is so because

(i) the pieces are in $B_1 \cup \ldots \cup B_n$, by (1);
(ii) each base is disjoint from all the columns to the left of it, by definition;
(iii) each base is disjoint from all the columns to the right of it, by definition (for the base) and by (1) (for the higher levels – since the images of the base to the right are contained in B_js even further to the right).

Thus, if two columns were to intersect, we could apply T^{-1} until one column intersected the base of the other; and we have seen that this does not happen.

Now it is a simple matter to make the estimate

$$\int_{B_1 \cup \ldots \cup B_n} f \, d\mu = \sum_{k=1}^{n} \int_{C_k} f \, d\mu = \sum_{k=1}^{n} \int_{B'_k \cup TB'_k \cup \ldots \cup T^{k-1}B'_k} f \, d\mu$$
$$= \sum_{k=1}^{n} \int_{B'_k} (f + fT + \ldots + fT^{k-1}) \, d\mu \geq 0,$$

since $f + fT + \ldots + fT^{k-1} > 0$, by definition, on $B_k \supset B'_k$.

Corollary 2.2 For each $\alpha \in \mathbb{R}$,
$$\int_{\{f^* > \alpha\}} f \, d\mu \geq \alpha \mu\{f^* > \alpha\}.$$
Proof: Letting $g = f - \alpha$, we see that $\{f^* > \alpha\} = \{g^* > 0\}$, so that
$$0 \leq \int_{\{g^* > 0\}} g \, d\mu = \int_{\{f^* > \alpha\}} (f - \alpha) d\mu,$$
and hence
$$\int_{\{f^* > \alpha\}} f \, d\mu \geq \alpha \mu\{f^* > \alpha\}$$

Now we are in a position to state and prove the fundamental theorem of ergodic theory, the Pointwise Ergodic Theorem (also known as just the Ergodic Theorem) of G. D. Birkhoff.

Theorem 2.3 The Ergodic Theorem (Birkhoff 1931) Let $(X, \mathscr{B}, \mu)$ be a probability space, $T: X \to X$ a m.p.t. and $f \in L^1(X, \mathscr{B}, \mu)$. Then

(1) $\lim_{n \to \infty} (1/n) \sum_{k=0}^{n-1} f(T^k x) = \bar{f}(x)$ exists a.e.;
(2) $\bar{f}(Tx) = \bar{f}(x)$ a.e.;
(3) $\bar{f} \in L^1$, and in fact $\|\bar{f}\|_1 \leq \|f\|_1$;
(4) if $A \in \mathscr{B}$ with $T^{-1}A = A$, then $\int_A f \, d\mu = \int_A \bar{f} \, d\mu$
    (this says that if $\mathscr{I}$ is the sub-σ-algebra of $\mathscr{B}$ consisting of all the T-invariant sets, then $\bar{f} = E(f|\mathscr{I})$ a.e.);
(5) $(1/n) \sum_{k=0}^{n-1} f T^k \to \bar{f}$ in L^1.

Remark 2.4 If X has infinite measure, the same statements hold, except that (4) and (5) apply only to invariant sets of finite measure.
Proof of the Ergodic Theorem: (1) For each $\alpha, \beta \in \mathbb{R}$ with $\alpha < \beta$, let
$$E_{\alpha,\beta} = \left\{ x \in X : \liminf_{n \to \infty} \frac{1}{n} \sum_{k=0}^{n-1} f(T^k x) < \alpha < \beta \right.$$
$$\left. < \limsup_{n \to \infty} \frac{1}{n} \sum_{k=0}^{n-1} f(T^k x) \right\}.$$

We will show that $\mu(E_{\alpha,\beta}) = 0$ for each α, β. Then the union over all rational α, β will also have measure 0, and hence the limit exists a.e.

Now $E_{\alpha,\beta}$ is an invariant subset of $\{f^* > \beta\}$, so by considering T restric-

2.2. The pointwise ergodic theorem

ted to $E_{\alpha,\beta}$, we see that the Maximal Ergodic Theorem implies that

$$\int_{E_{\alpha,\beta}} f\,\mathrm{d}\mu \geq \beta\mu(E_{\alpha,\beta}).$$

Next we consider $-f$. Since if $x \in E_{\alpha,\beta}$ there is an $n \geq 1$ with

$$\frac{1}{n}\sum_{k=0}^{n-1} f(T^k x) < \alpha,$$

we see that

$$E_{\alpha,\beta} \subset \{(-f)^* > -\alpha\}.$$

Hence, by the Maximal Ergodic Theorem,

$$\int_{E_{\alpha,\beta}} -f\,\mathrm{d}\mu \geq -\alpha\mu(E_{\alpha,\beta}),$$

or

$$\int_{E_{\alpha,\beta}} f\,\mathrm{d}\mu \leq \alpha\mu(E_{\alpha,\beta}).$$

Thus

$$\beta\mu(E_{\alpha,\beta}) \leq \int_{E_{\alpha,\beta}} f\,\mathrm{d}\mu \leq \alpha\mu(E_{\alpha,\beta}),$$

and, since $\alpha < \beta$, this is possible only if $\mu(E_{\alpha,\beta}) = 0$.

(2) It is clear that $\bar{f}T = \bar{f}$ a.e.

(3) We use Fatou's Lemma. Since

$$\left|\frac{1}{n}\sum_{k=0}^{n-1} fT^k\right| \leq \frac{1}{n}\sum_{k=0}^{n-1} |f|T^k,$$

we have $|\bar{f}| \leq \overline{|f|}$, and thus

$$\int |\bar{f}|\,\mathrm{d}\mu \leq \int \overline{|f|}\,\mathrm{d}\mu \leq \liminf_{n\to\infty} \int \frac{1}{n}\sum_{k=0}^{n-1} |f(T^k x)|\,\mathrm{d}\mu = \int |f|\,\mathrm{d}\mu < \infty.$$

(4) This is usually proved by means of the Maximal Ergodic Theorem as follows; we will also obtain this result as an immediate corollary of (5).

For each $n = 0, 1, 2, \ldots$ and $k = 0, \pm 1, \pm 2, \ldots$, let

$$A_{n,k} = \left\{x \in A : \frac{k}{2^n} \leq \bar{f}(x) < \frac{k+1}{2^n}\right\}.$$

The $A_{n,k}$ are invariant sets, and for each n we have $A = \bigcup_k A_{n,k}$.

Fix $\varepsilon > 0$. Then $f^* > k/2^n - \varepsilon$ on $A_{n,k}$, so

$$\int_{A_{n,k}} f\, d\mu \geq \left(\frac{k}{2^n} - \varepsilon\right)\mu(A_{n,k}).$$

Similarly, $(-f)^* > -(k+1)/2^n$ on $A_{n,k}$, so

$$\int_{A_{n,k}} -f\, d\mu \geq -\frac{k+1}{2^n}\mu(A_{n,k}).$$

Thus

$$\left(\frac{k}{2^n} - \varepsilon\right)\mu(A_{n,k}) \leq \int_{A_{n,k}} f\, d\mu \leq \frac{k+1}{2^n}\mu(A_{n,k}),$$

and letting $\varepsilon \to 0$ gives

$$\frac{k}{2^n}\mu(A_{n,k}) \leq \int_{A_{n,k}} f\, d\mu \leq \frac{k+1}{2^n}\mu(A_{n,k}).$$

Since obviously

$$\frac{k}{2^n}\mu(A_{n,k}) \leq \int_{A_{n,k}} \bar{f}\, d\mu \leq \frac{k+1}{2^n}\mu(A_{n,k}),$$

we have that

$$\left|\int_{A_{n,k}} f\, d\mu - \int_{A_{n,k}} \bar{f}\, d\mu\right| \leq \frac{1}{2^n}\mu(A_{n,k}).$$

Summing over k,

$$\left|\int_A f\, d\mu - \int_A \bar{f}\, d\mu\right| \leq \frac{1}{2^n}\mu(A),$$

and the conclusion follows upon letting $n \to \infty$.

(5) For bounded functions, the L^1 convergence would follow from the Bounded Convergence Theorem. The general case can then be proved by approximating by bounded functions (which are dense in L^1) and using (3). There is no problem in assuming that $f \geq 0$, since we may write $f = f^+ - f^-$ and deal with the two parts separately. If g is bounded and $0 \leq g \leq f$, then

$$\left\|\frac{1}{n}\sum_{k=0}^{n-1} fT^k - \bar{f}\right\|_1 \leq \left\|\frac{1}{n}\sum_{k=0}^{n-1} (fT^k - gT^k)\right\|_1$$
$$+ \left\|\frac{1}{n}\sum_{k=0}^{n-1} gT^k - \bar{g}\right\|_1 + \|\bar{g} - \bar{f}\|_1.$$

By (3), the third term is less than or equal to $\|g - f\|_1$, which can be made arbitrarily small by appropriate choice of g. Similarly, the first term is also less than or equal to $\|f-g\|_1$. But once g is fixed, the second term approaches 0 as $n \to \infty$, by the Bounded Convergence Theorem. This proves (5). (Notice that it is essential here that X have finite measure.)

We can also give in this vein an *alternative proof of* (4):

$$\left| \int_A f\,d\mu - \int_A \bar{f}\,d\mu \right| = \left| \int_A \left(\frac{1}{n} \sum_{k=0}^{n-1} fT^k - \bar{f} \right) d\mu \right|$$

$$\leq \int_A \left| \frac{1}{n} \sum_{k=0}^{n-1} fT^k - \bar{f} \right| d\mu = \left\| \frac{1}{n} \sum_{k=0}^{n-1} fT^k - \bar{f} \right\|_{L^1(A)} \to 0,$$

by (5).

Remark 2.5 It is not hard to extend the Ergodic Theorem from the case of point transformations to that of more general *Markov operators* (3.7) having subinvariant measures (see Foguel 1980 for the details).

Remark 2.6 The Maximal Ergodic Theorem (2.2.1) and parts (1), (2) and (4) of the Ergodic Theorem (2.2.3) hold under the weaker hypothesis that one of f^+ and f^- is in L^1 rather than $f \in L^1$.

Exercises
1. State and prove versions of the Maximal Ergodic Theorem and Pointwise Ergodic Theorem for one-parameter measure-preserving flows.
2. Prove Remark 2.6.
3. Show by example that (4) and (5) of the Ergodic Theorem need not hold if $\mu(X) = \infty$.
4. If $\mu(X) = \infty$ and $g \in L^\infty$, then $\{A_n g\}$ need not converge a.e., but it does converge in some sense.
5. Identify $\bar{f}(x)$ in case
   (a) $(X, \mathcal{B}, \mu, T)$ is $\mathcal{B}(p_0, p_1, \ldots, p_{n-1})$ and $f(\ldots x_{-1}, x_0, x_1 \ldots) = \chi_{\{i\}}(x_0)$.
   (b) $Tx = x + \alpha \bmod 1$, α irrational, and $f = \chi_I$ for some interval I.
   (c) $Tx = x + 1$ on $\mathbb{R}$, μ is Lebesgue measure, and $f \in L^1$.
6. Show that if $\sum_{k=0}^{n-1} f(T^k x) \to \infty$ a.e., then $\int_X f\,d\mu > 0$.

2.3. Recurrence

Let $(X, \mathcal{B}, \mu)$ be a probability space and $T: X \to X$ a measure-preserving transformation. In this section we discuss the problem of recurrence, the most basic question to be asked about the natures of orbits of points and measurable sets. We prove the recurrence theorems of

2. The fundamentals of ergodic theory

Poincaré, Khintchine, and Halmos and define the induced transformation on a set of positive measure.

The mathematically simple Poincaré Recurrence Theorem has become famous because of its physical and philosophical implications, some of which we will indicate below. It may be considered to be the most basic result in ergodic theory.

Definition 3.1 Let $B \in \mathscr{B}$. A point $x \in B$ is said to be *recurrent with respect to B* if there is a $k \geq 1$ for which $T^k x \in B$.

Theorem 3.2 *Poincaré Recurrence Theorem* (1899) For each $B \in \mathscr{B}$, almost every point of B is recurrent with respect to B.
Proof: Let F be the set of all those points of B which are not recurrent with respect to B; then

$$F = B - \bigcup_{k=1}^{\infty} T^{-k}B = B \cap T^{-1}(X - B) \cap T^{-2}(X - B) \cap \ldots .$$

Note that if $x \in F$, then $T^n x \notin F$ for each $n \geq 1$. Thus $F \cap T^{-n}F = \varnothing$ for $n \geq 1$, and hence $T^{-k}F \cap T^{-(n+k)}F = \varnothing$ for each $n \geq 1$ and each $k \geq 0$. Then the sets $F, T^{-1}F, T^{-2}F, \ldots$ are pairwise disjoint and each has measure $\mu(F)$. Since $\mu(X) < \infty$, $\mu(F)$ must be zero.

One may make the following physical interpretation of this result. As usual the space X is supposed to include all the possible states of some physical system, the σ-algebra $\mathscr{B}$ to consist of all *observable* states of the system, and the measure μ to specify the probability of each observable state. The physical system is evolving with respect to a discrete time (i.e., we make measurements on it, say, once a second), and $T: X \to X$ is the transformation which carries each state of the system into its succeeding state. In a situation of equilibrium, the transformation T preserves the measure μ, so that the probability of an observable state does not change with the time.

Under this interpretation Poincaré's Recurrence Theorem says that if at time zero the physical system is in some observable state B, then almost surely the system will eventually return to this observable state. For example, if we insert a partition in a box and pump out all the air on one side of the partition, our system of air molecules is in a state for which all the molecules are in half of the box. Poincaré's Recurrence Theorem asserts that if we remove the partition and wait long enough, the molecules will almost surely once again congregate in their original half of the box. The fact that this theorem implies that such an unlikely event will *almost*

2.3. Recurrence

surely happen has worried many people for a long time. Zermelo pointed out the apparent incompatibility of this prediction with such important conclusions of thermodynamics as the Second Law and Boltzmann's *H*-Theorem. Apparently much of the controversy is based on various misunderstandings, such as the fact that the *H*-Theorem involves not a single orbit (or history) but a quantity calculated by performing an average over all possible initial conditions. We also face the difficulty that the expected return times are even larger than enormous. Consider the following simple example, due to the Ehrenfests (1957), which involves only a relatively small number of particles.

Let us suppose that there are two urns, one of them containing 100 balls numbered from 1 to 100, and the other being empty. There is also a hat containing 100 slips of paper numbered from 1 to 100. Once a second we draw a slip of paper from the hat, read the number on it, replace it in the hat, and move the ball bearing that number from whichever urn it is in to the other urn. According to the Second Law of Thermodynamics, as well as our naive intuition, the system will settle down towards the statistical equilibrium state in which there are 50 balls in each urn. Of course there will continue to be random fluctuations about the 50–50 division, but it appears highly unlikely that a fluctuation could be so large that all 100 balls would return to the urn from which they started. The Recurrence Theorem says that although such a fluctuation may be highly unlikely, still it will occur almost surely.

The Recurrence Theorem really applies to this situation because this sequence of experiments can be represented by a *Markov shift*. We can describe the state of our physical system at a given time $k = 0, 1, 2 \ldots$ by specifying the number $\omega_k \in \{0, 1, 2, \ldots, 100\}$ of balls in the first urn. If at time $k = 0$ there are ω_0 balls in the first urn and we proceed to draw numbers from the hat and so forth, the system passes through the states $\omega_0, \omega_1, \omega_2, \ldots$ in succession, subject to the conditions

(1)
$$\omega_k \in \{0, 1, \ldots, 100\}$$
$$|\omega_k - \omega_{k+1}| = 1$$

for all $k = 0, 1, 2, \ldots$. Matters can be simplified considerably if we assume that the experiment began infinitely long ago and continues infinitely far into the future.

Let $\Omega = \prod_{i=-\infty}^{\infty} \{0, 1, 2, \ldots, 100\}$, let $\sigma : \Omega \to \Omega$ be the shift transformation defined by $(\sigma\omega)(n) = \omega(n+1)$ for $n \in \mathbb{Z}$, and let $\mathscr{B}$ be the σ-algebra generated by the finite segments of histories

$$B_k(i_1, i_2, \ldots, i_n) = \{\omega \in \Omega : \omega(j+k) = i_j, j = 1, 2, \ldots, n\}.$$

2. The fundamentals of ergodic theory

We wish now to construct a measure μ on $(\Omega, \mathscr{B})$ such that $\mu(B_k(i_1,\ldots,i_n))$ gives the probability of observing the sequence $(i_1,\ldots,i_n)$ of successive states of the system beginning at time k. The a priori probability p_i of observing the state i at time k is independent of k and is

$$p_i = \frac{1}{2^{100}} \binom{100}{i}.$$

Let p denote the vector $(p_0, p_1, \ldots, p_{100})$.

For each $i, j = 0, 1, \ldots, 100$, let $a_{i,j}$ denote the conditional probability of the state j given the state i. If we are in state i, then there are i balls in the first urn and $100-i$ in the second. Drawing a number m from the hat, the probability that ball number m is in the first (second) urn is $i/100$ $((100-i)/100)$. Now from state i we can only move to state $i+1$ or $i-1$. Thus

$$a_{i,i-1} = \frac{i}{100}$$

$$a_{i,i+1} = \frac{100-i}{100}$$

$$a_{i,j} = 0 \text{ if } j \text{ is neither } i+1 \text{ nor } i-1.$$

We have the *transition matrix*

$$A = \begin{bmatrix} 0 & 1 & 0 & 0 & 0 & \cdots \\ \frac{1}{100} & 0 & \frac{99}{100} & 0 & 0 & \cdots \\ 0 & \frac{2}{100} & 0 & \frac{98}{100} & 0 & \cdots \\ 0 & 0 & \frac{3}{100} & 0 & \frac{97}{100} & \cdots \\ \vdots & \vdots & \vdots & \vdots & \vdots & \end{bmatrix}$$

Then the probability of observing a sequence $(i_1,\ldots,i_n)$ beginning at time k is independent of k and is given by

$$\mu(B_k(i_1,\ldots,i_n)) = p_{i_1} a_{i_1,i_2} \cdots a_{i_{n-1},i_n}$$

Since $pA = p$, the conditions of Kolmogorov's Theorem are satisfied and μ extends to a measure on all of $\mathscr{B}$ such that $\sigma: X \to X$ is a m.p.t.

Let $E = \{\omega \in \Omega : \omega(0) = 0\}$, so E consists of those performances of the experiment in which we are interested. The measure of E is $p_0 > 0$. Accord-

2.3. Recurrence

ing to the Recurrence Theorem,

$$\{\omega \in E : \text{there is } k > 0 \text{ such that } \sigma^k \omega \in E\}$$

has measure equal to $\mu(E)$. That is, with probability 1 the system will recur again and again to the state in which there are no balls in the first urn. Of course, the expected time is astronomical: by Kac's Theorem (2.4.6), it is $1/\mu(E) = 2^{100}$ drawings. For real systems with millions of particles, the expected return times will be even more absurd.

For recurrence of *sets* a sharper, quantitative statement can be made. A set $E \subset \mathbb{Z}$ is called *relatively dense* if it has *bounded gaps*, in that there is a positive integer K such that

$$E \cap \{j, j+1, \ldots, j+K-1\} \neq \emptyset \text{ for each } j \in \mathbb{Z}.$$

The following theorem says that the images of a measurable set under the iterates of T come back and overlap the set fairly regularly – namely, with bounded gaps.

Theorem 3.3 (Khintchine 1934) For any $B \in \mathcal{B}$ and any $\varepsilon > 0$, the set

$$E_\varepsilon = \{k \in \mathbb{Z} : \mu(T^k B \cap B) \geq \mu(B)^2 - \varepsilon\}$$

is relatively dense.

Proof: We direct our attention to the Hilbert space $L^2(X, \mathcal{B}, \mu)$ and apply the Mean Ergodic Theorem. Let $U : L^2 \to L^2$ be the unitary operator induced by T. If $B \in \mathcal{B}$ and χ_B is the characteristic function of B, then $\chi_{T^k B} = U^{-k} \chi_B$. Thus

$$\mu(T^k B \cap B) = \int_X \chi_B \chi_{T^k B} \, d\mu = \int_X \chi_B U^{-k} \chi_B \, d\mu = (\chi_B, U^{-k} \chi_B).$$

Applying the Mean Ergodic Theorem to $f = \chi_B$, given $\varepsilon > 0$ there is an n such that

$$\left\| \frac{1}{n} \sum_{k=0}^{n-1} U^k f - Pf \right\| < \frac{\varepsilon}{\|f\| + 1},$$

and hence for each $j \in \mathbb{Z}$

$$\left\| \frac{1}{n} \sum_{k=0}^{n-1} U^{k+j} f - Pf \right\| < \frac{\varepsilon}{\|f\| + 1}, \text{ or}$$

$$\left\| \frac{1}{n} \sum_{k=j}^{n+j-1} U^k f - Pf \right\| < \frac{\varepsilon}{\|f\| + 1} \text{ for all } j \in \mathbb{Z}.$$

Because of the orthogonal decomposition $L^2 = \mathcal{M} \oplus \mathcal{N}$ in the Mean

Ergodic Theorem, $(Pf, Pf) = (Pf, f)$; also, since $1 \in \mathcal{M}$, $(f, 1) = (Pf, 1)$; consequently

$$(f, 1)^2 = (Pf, 1)^2 \leq \|Pf\|^2 (1,1) = (Pf, Pf) = (Pf, f).$$

Then

$$\left| \left(\frac{1}{n} \sum_{k=j}^{n+j-1} U^k f - Pf, f \right) \right| \leq \left\| \frac{1}{n} \sum_{k=j}^{n+j-1} U^k f - Pf \right\| \|f\| < \varepsilon \text{ for all } j \in \mathbb{Z},$$

so

$$\frac{1}{n} \sum_{k=j}^{n+j-1} \mu(T^k B \cap B) = \frac{1}{n} \sum_{k=j}^{n+j-1} (U^k f, f) \geq (Pf, f) - \varepsilon \geq (f, 1)^2 - \varepsilon$$

$$= \mu(B)^2 - \varepsilon \text{ for all } j \in \mathbb{Z}.$$

Thus the interval $[j, n+j-1]$ must contain at least one k for which $\mu(T^k B \cap B) \geq \mu(B)^2 - \varepsilon$, and this shows that E_ε is relatively dense.

One may study recurrence also in more general situations, even where there is no finite invariant measure and where T is merely a set transformation rather than a point transformation. Let $(X, \mathcal{B})$ be a measurable space, $\mathcal{I} \subset \mathcal{B}$ a σ-ideal with $T^{-1} \mathcal{I} \subset \mathcal{I}$, and $T: X \to X$ a (possibly noninvertible) measurable transformation. For each $B \in \mathcal{B}$, if

$$B^* = \bigcup_{k=1}^{\infty} T^{-k} B,$$

then $B \setminus B^*$ is the set of all those points of B which never return to B. We say that the transformation T is *recurrent* in case $B \setminus B^* \in \mathcal{I}$ for all $B \in \mathcal{B}$.

For each $B \in \mathcal{B}$,

$$B \setminus \bigcap_{n=0}^{\infty} \bigcup_{j \geq n} T^{-j} B$$

is the set of all those points of B which do not return to B infinitely many times. If for each $B \in \mathcal{B}$ this set is in $\mathcal{I}$, then we say that T is *infinitely recurrent*. Poincaré's Recurrence Theorem implies that a m.p.t. on a finite measure space is infinitely recurrent, where $\mathcal{I}$ is the σ-ideal of sets of measure zero.

A set $B \in \mathcal{B}$ is called *wandering* if the sets $B, T^{-1}B, T^{-2}B, \ldots$ are pairwise disjoint. T is called *conservative* if every wandering set is in $\mathcal{I}$. Since $B \setminus B^*$ is wandering for each $B \in \mathcal{B}$, it is true that T conservative implies T recurrent.

Finally, T is called *incompressible* if whenever $B \in \mathcal{B}$ and $T^{-1}B \subset B$, then $B \setminus T^{-1}B \in \mathcal{I}$.

2.3. Recurrence

Theorem 3.4 (*Halmos* 1947) Let X, $\mathscr{B}$, $\mathscr{I}$, and T be as above. Then the following statements are equivalent: (1) T is incompressible. (2) T is recurrent. (3) T is conservative. (4) T is infinitely recurrent.

Proof (*Wright* 1961b): For each $B \in \mathscr{B}$, continue to let $B^* = \bigcup_{k=1}^{\infty} T^{-k} B$. Note that each set $B \backslash B^*$ is wandering, and conversely if B is wandering then $B \cap B^* = \varnothing$ so $B = B \backslash B^*$.

(1) $\Rightarrow$ (2): Let $B \in \mathscr{B}$ and $E = B \cup B^*$. Then $T^{-1} E = B^* \subset E$, so $E \backslash T^{-1} E \in \mathscr{I}$. But $E \backslash T^{-1} E = (B \cup B^*) \backslash B^* = B \backslash B^*$, so $B \backslash B^* \in \mathscr{I}$.

(2) $\Rightarrow$ (3): If B is a wandering set, then $B = B \backslash B^* \in \mathscr{I}$, since T is recurrent.

(3) $\Rightarrow$ (1): Suppose $B \in \mathscr{B}$ and $T^{-1} B \subset B$. Then $B^* = T^{-1} B$, so $B \backslash T^{-1} B = B \backslash B^* \in \mathscr{I}$, since $B \backslash B^*$ is wandering and T is conservative.

(4) $\Rightarrow$ (2): Obvious.

(2) $\Rightarrow$ (4): If T is recurrent and $B \in \mathscr{B}$, then $B \backslash B^* \in \mathscr{I}$. We need to show that $B \backslash \bigcap_{n=0}^{\infty} T^{-n} B^* \in \mathscr{I}$. Now

$$B \backslash \bigcap_{n=0}^{\infty} T^{-n} B^* = B \cap \left(X \backslash \bigcap_{n=0}^{\infty} T^{-n} B^* \right) = B \cap \bigcup_{n=0}^{\infty} (X \backslash T^{-n} B^*)$$
$$= B \cap \left[(X \backslash B^*) \cup (X \backslash T^{-1} B^*) \cup (X \backslash T^{-2} B^*) \cup \ldots \right]$$
$$= B \cap \left[(X \backslash B^*) \cup (B^* \backslash T^{-1} B^*) \cup (T^{-1} B^* \backslash T^{-2} B^*) \cup \ldots \right]$$
$$= (B \backslash B^*) \cup \left[B \cap \bigcup_{n=0}^{\infty} (T^{-n} B^* \backslash T^{-(n+1)} B^*) \right].$$

Now $T^{-1} T^{-n} B^* = T^{-(n+1)} B^* \subset T^{-n} B^*$ and T is recurrent, hence incompressible, so $T^{-n} B^* \backslash T^{-(n+1)} B^* \in \mathscr{I}$ for each $n = 0, 1, 2, \ldots$ Consequently $B \backslash \bigcap_{n=0}^{\infty} T^{-n} B^* \in \mathscr{I}$ and T is infinitely recurrent.

The phenomenon of recurrence makes possible an interesting construction of m.p.t.s which is very useful for producing examples with a wide variety of properties. Let $T: X \to X$ be a m.p.t. and $A \subset X$ a measurable subset of X of positive measure. The integer

$$n_A(x) = \inf \{ n \geq 1 : T^n x \in A \}$$

is defined (and finite) for a.e. $x \in A$. A is a probability space with measure $\mu_A = \mu / \mu(A)$. We define the *induced* or *derivative transformation* $T_A : A \to A$ by

$$T_A(x) = T^{n_A(x)} x$$

for a.e. $x \in A$ (Kakutani 1943). Of course n_A and T_A are measurable, and in fact T_A *is a measure-preserving transformation*. This is all checked most easily with the aid of the diagram below, in which

$$A_n = \{ x \in A : n_A(x) = n \}.$$

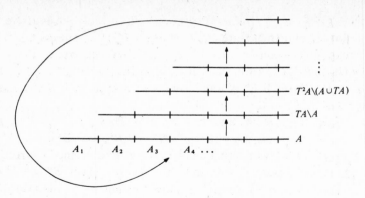

The action of T on this part of X is described as follows: a point not on the top floor is moved directly up, and the top floor is moved back down to the first floor. Under T_A, each point of A is mapped to the point where it first returns to A.

This idea of a derivative transformation can also be reversed, so as to define an *induced* or *primitive transformation* $\tilde{T}\colon \tilde{X} \to \tilde{X}$ on a space $\tilde{X}$ larger than X. For example, if $A \subset X$ is a measurable set with positive measure, let A' be a distinct copy of A. The disjoint union $\tilde{X} = X \cup A'$ is made a probability space in the obvious way. We think of A' as sitting over A.

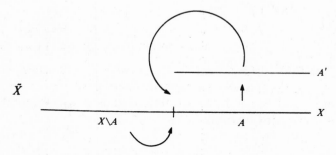

The action of $\tilde{T}$ is described as follows. Since A' is a copy of A, there is a one-to-one onto map $x \to x'$ from A to A'. We let

$$\tilde{T}x = \begin{cases} Tx & \text{if } x \in X \setminus A \\ x' & \text{if } x \in A \\ Ty & \text{if } x \in A' \text{ and } x = y' \text{ for } y \in A. \end{cases}$$

The action of $\tilde{T}$ is described in the same way as that of T above. Again $\tilde{T}$ is a *measure-preserving transformation*. Of course this process can be repeated so as to construct skyscrapers with $3, 4, \ldots, \infty$ floors as well.

2.4. Ergodicity

Exercises

1. Let $T: X \to X$ be an invertible, measurable, *nonsingular* transformation on a σ-finite measure space $(X, \mathscr{B}, m)$, in that T preserves the σ-ideal of null sets of m.

   A set $W \in \mathscr{B}$ of positive measure is called *weakly wandering* if there is a sequence $n_k \to \infty$ such that the sets $T^{n_k} W$ are all pairwise disjoint. Show that if T has a weakly wandering set, then there does not exist a finite invariant measure equivalent to m. (The converse, due to Hajian and Kakutani (1964), is also true.)

2. Let T be as in (1). Show that X has a decomposition into disjoint, measurable, invariant *conservative and dissipative parts*, $X = C \cup D$, in the following sense: $T|C$ is conservative and $D = \{T^n W : n \in \mathbb{Z}\}$ for some wandering set W. (Hint: We may assume that $m(X) < \infty$. Take W_n with $m(W_n) \nearrow \sup\{m(W): W \text{ is wandering}\}$ and let C be the complement of the invariant set generated by the W_n.)

3. Let T be as in (1). Show that T is conservative if and only if $\sum_{k=0}^{\infty} uT^k$ takes only the two values 0 and ∞ a.e. for each nonnegative $u \in L^\infty(X, \mathscr{B}, m)$.

4. Verify that the induced transformations T_A and $\tilde{T}$ really are m.p.t.s when X is.

5. Describe the action of T_A in case
   (a) $X = [0, 1)$, $Tx = x + \alpha \pmod 1$, and $A = [0, \tfrac{1}{2})$,
   (b) $X = \{0, 1\}^{\mathbb{Z}}$, σ is the shift, and $A = \{x: x(0) = 0\}$.

6. Prove directly (i.e. without using 3.3 or the Mean or Pointwise Ergodic Theorems) that if T is a m.p.t. on a finite measure space and $\mu(A) > 0$, then $\{n \geq 1 : \mu(T^{-n} A \cap A) > 0\}$ has bounded gaps. (Hint: If not, there is a sequence $\{n_i\}$ such that the sets $T^{-n_i} A$ are essentially disjoint.)

7. If $\mu(A) > 0$, then almost every point of A returns to A with a positive limiting frequency. Need almost every point of A return to A with bounded gap?

2.4. Ergodicity

From an abstract point of view, the qualitative ideas of recurrence discussed in the preceding section lead naturally to quantitative considerations. Granted that points almost surely return to subsets of positive measure, how often do these recurrences take place, and how much time does (the orbit of) each point spend in the vicinity of a given subset? Considering again the example of the Ehrenfests, according to intuition

2. The fundamentals of ergodic theory

the two-urn system should spend most of its time near the highest probability situation (an even division), with only very infrequent and brief diversions to low-probability states. Thus the high probability states at least *seem* attractive and stable, with high probability and on a reasonable time scale.

Historically and physically the idea of ergodicity came from Boltzmann's *ergodic hypothesis*. It was desirable that the time mean of a physical variable should coincide with its space mean, i.e. that the long-term time average along a single history (or orbit) should equal the average over all possible initial conditions, or, equivalently, the average at any other single moment over all possible histories. (The word 'ergodic' comes from the Greek for 'energy path'.) In order to arrive at this conclusion, Boltzmann hypothesized that each orbit (under the action of $\mathbb{R}$) filled out *all* of the phase space (in his case a surface of constant energy). It did not take long to realize that such a condition was topologically impossible, and so it was replaced by the *quasi-ergodic hypothesis*: each orbit is dense in each surface of constant energy. (This is *minimality*; see Section 4.2). However, in spite of many attempts to prove that it did, this weaker hypothesis does not imply the equality of space means and time means. (This follows from the fact that a minimal system need not be uniquely ergodic – see Section 4.2 –, as was first shown by Markov – see Nemytskii and Stepanov 1960.)

In this section we will establish several correct and mathematically easy necessary and sufficient conditions for the equality of space means and time means. The most basic of these is *metric indecomposability*; we will in fact take this as the definition of *ergodicity*. Physicists have complained that none of these conditions is easy to verify in practice. Indeed, it was only recently (1963) that Sinai was able to prove the ergodicity of (a simplified version of) the hard sphere gas, the most fundamental system of interest in statistical mechanics. Practical difficulties aside, the property is interesting in itself and worthy of close examination; and since every system can be decomposed into ergodic systems (see Jacobs 1960 pp 85 ff) the assumption of ergodicity, if necessary, is frequently harmless.

Recall that a set $B \in \mathscr{B}$ is called *invariant* if $\mu(T^{-1}B \triangle B) = 0$. The transformation T (or, more properly, the system $(X, \mathscr{B}, \mu, T)$) is called *ergodic* or *metrically transitive* if every invariant set has measure 0 or 1.

Proposition 4.1 $(X, \mathscr{B}, \mu, T)$ is ergodic if and only if every invariant ($f \circ T = f$ a.e.) measurable function on X is constant a.e.

Proof: Suppose every invariant measurable function is constant a.e., and

2.4. Ergodicity

let $E \in \mathscr{B}$ be an invariant set. Then χ_E is constant a.e., so χ_E is either 0 a.e. or 1 a.e. Thus $\mu(E)$ is 0 or 1.

Conversely, suppose $(X, \mathscr{B}, \mu, T)$ is ergodic and f is an invariant measurable function. Then for each $r \in R$, $E_r = \{x \in X : f(x) > r\}$ is measurable and invariant, hence has measure 0 or 1. But if f is not constant a.e., there exists an $r \in R$ such that $0 < \mu(E_r) < 1$. Therefore f must be constant a.e.

Theorem 4.2 $(X, \mathscr{B}, \mu, T)$ is ergodic if and only if 1 is a simple eigenvalue of the transformation U induced on $L^2(X, \mathscr{B}, \mu)$ (complex) by T.

Moreover, if $(X, \mathscr{B}, \mu, T)$ is ergodic, then every eigenvalue of U is simple and the set of all eigenvalues of U is a subgroup of the circle group $\mathbb{K} = \{z \in \mathbb{C} : |z| = 1\}$.

Conversely, given a subgroup G of the circle group, there is an ergodic system $(X, \mathscr{B}, \mu, T)$ such that G is the group of eigenvalues of the unitary operator on $L^2(X, \mathscr{B}, \mu)$ determined by T.

Proof: 1 is always an eigenvalue of U, since constant functions are invariant. For 1 to be a simple eigenvalue means that $Uf = f$ implies f is constant a.e. Thus the first statement is obvious.

Since U is unitary, every eigenvalue of U has absolute value 1. (For if $Uf = \lambda f$, then $f = \lambda U^* f$, so $(Uf, f) = \lambda (f, f)$ and $(Uf, f) = (f, U^* f) = 1/\bar{\lambda} (f, f)$. Thus $\lambda \bar{\lambda} = 1$.) If f is an eigenfunction with eigenvalue λ, then $U|f| = |Uf| = |\lambda f| = |f|$, so $|f|$ is a nonzero constant a.e., since T is ergodic. Therefore $f \neq 0$ a.e. Now if g is another eigenfunction with eigenvalue η, then $U(g/f) = (\eta/\lambda)(g/f)$, so g/f is an eigenfunction with eigenvalue η/λ; therefore the eigenvalues of U form a subgroup of $\mathbb{K}$. If we take $\eta = \lambda$, we see that g/f is constant a.e.; therefore each eigenvalue is simple.

Finally, suppose we are given a subgroup G of $\mathbb{K}$. If we give G the discrete topology, then its character group $\hat{G}$ is a compact abelian group and hence has a unique normalized Haar measure μ. Define $\phi \in \hat{G}$ by $\phi(g) = g$ for all $g \in G$, and $T : \hat{G} \to \hat{G}$ by $T\gamma = \phi \cdot \gamma$. Equipping $\hat{G}$ with the σ-algebra $\mathscr{B}$ of Borel sets, we obtain a probability space $(\hat{G}, \mathscr{B}, \mu)$ and a m.p.t. $T : \hat{G} \to \hat{G}$.

Note that if $f : \hat{G} \to \mathbb{K}$ is a character of $\hat{G}$, then for each $\gamma \in \hat{G}$, $(Uf)(\gamma) = f(\phi \gamma) = f(\phi) f(\gamma)$, so f is an eigenfunction of U with eigenvalue $f(\phi)$. We will see that all eigenfunctions (up to constant multiples) arise in this way, and that each eigenvalue is simple.

The map $G \to \hat{\hat{G}}$ according to $g \to f_g$, where $f_g(\gamma) = \gamma(g)$, is well known to be an isomorphism onto. The eigenvalue corresponding to $f_g \in \hat{\hat{G}}$ is $f_g(\phi) = \phi(g) = g$. Thus G is contained in the group of eigenvalues of U.

Now the elements of $\hat{\hat{G}}$ form a complete orthonormal set in $L^2(\hat{G}, \mathcal{B}, \mu)$. Thus, given an eigenfunction f with eigenvalue λ, we have

$$f = \sum_{g \in G} a_g f_g \quad (a_g \in \mathbb{C}),$$

and thus

$$f(\phi\gamma) = \sum_{g \in G} a_g f_g(\phi\gamma) = \sum a_g f_g(\phi) f_g(\gamma) = \sum a_g g\gamma(g)$$
$$= \lambda f(\gamma) = \sum_{g \in G} \lambda a_g f_g(\gamma) = \sum \lambda a_g \gamma(g),$$

so $a_g g = \lambda a_g$ for all $g \in G$, and hence $a_g = 0$ whenever $g \neq \lambda$. Therefore $g = a_\lambda f_\lambda$, and we see that each eigenfunction is a constant multiple of a character of $\hat{G}$. The foregoing also shows that the group of eigenvalues coincides with G and each eigenvalue is simple.

Remark 4.3 A system $(X, \mathcal{B}, \mu, T)$, such as the system $(\hat{G}, \mathcal{B}, \mu, T)$ constructed in the preceding proof, which has the property that the eigenfunctions of the induced unitary operator U on $L^2(X, \mathcal{B}, \mu)$ span $L^2(X, \mathcal{B}, \mu)$, is said to have *discrete spectrum*.

Theorem 4.4 $(X, \mathcal{B}, \mu, T)$ is ergodic if and only if for each $f \in L^1(X, \mathcal{B}, \mu)$ the time mean of f equals the space mean of f a.e.:

$$\bar{f}(x) = \lim_{n \to \infty} \frac{1}{n} \sum_{k=0}^{n-1} f(T^k x) = \int_X f \, d\mu \quad \text{a.e.}$$

Proof: Suppose $(X, \mathcal{B}, \mu, T)$ is ergodic. Then $\bar{f}$, being invariant, must be constant a.e. Moreover, from the Ergodic Theorem we also have

$$\int_X f \, d\mu = \int_X \bar{f} \, d\mu = \bar{f}(x) \quad \text{a.e.}$$

Conversely, suppose that for each $f \in L^1(X, \mathcal{B}, \mu)$, $\bar{f}$ is constant a.e. If f is an invariant function in $L^1(X, \mathcal{B}, \mu)$, then

$$\frac{1}{n} \sum_{k=0}^{n-1} f(T^i x) = f(x) \quad \text{a.e.,}$$

so $f = \bar{f}$ a.e. Thus f is constant a.e., and Proposition 2.4.1 implies that $(X, \mathcal{B}, \mu, T)$ is ergodic.

For a measurable set E and a point $x \in X$, we define the *mean sojourn time* of x in E to be

$$\lim_{n \to \infty} \frac{1}{n} \sum_{k=0}^{n-1} \chi_E(T^k x) = \bar{\chi}_E(x).$$

2.4. Ergodicity

If T is ergodic, then $\bar{\chi}_E = \mu(E)$ a.e. Conversely, if the preceding theorem holds for characteristic functions of measurable sets, then, for an invariant set $E \in \mathscr{B}$, $\bar{\chi}_E = \chi_E$ a.e. is the constant $\mu(E)$ a.e., so $\mu(E)$ must be 0 or 1. Thus *T is ergodic if and only if the mean sojourn time in a measurable set equals the measure of the set for almost all points of X*. The theory of Krylov and Bogoliouboff (1937) (see also Oxtoby 1952) reverses this situation by starting with a measurable transformation and trying to define an ergodic invariant measure by

$$\mu(E) = \lim_{n \to \infty} \frac{1}{n} \sum_{k=0}^{n-1} \chi_E(T^k x)$$

when the limit exists.

Proposition 4.5 $(X, \mathscr{B}, \mu, T)$ is ergodic if and only if for each $f, g \in L^2(X, \mathscr{B}, \mu)$ we have

$$\lim_{n \to \infty} \frac{1}{n} \sum_{k=0}^{n-1} (U^k f, g) = (f, 1)\overline{(g, 1)}.$$

Proof: If $(X, \mathscr{B}, \mu, T)$ is ergodic, then

$$\lim_{n \to \infty} \frac{1}{n} \sum_{k=0}^{n-1} (U^k f, g) = (\bar{f}, g) = \left(\int_X f \, d\mu, g \right)$$
$$= ((f, 1), g) = (f, 1)(1, g) = (f, 1)\overline{(g, 1)}.$$

Conversely, given $f \in L^2(X, \mathscr{B}, \mu) \subset L^1(X, \mathscr{B}, \mu)$, suppose $(\bar{f}, g) = (f, 1)\overline{(g, 1)}$ for all $g \in L^2(X, \mathscr{B}, \mu)$. Then we must have $\bar{f} = (f, 1) = \int_X f \, d\mu$ a.e., so T is ergodic by the preceding theorem.

In case $T: X \to X$ is ergodic, the Kakutani skyscraper picture we associated with an induced transformation $T_A: A \to A$ on a subset of positive measure is actually a picture of the action of T on (almost) all of X, since almost every point of X must eventually enter A.

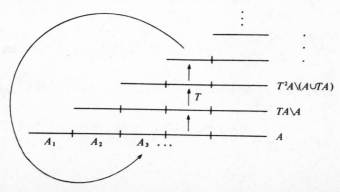

46   2. *The fundamentals of ergodic theory*

Here $n_A(x) = \inf\{n > 0 : T^n x \in A\}$ and $A_n = \{x \in A : n_A(x) = n\}$. This diagram makes possible a calculation of the *expected recurrence time*

$$\frac{1}{\mu(A)} \int_A n_A \, d\mu$$

of a point of A to A: it is $1/\mu(A)$. We give three proofs of this interesting fact (partly because the question of expected *multiple* recurrence times – see 4.3 – is open).

Theorem 4.6 (*Kac* 1947) Let T be an ergodic m.p.t. on a Lebesgue space $(X, \mathscr{B}, \mu)$. If $A \in \mathscr{B}$ with $\mu(A) > 0$, then

$$\int_A n_A \, d\mu = 1.$$

Proofs: (1) (Wright 1961a) Referring to the diagram above,

$$\int_A n_A \, d\mu = \sum_{n=1}^{\infty} n\mu(A_n) = \mu(X) = 1,$$

since the column $A_n \cup TA_n \cup \ldots \cup T^{n-1}A_n$ consists of n pieces, each of measure $\mu(A_n)$, and the union of the columns is X.

(2) (Kac 1947) For each $k = 0, 1, 2, \ldots$ let $\tilde{A}_k = X \setminus T^{-k}A$. Define

$$m_0 = 1 \quad \text{and}$$
$$m_k = \mu(\tilde{A}_0 \cap \tilde{A}_1 \cap \ldots \cap \tilde{A}_{k-1}) \quad \text{for } k \geq 1.$$

Then for each $k \geq 0$,

$$\mu(A \cap \tilde{A}_1 \cap \ldots \cap \tilde{A}_k \cap T^{-(k+1)}A)$$
$$= \mu(A \cap \tilde{A}_1 \cap \ldots \cap \tilde{A}_k) - \mu(A \cap \tilde{A}_1 \cap \ldots \cap \tilde{A}_k \cap \tilde{A}_{k+1})$$
$$= \mu(\tilde{A}_1 \cap \ldots \cap \tilde{A}_k) - \mu(\tilde{A}_0 \cap \ldots \cap \tilde{A}_k)$$
$$\quad - [\mu(\tilde{A}_1 \cap \ldots \cap \tilde{A}_{k+1}) - \mu(\tilde{A}_0 \cap \ldots \cap \tilde{A}_{k+1})]$$

(and, because T is measure-preserving)

$$= \mu(\tilde{A}_0 \cap \ldots \cap \tilde{A}_{k-1}) - \mu(\tilde{A}_0 \cap \ldots \cap \tilde{A}_k)$$
$$\quad - [\mu(\tilde{A}_0 \cap \ldots \cap \tilde{A}_k) - \mu(\tilde{A}_0 \cap \ldots \cap \tilde{A}_{k+1})]$$
$$= m_k - m_{k+1} - m_{k+1} + m_{k+2} = m_k - 2m_{k+1} + m_{k+2}.$$

Now

$$\int_A n_A \, d\mu = \sum_{k=0}^{\infty} (k+1)\mu(A \cap \tilde{A}_1 \cap \ldots \cap \tilde{A}_k \cap T^{-(k+1)}A)$$
$$= \sum_{k=0}^{\infty} (k+1)(m_k - 2m_{k+1} + m_{k+2})$$

2.4. Ergodicity

$$= \lim_{n\to\infty} \sum_{k=0}^{n} (k+1)(m_k - 2m_{k+1} + m_{k+2})$$

$$= \lim_{n\to\infty} \left[\sum_{k=0}^{n} (k+1)m_k - \sum_{k=1}^{n+1} 2km_k + \sum_{k=2}^{n+2} (k-1)m_k \right]$$

$$= \lim_{n\to\infty} \left[m_0 + 2m_1 - 2m_1 + \sum_{k=2}^{n} [(k+1) - 2k \right.$$

$$\left. + (k-1)]m_k - 2(n+1)m_{n+1} + nm_{n+1} + (n+1)m_{n+2} \right]$$

$$= 1 + \lim_{n\to\infty} \left[-2m_{n+1} - nm_{n+1} + (n+1)m_{n+2} \right]$$

$$= 1 + \lim_{n\to\infty} \left[(n+1)(m_{n+2} - m_{n+1}) - m_{n+1} \right]$$

$$= 1 - \lim_{n\to\infty} \left[n(m_n - m_{n+1}) + m_n \right].$$

Since the partial sums increase, $\{n(m_n - m_{n+1}) + m_n\}$ is decreasing, and hence the limit exists. But

$$\bigcap_{k=0}^{\infty} \tilde{A}_k$$

is an invariant subset of $\tilde{A}$, and $\mu(A) > 0$, so by ergodicity of T we must have

$$\lim_{n\to\infty} m_n = \mu\left(\bigcap_{k=0}^{\infty} \tilde{A}_k \right) = 0.$$

The series

$$\sum_{n=0}^{\infty} (m_n - m_{n+1})$$

has kth partial sum $m_0 - m_{k+1}$, and hence the series is convergent and has sum $m_0 = 1$. Moreover, since $\{m_n\}$ is a decreasing sequence, this series has positive terms. Therefore $\lim_{n\to\infty} n(m_n - m_{n+1}) = 0$, because otherwise (since the limit exists) we could conclude that the series $\sum (1/n)$ is convergent. It follows that

$$\int_A n_A \, d\mu = 1.$$

(3) Since the derivative transformation $T_A : A \to A$ is ergodic (Exercise 1),

$$\frac{1}{\mu(A)} \int_A n_A \, d\mu = \lim_{k\to\infty} \frac{1}{k} \sum_{j=0}^{k-1} n_A(T_A^j x) \quad \text{a.e. on } A.$$

For each k let $N_k(x) = n_A(x) + n_A(T_A x) + \ldots + n_A(T_A^{k-1} x)$, so that

$$\frac{N_k(x)}{k} \to \frac{1}{\mu(A)} \int_A n_A \, d\mu \quad \text{a.e. as } k \to \infty.$$

However, $N_k(x)$ is the number of T-steps required to produce k entries to A; equivalently, k counts the number of visits of $Tx, T^2 x, \ldots T^{N_k(x)} x$ to A:

$$
\begin{array}{cccc}
& T_A x & T_A^2 x & T_A^k x \\
x, Tx, T^2 x, \ldots T^{j_1} x, \ldots & T^{j_2} x, & \ldots & \ldots T^{N_k(x)}(x) \\
\underbrace{} & \underbrace{} & & \underbrace{} \\
n_A(x) & n_A(T_A x) & \ldots & n_A(T_A^{k-1} x)
\end{array}
$$

Thus $k = \sum_{j=0}^{N_k(x)-1} \chi_A(T^j x)$ for $x \in A$, and so by ergodicity of T

$$\frac{k}{N_k(x)} \to \int \chi_A \, d\mu = \mu(A) \quad \text{a.e.}$$

The Kakutani skyscraper decomposition of an ergodic transformation leads to a lemma on which are based many of the important constructions of ergodic theory. With a bit more work, the lemma can be shown to hold actually for all *aperiodic* transformations – those whose periodic points form a set of measure 0 (see Halmos 1956).

Lemma 4.7 *Kakutani–Rokhlin Lemma (Kakutani 1943, Rokhlin 1948)* Let $T: X \to X$ be an ergodic m.p.t. on a nonatomic measure space $(X, \mathcal{B}, \mu)$, n a positive integer, and $\varepsilon > 0$. Then there is a measurable set $A \subset X$ such that $A, TA, \ldots T^{n-1} A$ are pairwise disjoint and cover X up to a set of measure less than ε.

Proof: Select a measurable set $B \subset X$ of small measure (exactly how small to be determined later). We use the set B to form a Kakutani skyscraper decomposition of X. That is, for each $k = 1, 2, \ldots$ let

$$B_k = \{x \in B : n_B(x) = k\},$$

so that X decomposes (up to a set of measure 0) into the union of disjoint columns $B_k \cup TB_k \cup \ldots \cup T^{k-1} B_k$, $k = 1, 2, \ldots$

In order to form the set A, we ignore the first $n - 1$ columns from the left; and beginning with the nth column from the left we select the base of each column and every nth one above it, moving up the column, and stopping before we arrive at a point n levels from the top of the column. Analytically,

2.4. Ergodicity

```
                                        ——— T^{k-1}B_k
                   ——— T²B_3            ⋮
       ——— TB_2    ——— TB_3    ...  ——— TB_k    ...
——— B_1 ——— B_2    ——— B_3    ...  ——— B_k     ...
```

we may write

$$A = \bigcup_{k \geq n} \bigcup_{j=0}^{[(k-n-1)/n]} T^{jn} B_k.$$

Then clearly $A, TA, \ldots T^{n-1}A$ are pairwise disjoint and cover X except maybe for a set of measure at most

$$n \sum_{k=1}^{\infty} \mu(B_k) \leq n\mu(B).$$

Thus we only need initially to choose $\mu(B) < \varepsilon/n$.

Here are some examples of ergodic transformations.

(1) *Bernoulli shifts are ergodic.* This follows from Exercise 4(b) (p. 56), since the condition

$$\frac{1}{n} \sum_{k=0}^{n-1} \mu(T^{-k}A \cap B) \to \mu(A)\mu(B)$$

is easily verified when A and B are cylinder sets. In the next section we will see that essentially the same argument proves that in fact Bernoulli shifts are strongly mixing.

(2) *Irrational rotations of the circle are ergodic.* Let $Tx = x + \alpha \pmod{1}$ for $x \in [0, 1)$, where α is irrational. If $f \in L^2(X, \mathscr{B}, m)$, then f has a Fourier expansion

$$f(x) = \sum_{-\infty}^{\infty} a_n e^{2\pi i n x}$$

in L^2. If $Uf = f$, we find that

$$Uf(x) = f(x + \alpha) = \sum_{-\infty}^{\infty} a_n e^{2\pi i n \alpha} e^{2\pi i n x} = \sum_{-\infty}^{\infty} a_n e^{2\pi i n x},$$

so that $a_n = a_n e^{2\pi i n \alpha}$ for all n. Since $e^{2\pi i n \alpha} \neq 1$ unless $n = 0$, we must have $a_n = 0$ for $n \neq 0$. Hence $f(x) = a_0$ in L^2. By Proposition 2.4.1, T is ergodic.

Thus if $f \in L^1(X, \mathscr{B}, m)$, we have

$$\frac{1}{n} \sum_{k=0}^{n-1} f(x + k\alpha) \to \int f \, dm \quad \text{a.e.}$$

50   2. The fundamentals of ergodic theory

H. Weyl (1916) showed that in fact *if f is Riemann integrable, then this formula holds for all x*. His way to see this was as follows (see also Section 4.2).

Translating f if necessary, it is enough to consider the case $x = 0$. If f is Riemann integrable, there are trigonometric polynomials

$$p_1(x) = \sum_{n=-n_1}^{n_1} c_n e^{2\pi i n x}, \qquad p_2(x) = \sum_{n=-n_2}^{n_2} d_n e^{2\pi i n x}$$

such that

$$p_1 < f < p_2 \text{ on } [0, 1)$$

and

$$\int_0^1 (p_2 - f)\,dm \quad \text{and} \quad \int_0^1 (f - p_1)\,dm$$

are both arbitrarily small. Now

$$\frac{1}{N} \sum_{k=0}^{N-1} p_1(k\alpha) \to c_0 = \int_0^1 p_1\,dm,$$

and similarly for p_2, because for $n \neq 0$

$$\frac{1}{N} \sum_{k=0}^{N-1} e^{2\pi i n k \alpha} = \frac{1}{N} \frac{1 - e^{2\pi i N n \alpha}}{1 - e^{2\pi i n \alpha}} \to 0.$$

Then

$$\frac{1}{N} \sum_{k=0}^{N-1} p_1(k\alpha) < \frac{1}{N} \sum_{k=0}^{N-1} f(k\alpha) < \frac{1}{N} \sum_{k=0}^{N-1} p_2(k\alpha) \qquad \text{for all } N$$

implies that $(1/N)\sum_{k=0}^{N-1} f(k\alpha)$ is arbitrarily close to $\int_0^1 f\,dm$ for large N.

This fact has an interesting application, the solution of *Gelfand's problem*: find the frequency of the number 7 in the sequence 1, 2, 4, 8, 1, 3, 6, ... of first digits (in decimal notation) of the numbers 2^n, $n = 0, 1, 2, ...$ That is, if for each $k = 0, 1, ...\, 9$, $p_k(n)$ denotes the number of ks in the first n terms of the above sequence, Gelfand's problem is to evaluate, if it exists,

$$p_7 = \lim_{n \to \infty} \frac{p_7(n)}{n}.$$

Let $\alpha = \log_{10} 2$ and $I_k = [\log_{10} k, \log_{10}(k+1))$ for $k = 1, 2, ...\, 9$. Then the first digit of 2^j is k if and only if $k \cdot 10^r \leq 2^j < (k+1) \cdot 10^r$ for some r, i.e., if and only if $j\alpha \in I_k$. Thus

$$p_7(n) = \sum_{j=0}^{n-1} \chi_{I_7}(j\alpha)$$

2.4. Ergodicity

and

$$p_7 = \lim_{n\to\infty} \frac{1}{n} \sum_{j=0}^{n-1} \chi_{I_7}(j\alpha) = m(I_7) = \log_{10}\frac{8}{7}.$$

(3) *Translations of the torus.* The same way as in (2), we can determine which translations of the n-torus

$$\mathbb{K}^n = [0,1)^n = \{x = (x_1, x_2, \ldots x_n) : 0 \leq x_i < 1 \text{ for } i = 1, 2, \ldots n\}$$

are ergodic. $\mathbb{K}^n$ is a compact abelian group under addition modulo 1. Fix $\alpha = (\alpha_1, \alpha_2, \ldots \alpha_n) \in \mathbb{K}^n$ and define $T: \mathbb{K}^n \to \mathbb{K}^n$ by $Tx = x + \alpha \pmod{1}$ for all $x \in \mathbb{K}^n$. Let $\mathscr{B}$ be the σ-algebra of Borel subsets of $\mathbb{K}^n$, and let μ denote the product Lebesgue measure on $\mathbb{K}^n$. Then μ is just normalized Haar measure, and T is a m.p.t.

$(\mathbb{K}^n, \mathscr{B}, \mu, T)$ is ergodic if and only if invariant L^2 functions are constant a.e. Suppose then that f is in the complex L^2 space $L^2(\mathbb{K}^n, \mathscr{B}, \mu)$. Now $f - f \circ T$ is zero a.e. if and only if its Fourier transform $(f - f\circ T)\hat{\,}$ is zero as a function on $\hat{\mathbb{K}}^n = \mathbb{Z}^n$. For each $k \in \mathbb{Z}^n$,

$$(f - f \circ T)\hat{\,}(k) = \int_{\mathbb{K}^n} e^{-2\pi i k \cdot x}(f(x) - f(Tx))\,d\mu(x)$$

$$= \int_{\mathbb{K}^n} e^{-2\pi i k \cdot x} f(x)\,d\mu(x) - \int_{\mathbb{K}^n} e^{-2\pi i k \cdot x} e^{2\pi i k \cdot \alpha} f(x)\,d\mu(x)$$

$$= (1 - e^{2\pi i k \cdot \alpha}) \int_{\mathbb{K}^n} e^{-2\pi i k \cdot x} f(x)\,d\mu(x) = (1 - e^{2\pi i k \cdot \alpha}) \hat{f}(k).$$

Suppose that α is *rationally independent* in the sense that, for $k \in \mathbb{Z}^n$, $k \cdot \alpha \in \mathbb{Z}$ if and only if $k = 0$. If $f \in L^2$ is invariant and α is rationally independent, then the above calculation shows that $\hat{f}(k) = 0$ for all $k \neq 0$. This implies that f equals the constant $\hat{f}(0)$ a.e.

Conversely, if $(\mathbb{K}^n, \mathscr{B}, \mu, T)$ is ergodic, then α must be rationally independent. For if there is a nonzero $k \in \mathbb{Z}^n$ such that $k \cdot \alpha \in \mathbb{Z}$, then $f(x) = e^{2\pi i k \cdot x}$ is a nonconstant invariant function and $(\mathbb{K}^n, \mathscr{B}, \mu, T)$ is not ergodic. Thus $(\mathbb{K}^n, \mathscr{B}, \mu, T)$ is ergodic if and only if α is rationally independent.

The system $(\mathbb{K}^n, \mathscr{B}, \mu, T)$ has discrete spectrum, as does any translation on a compact abelian group.

(4) *An irreducible Markov shift is ergodic.* Let $(\Omega, \mathscr{B}, \mu, \sigma)$ be a Markov shift on the alphabet $\{1, 2, \ldots N\}$ determined by an $N \times N$ stochastic matrix $A = (a_{ij})$ with left-fixed probability row vector p (see 1.2 and 2.3). Recall that $A^k = (a_{ij}^k)$ gives the k-step transition probabilities:

$$a_{ij}^k = Pr\{\omega_k = j | \omega_0 = i\}.$$

2. The fundamentals of ergodic theory

Assume that A is *irreducible*, in that for each i and j there is a k for which $a_{ij}^k > 0$. Let

$$E_i = \{\omega \in \Omega : \omega_0 = i\}, \qquad i = 1, 2, \ldots N.$$

The Ergodic Theorem yields the a.e. existence of the limits

$$\lim_{n \to \infty} \frac{1}{n} \sum_{k=0}^{n-1} \chi_{E_i}(\sigma^k x)$$

and hence of

$$q_{ij} = \frac{1}{p_i} \int \left[\lim_{n \to \infty} \frac{1}{n} \sum_{k=0}^{n-1} \chi_{E_i}(\sigma^k x) \chi_{E_j}(x) \right] d\mu(x)$$

$$= \frac{1}{p_i} \lim_{n \to \infty} \frac{1}{n} \sum_{k=0}^{n-1} \mu(\sigma^{-k} E_i \cap E_j) = \lim_{n \to \infty} \frac{1}{n} \sum_{k=0}^{n-1} a_{ij}^k.$$

The matrix $Q = (q_{ij})$ is stochastic and satisfies

$$Q = QA = AQ, \qquad Q^2 = Q,$$

since $Q = \lim_{n \to \infty} (1/n) \sum_{k=0}^{n-1} A^k$.

I claim that *if A is irreducible, then all the entries of Q are positive and all the rows of Q are identical, each row being equal to p*.

From $Q = QA$, we see that, for fixed i and j,

$$q_{ij} = \sum_k q_{ik} a_{kj}^n \geq q_{ik} a_{kj}^n \quad \text{for each } k \text{ and } n.$$

Thus if

$$\mathscr{E}_i = \{j : q_{ij} > 0\},$$

then if $k \in \mathscr{E}_i$ and $a_{kj}^n > 0$, we have $j \in \mathscr{E}_i$. Moreover, $\mathscr{E}_i \neq \varnothing$. So by irreducibility $\mathscr{E}_i = \{1, 2, \ldots N\}$ for each i, that is to say, $q_{ij} > 0$ for all i and j.

To see that all the rows of Q are identical, suppose that for some $j_0 k_0$

$$q_{j_0 k_0} < q = \max_i q_{i k_0}.$$

Then from $Q^2 = Q$ we have

$$q_{i k_0} = \sum_j q_{ij} q_{j k_0} < q \sum_j q_{ij} = q$$

for all i, which is impossible.

Now

$$(pQ)_j = \sum_i p_i q_{ij} = \lim_{n \to \infty} \frac{1}{n} \sum_k \sum_i p_i a_{ij}^k$$

$$= \lim_{n \to \infty} \frac{1}{n} \sum_k (pA^k)_j = p_j;$$

2.4. Ergodicity

since q_{ij} is independent of i, we have $q_{ij} = p_j$ for all i, j.

To prove that σ is ergodic, it is enough to show that

$$\frac{1}{n} \sum_{k=0}^{n-1} \mu(\sigma^{-k} E \cap F) \to \mu(E)\mu(F)$$

for each pair of cylindrical sets

$$E = \{\omega \in \Omega : \omega_r \omega_{r+1} \ldots \omega_{r+l} = i_0 i_1 \ldots i_l\}$$
$$F = \{\omega \in \Omega : \omega_s \omega_{s+1} \ldots \omega_{s+m} = j_0 j_1 \ldots j_m\},$$

where $i_0, \ldots i_l, j_0, \ldots j_m \in \{1, 2, \ldots N\}$. If k is large enough, then

$$(\{r, r+1, \ldots r+l\} + k) \cap \{s, s+1, \ldots s+m\} = \emptyset,$$

and we have

$$\mu(\sigma^{-k} E \cap F) = (p_{j_0} a_{j_0 j_1} \ldots a_{j_{m-1} j_m})(a_{j_m i_0}^{k-m} a_{i_0 i_1} \ldots a_{i_{l-1} i_l}).$$

Since

$$\frac{1}{n} \sum_{k=0}^{n-1} a_{j_m i_0}^{k-m} \to p_{i_0} \quad \text{as } n \to \infty,$$

the ergodicity of σ follows.

According to Exercise 8, the converses of all these implications also hold. Thus *for a Markov shift with all $p_i > 0$ the following statements are equivalent*:

(1) σ is ergodic.
(2) A is irreducible.
(3) q_{ij} is independent of i.
(4) $q_{ij} = p_j$ for all i and j.
(5) $q_{ij} > 0$ for all i and j.

(5) *Ergodicity of skew products* (Anzai 1951). Let T be a m.p.t. on a probability space $(X, \mathscr{B}, \mu)$, and G a compact abelian group with Haar measure v. Let $f : X \to G$ be a measurable function, and on the product probability space $(X \times G, \mathscr{B}_X \otimes \mathscr{B}_G, \mu \times v)$ define the *skew product transformation* T_f by

$$T_f(x, g) = (Tx, g \cdot f(x)).$$

Recall that T_f is measurable, invertible, and measure-preserving.

Theorem 4.8 Suppose $(X, \mathscr{B}, \mu, T)$ is ergodic. Then the skew product transformation $T_f : X \times G \to X \times G$ is not ergodic if and only if there is a nontrivial character $\tau \in \hat{G}$ and a measurable map η of X into the unit circle $\mathbb{K}$ such that

$$\tau(f(x)) = \eta(x)\eta(Tx)^{-1} \quad \text{for almost all } x \in X.$$

Proof: Suppose there is such a $\tau \in \hat{G}$ and such a map $\eta: X \to \mathbb{K}$. Define $h: X \times G \to \mathbb{K}$ by $h(x, g) = \eta(x)\tau(g)$, so h is clearly measurable. The function h is T_f-invariant because

$$h(Tx, gf(x)) = \eta(Tx)\tau(gf(x)) = \eta(Tx)\tau(g)\tau(f(x)) = \eta(x)\tau(g) = h(x, g) \quad \text{a.e.}$$

Moreover, h cannot be constant a.e. since $(\tau, 1) = 0$ in $L^2(G, \mathcal{B}_G, \nu)$ and hence

$$\iint_{X \times G} h(x, g) d\mu(x) d\nu(g) = \int_X \left(\int_G \eta(x)\tau(g) d\nu(g) \right) d\mu(x) = 0.$$

Thus h could be constant only if $h = 0$ a.e., but $|h| = 1$.

Conversely, suppose the skew product is not ergodic and let $h: X \times G \to \mathbb{C}$ be a nonconstant invariant function in L^2. For each $\tau \in \hat{G}$ define $\phi_\tau: X \to \mathbb{C}$ by

$$\phi_\tau(x) = \int_G h(x, g)\tau(g)^{-1} d\nu(g).$$

Then

$$\phi_\tau(Tx)\tau(f(x)) = \int_G h(Tx, g)\tau(g)^{-1}\tau(f(x)) d\nu(g)$$

$$= \int_G h(x, gf(x)^{-1})\tau(g)^{-1}\tau(f(x)) d\nu(g)$$

$$= \int_G h(x, g)\tau(gf(x))^{-1}\tau(f(x)) d\nu(g)$$

$$= \int_G h(x, g)\tau(g)^{-1} d\nu(g) = \phi_\tau(x) \quad \text{a.e.,}$$

so $|\phi_\tau(Tx)| = |\phi_\tau(x)|$ a.e. Since T is ergodic, $|\phi_\tau|$ must be a constant C_τ a.e.

There exists a nontrivial $\tau \in \hat{G}$ for which $C_\tau \neq 0$. For if not, then for almost all x the function $h(x, \cdot) \in L^2(G)$ has each of its Fourier coefficients $\int h(x, \cdot)\tau^{-1} d\nu$ equal to zero for $\tau \neq 1$ and consequently $h(x, g)$ is independent of g. Since h is invariant and T is ergodic, this would imply that h is constant. Choose a nontrivial $\tau \in \hat{G}$ such that $C_\tau \neq 0$, and let

$$\eta(x) = \frac{\phi_\tau(x)}{C_\tau},$$

so $\eta: X \to \mathbb{K}$ is measurable. Then

$$\eta(x)\eta(Tx)^{-1} = \frac{\phi_\tau(x)}{C_\tau} \cdot \frac{C_\tau}{\phi_\tau(Tx)} = \frac{\phi_\tau(x)}{\phi_\tau(x)\tau(f(x))^{-1}} = \tau(f(x)),$$

as required.

2.4. Ergodicity

Anzai originally considered skew products in the case that $X = G = \mathbb{K} = 1$-torus. Actually, this example goes back to von Neumann, who showed that the skew product $(x, y) \to (x + \alpha, x + y)$ (α irrational) on $\mathbb{K}^2$ has the same spectral type (i.e., the induced operators in L^2 are unitarily equivalent) as the cross product of the shift transformation on an infinite torus with the translation $x \to x + \alpha$ of $\mathbb{K}$, but these two systems are not isomorphic (see Anzai 1951). In Anzai's situation, the transformation $T:\mathbb{K}^2 \to \mathbb{K}^2$ is given by $T(x, y) = (x + \alpha, y + f(x))$, where α is irrational and $f: \mathbb{K} \to \mathbb{K}$ is measurable, and the condition for nonergodicity becomes the following: there is an integer $p \neq 0$ and a measurable map $\eta: \mathbb{K} \to \mathbb{K}$ such that

$$pf(x) = \eta(x) - \eta(Tx) \quad \text{a.e.}$$

Obviously the idea of a skew product generalizes further, for example to infinite products. We might consider the map $T: \mathbb{K}^\infty \to \mathbb{K}^\infty$ given by, say, $T(x_1, x_2, \ldots) = (x_1 + \alpha, x_1 + x_2, x_2 + x_3, \ldots)$. Skew products are related to cohomology and representations of groups and to the virtual groups of Mackey. They also provide a useful tool for the analysis of the structure of a given m.p.t., as in structure theorems of Furstenberg (1963) and Zimmer (1975, 1976a, b).

Recall that a given system $(X, \mathscr{B}, \mu, T)$ is said to have *discrete spectrum* in case the eigenfunctions of the unitary operator U induced on L^2 by T span L^2. There are two generalizations of this concept. Let $D_0 \subset L^2(X, \mathscr{B}, \mu)$ consist of the constant functions of absolute value 1, and for each $i = 1, 2, \ldots$ let $D_i = \{f \in L^2(X, \mathscr{B}, \mu): |f| = 1$ a.e. and $Uf/f \in D_{i-1}\}$. The elements of $D = \bigcup_{i=0}^{\infty} D_i$ are called *quasi-eigenfunctions* of U, and T is said to have *quasi-discrete spectrum* in case D spans $L^2(X, \mathscr{B}, \mu)$. Again, let $H_0 \subset L^2(X, \mathscr{B}, \mu)$ denote the constant functions of absolute value 1. For each ordinal i, E_i will denote the closed linear span in $L^2(X, \mathscr{B}, \mu)$ of H_i. Assuming that H_j and E_j have been defined for all $j < i$, we define

$$H_i = \bigcup_{j < i} H_j \text{ if } i \text{ is a limit ordinal, and}$$

$$H_i = \{f \in L^2(X, \mathscr{B}, \mu): |f| = 1 \text{ a.e. and } Uf/f \in E_{i-1}\}$$

if i is a successor ordinal. The elements of $H = \bigcup_i H_i$ are called *generalized eigenfunctions of* U, and T is said to have *generalized discrete spectrum* in case H spans $L^2(X, \mathscr{B}, \mu)$ (Abramov 1962, Parry 1969a, Petersen 1971 et al.). These classes of transformations may be considered the next simplest, after the discrete spectrum transformations.

2. The fundamentals of ergodic theory

Exercises

1. Prove that if T is ergodic, then so are the induced transformations T_A and $\tilde{T}$ (see Section 2.3). ($\tilde{T}$ is metrically indecomposable even if $\tilde{\mu}(\tilde{X}) = \infty$.)

2. Prove that if T is ergodic, $f \geq 0$, and
$$\varlimsup_{n \to \infty} \frac{1}{n} \sum_{k=0}^{n-1} f(T^k x) < \infty \text{ a.e.,}$$
then $f \in L^1$.

3. Which of the equivalent characterizations of ergodicity fail when X has infinite measure?

4. (a) Prove that T is ergodic if and only if
$$\frac{1}{n} \sum_{k=0}^{n-1} \mu(T^k A \cap B) \to \mu(A)\mu(B) \quad \text{for all } A, B \in \mathscr{B}.$$

   (b) Prove that T is ergodic if and only if (a) holds for all A, B in a semialgebra which generates the σ-algebra $\mathscr{B}$.

5. Let T be ergodic and $\nu \ll \mu$ a measure on $(X, \mathscr{B})$ such that $\nu T^{-1} \leq \nu$. Show that $\nu T^{-1} = \nu$ and ν is a constant multiple of μ.

6. Let $X = [0, 1]$ with Lebesgue measure m. Then T (preserving m) is ergodic on X if and only if
$$\frac{1}{n} \sum_{k=0}^{n-1} f(T^k x) \to \int f \, dm \text{ a.e.}$$
for each continuous function f.
(Note: It is not true that ergodicity is characterized by the absence of *continuous* invariant functions.)

7. Let $U: L^2(X, \mathscr{B}, \mu) \to L^2(X, \mathscr{B}, \mu)$ be the unitary operator associated with an ergodic m.p.t. on a nonatomic probability space $(X, \mathscr{B}, \mu)$. Then every point of the unit circle is an *approximate eigenvalue* of U; i.e., given λ with $|\lambda| = 1$ there are $f_n \in L^2$ with $\|f_n\|_2 = 1$ for all n and $\|Uf_n - \lambda f_n\| \to 0$. Consequently the spectrum of U is the entire unit circle.

8. Suppose that $(\Omega, \mathscr{B}, \mu, \sigma)$ is a Markov shift determined by a given stochastic matrix A and fixed probability vector p with all $p_i > 0$. Prove that if $(\Omega, \mathscr{B}, \mu, \sigma)$ is ergodic, then A is irreducible.

9. Prove that for any ergodic m.p.t. on a nonatomic space there is a set A of positive measure for which the return time n_A is unbounded.

10. Prove that if T has discrete spectrum, then there is a sequence of

integers $n_k \nearrow \infty$ with $T^{n_k} \to I$ in the strong operator topology on L^2, i.e., $\|T^{n_k}f - f\|_2 \to 0$ for all $f \in L^2$.

2.5. Strong mixing

While every m.p.t. on a space of finite measure is recurrent, only certain ones, namely those that move almost all points throughout the space, are ergodic. The concept of ergodicity can be strengthened in several ways to define classes of transformations which satisfy even more stringent quantitative recurrence conditions. Recall that a m.p.t. $T: X \to X$ is ergodic if and only if

(1) $\quad \dfrac{1}{n}\sum_{k=0}^{n-1} \mu(T^{-k}A \cap B) \to \mu(A)\mu(B) \quad$ for all $A, B \in \mathscr{B}$.

We say that T is *weakly mixing* if

(2) $\quad \dfrac{1}{n}\sum_{k=0}^{n-1} |\mu(T^{-k}A \cap B) - \mu(A)\mu(B)| \to 0 \quad$ for all $A, B \in \mathscr{B}$,

and *strongly mixing* if

(3) $\quad |\mu(T^{-k}A \cap B) - \mu(A)\mu(B)| \to 0 \quad$ for all $A, B \in \mathscr{B}$.

Thus T is strongly mixing if and only if
$$\mu(T^{-k}A \cap B) \to \mu(A)\mu(B),$$
i.e.,
$$\frac{\mu(T^{-k}A \cap B)}{\mu(B)} \to \mu(A) \quad \text{for all } A, B \in \mathscr{B} \text{ with } \mu(B) > 0.$$

This last condition says that eventually T distributes A fairly evenly throughout the space X: for large k, the proportion of $T^{-k}A$ that lies in B, namely $\mu(T^{-k}A \cap B)/\mu(B)$, is approximately the same as the relative size of A in X, namely $\mu(A)$.

The following observations are important even though they are evident.

Proposition 5.1. Strong mixing implies weak mixing and weak mixing implies ergodicity.

Proposition 5.2. T is strongly mixing if and only if
$$(U^n f, g) \to (f, 1)\overline{(g, 1)} \quad \text{for all } f, g \in L^2.$$

Thus T is strongly mixing if and only if $U^n f \to \int f \, d\mu$ weakly in $L^2(X, \mathscr{B}, \mu)$ for each $f \in L^2(X, \mathscr{B}, \mu)$. Too bad about that.

2. The fundamentals of ergodic theory

Proposition 5.3 T is strongly (respectively weakly) mixing if and only if (3) (respectively (2)) holds for all A, B in a semialgebra which generates $\mathscr{B}$, or for all A, B in a set dense in $\mathscr{B}$.

Proposition 5.4 Ergodicity, weak mixing, and strong mixing are isomorphism invariants of T. (Actually they are spectral invariants.)

Theorem 5.5 (*Rényi* 1958) T is strongly mixing if and only if for each $A \in \mathscr{B}$
$$\lim_{n \to \infty} \mu(T^{-n}A \cap A) = \mu(A)^2.$$

Proof: We prove that if $\lim_{n \to \infty} \mu(T^{-n}A \cap A) = \mu(A)^2$ for all $A \in \mathscr{B}$, then $(X, \mathscr{B}, \mu, T)$ is strongly mixing; the converse is obvious. Fix $A \in \mathscr{B}$ and let M be the closed linear subspace of $L^2(X, \mathscr{B}, \mu)$ generated by the constant functions together with $\{U^k \chi_A : k \in \mathbb{Z}\}$. Since

$$\lim_{n \to \infty} (U^n \chi_A, 1) = \lim_{n \to \infty} \mu(A) = (\chi_A, 1)(1, 1) \quad \text{and}$$

$$\lim_{n \to \infty} (U^n \chi_A, U^k \chi_A) = \lim_{n \to \infty} (U^{n-k} \chi_A, \chi_A) = \lim_{n \to \infty} \mu(T^{-n+k}A \cap A)$$

$$= \mu(A)^2 = (\chi_A, 1)(U^k \chi_A, 1),$$

we have

$$\lim_{n \to \infty} (U^n \chi_A, f) = (\chi_A, 1)(\overline{f, 1}) \quad \text{for all } f \in M.$$

Given $f \in L^2(X, \mathscr{B}, \mu)$, write $f = f_1 + f_2$, where $f_1 \in M$ and $f_2 \in M^\perp$. Then

$$\lim_{n \to \infty} (U^n \chi_A, f) = \lim_{n \to \infty} (U^n \chi_A, f_1) + \lim_{n \to \infty} (U^n \chi_A, f_2)$$

$$= (\chi_A, 1)(\overline{f_1, 1}) = (\chi_A, 1)(\overline{f, 1}),$$

since $(f_2, 1) = 0$. Thus in particular, for $B \in \mathscr{B}$ and $f = \chi_B$,

$$\lim_{n \to \infty} \mu(T^{-n}A \cap B) = \lim_{n \to \infty} (U^n \chi_A, \chi_B) = (\chi_A, 1)(\chi_B, 1) = \mu(A)\mu(B);$$

therefore $(X, \mathscr{B}, \mu, T)$ is strongly mixing.

Here are some examples of strongly mixing transformations.

(1) *Bernoulli shifts are strongly mixing.* This is true even for shifts based on infinite alphabets. Let

$$(\Omega, \mathscr{F}, \mu) = \prod_{-\infty}^{\infty} (X, \mathscr{B}, P),$$

where $(X, \mathscr{B}, P)$ is a probability space, and let $\sigma: \Omega \to \Omega$ be the shift. It is clear that if $A, B \in \mathscr{F}$ are cylindrical sets (i.e., of the form $\{\omega \in \Omega : \omega_{i_1} \in E_1, \ldots, \omega_{i_r} \in E_r\}$, where $E_1, \ldots E_r \in \mathscr{B}$), then

$$\mu(\sigma^{-n}A \cap B) \to \mu(A)\mu(B),$$

2.5. Strong mixing

since if n is large enough the sets of indices determining $\sigma^{-n}A$ and B do not overlap. By Proposition 5.3, σ is strongly mixing.

(2) *Mixing for Markov shifts.* An $N \times N$ stochastic matrix A is called *aperiodic* in case there do not exist an $i = 1, 2, \ldots, N$ and an integer $m > 1$ such that $a_{ii}^k > 0$ implies m divides k.

Theorem 5.6 Let $(\Omega, \mathscr{F}, \mu, \sigma)$ be the Markov shift determined by an $N \times N$ stochastic matrix A which fixes a positive probability vector p (see Sections 1.2.B and 2.4). The following statements are equivalent:

(1) $(\Omega, \mathscr{F}, \mu, \sigma)$ is strongly mixing;
(2) $\lim_{k \to \infty} a_{ij}^k = p_j$ for all i and j;
(3) A is irreducible and aperiodic.

Proof: (1) $\Rightarrow$ (2):

$$a_{ij}^k = Pr\{\omega_k = j | \omega_0 = i\}$$
$$= \frac{\mu(T^{-k}\{\omega_0 = j\} \cap \{\omega_0 = i\})}{\mu\{\omega_0 = i\}} \to \mu\{\omega_0 = j\} = p_j.$$

(2) $\Rightarrow$ (3): This is clear, since all p_j are positive.

(3) $\Rightarrow$ (1): If A is aperiodic and irreducible, then σ^n is ergodic for each n, and since $(1/n) \sum_{k=0}^{n-1} a_{ij}^k \to p_j > 0$, we must have $\limsup_{k \to \infty} a_{ij}^k > 0$ for all i, j. Now if, for example,

$$E = \{\omega \in \Omega : \omega_0 \omega_1 \ldots \omega_r = i_0 i_1 \ldots i_r\}$$

and

$$F = \{\omega \in \Omega : \omega_0 \omega_1 \ldots \omega_s = j_0 j_1 \ldots j_s\},$$

then for large enough k,

$$\mu(E \cap T^{-k}F) = (p_{i_0} a_{i_0 i_1} \ldots a_{i_{r-1} i_r})(a_{i_r j_0}^{k-r} a_{j_0 j_1} \ldots a_{j_{s-1} j_s})$$
$$= \mu(E)\mu(F) \frac{a_{i_r j_0}^{k-r}}{p_{j_0}},$$

and

$$\limsup_{k \to \infty} \mu(T^{-k}E \cap F) \leq c\mu(E)\mu(F),$$

where

$$c = \sup_{i,j} \limsup_{k \to \infty} \frac{a_{ij}^k}{p_j}.$$

By Ornstein's characterization (see Exercise 2.6.6), σ is strongly mixing.

Friedman and Ornstein (1970) proved that in fact *every strongly mixing Markov shift is Bernoulli*. This means that there is another measurement to make on the experiment, the records of whose outcomes will be indistinguishable from the histories produced by a certain Bernoulli shift: There is a finite measurable independent partition $\{A_1, A_2,...,A_n\}$ of Ω which generates $\mathscr{F}$, so that the coding $\Omega \xrightarrow{\phi} \prod_{-\infty}^{\infty} \{1, 2,...n\}$ by $(\phi\omega)_j = i$ if and only if $\phi^j\omega \in A_i$ gives an isomorphism of the Markov shift with the Bernoulli scheme on n symbols with weights $\mu(A_1), \mu(A_2),...\mu(A_n)$.

(3) *Automorphisms of compact abelian groups* (Halmos 1943). Let X be a compact abelian group, $\mathscr{B}$ the Borel subsets of X, μ normalized Haar measure on X, and $T: X \to X$ a continuous automorphism. Then T is a m.p.t. For, define a measure γ on X by $\gamma(A) = \mu(T^{-1}A)$ for each $A \in \mathscr{B}$; then for all $A \in \mathscr{B}$ and all $x \in X$, $\gamma(xA) = \mu[T^{-1}(xA)] = \mu[(T^{-1}x)(T^{-1}A)] = \mu(T^{-1}A) = \gamma(A)$, so because of the uniqueness of Haar measure γ is a constant multiple of μ. But $\gamma(X) = \mu(T^{-1}X) = \mu(X)$, so $\gamma = \mu$. Again we wish to find conditions for $(X, \mathscr{B}, \mu, T)$ to be ergodic.

The character group $\hat{X}$ of X forms a complete orthonormal set in $L^2(X, \mathscr{B}, \mu)$, and the unitary operator U induced on L^2 by T maps $\hat{X}$ into itself.

Theorem 5.7 If $(X, \mathscr{B}, \mu, T)$ is ergodic, then the only finite orbit of U on $\hat{X}$ is that of the trivial character. Conversely, if U on $\hat{X}$ has no finite orbits other than that of the trivial character, then $(X, \mathscr{B}, \mu, T)$ is strongly mixing and hence ergodic.

Proof: Suppose $\xi \in \hat{X}$ is a nontrivial character with a finite orbit under U. Let n be the smallest positive integer such that $U^n \xi = \xi$. Then $\psi = \xi + U\xi + ... + U^{n-1}\xi$ is an invariant function. Since $1, \xi, U\xi,...,U^{n-1}\xi$ are distinct elements of $\hat{X}$ they are linearly independent, and therefore ψ is not constant. Consequently $(X, \mathscr{B}, \mu, T)$ is not ergodic.

Conversely, suppose now that no element of $\hat{X}$ other than 1 has a finite orbit under U. Let $f, g \in \hat{X}$ and suppose at least one of f, g is not 1. Then $U^n f \neq g$ for all large n, and consequently

$$\lim_{n \to \infty} (U^n f, g) = 0 = (f, 1)\overline{(g, 1)}.$$

In case $f = g = 1$, the same formula still holds. If $f, g \in L^2$ are arbitrary, the fact that $\lim_{n \to \infty} (U^n f, g) = (f, 1)\overline{(g, 1)}$ follows from approximation of f and g by linear combinations of elements of $\hat{X}$ (see Proposition 1.4.2).

We consider now the special case where X is the n-torus $\mathbb{K}^n = \mathbb{R}^n/\mathbb{Z}^n$. It is known that a continuous automorphism $T: \mathbb{K}^n \to \mathbb{K}^n$ is given by an $n \times n$ matrix with integer entries and determinant ± 1. We claim that T on $\mathbb{K}^n$ is ergodic *if and only if the matrix A representing T has no roots of unity among its eigenvalues*.

2.5. Strong mixing

According to this Theorem, T on $\mathbb{K}^n$ is ergodic if and only if U on $\hat{\mathbb{K}}^n = \mathbb{Z}^n$ has no finite orbits other than that of the trivial character. If $f \in \hat{\mathbb{K}}^n$ is nontrivial, there is a nonzero $k \in \mathbb{Z}^n$ such that

$$f(x) = e^{2\pi i k \cdot x} \text{ for all } x \in \mathbb{K}^n.$$

Then

$$U^p f(x) = e^{2\pi i k \cdot T^p x} = e^{2\pi i k \cdot x A^p}$$
$$= e^{2\pi i k (A^{tr})^p \cdot x},$$

so if $U^p f = f$, then $k(A^{tr})^p = k$ and A^p has 1 among its eigenvalues. Then we have $A^p v = v$ for some $v \in \mathbb{R}^n$, and hence

$$0 = (A^p - I)v = (A - \xi_1 I)[(A - \xi_2 I)\ldots(A - \xi_p I)v],$$

where $\xi_1, \ldots \xi_p$ are the pth roots of unity. Thus either $(A - \xi_2 I)\ldots(A - \xi_p I)v$ is an eigenvector of A with eigenvalue ξ_1, or else it is zero. It is clear that we may continue this process to find that at least one ξ_i is an eigenvalue of A. Thus if U on $\hat{\mathbb{K}}^n$ has a finite orbit, then A has a root of unity among its eigenvalues.

Conversely, suppose A has a root of unity among its eigenvalues. Then there will be $k \in \mathbb{Z}^n$ and $p \in \mathbb{Z}^+$ such that $k(A^{tr})^p = k$. It is clear that then $f \in \hat{\mathbb{K}}^n$ defined by $f(x) = e^{2\pi i k \cdot x}$ will have a finite orbit under U.

Katznelson (1971) proved that any ergodic automorphism of $\mathbb{K}^n$ is isomorphic to a Bernoulli scheme. This was extended to compact abelian groups by Lind (1977) and Miles and Thomas (1978), and to certain ergodic automorphisms of nilmanifolds by Dani (1976).

There *are* some non-Bernoulli strongly mixing transformations, for example those with entropy 0 (see Chapter 5). We will mention now two other properties that lie between strong mixing and being Bernoulli.

Let N be a finite or infinite cardinal number. We say that a m.p.t. T has *Lebesgue spectrum of multiplicity N* in case there is a set Λ of cardinality N and a set of functions

$$\{f_{\lambda, j} : \lambda \in \Lambda, j \in \mathbb{Z}\}$$

which, together with 1, form an orthonormal basis for $L^2(X, \mathcal{B}, \mu)$, and such that

$$U_T f_{\lambda, j} = f_{\lambda, j+1} \text{ for all } \lambda \in \Lambda, j \in \mathbb{Z}.$$

It can be proved that *T has Lebesgue spectrum if and only if its maximal spectral type is absolutely continuous with respect to Lebesgue measure on the unit circle* (hence the name).

Example 5.8 Bernoulli schemes have countable Lebesgue spectrum
The idea will be clear from considering $\mathscr{B}(\frac{1}{2},\frac{1}{2})$, the shift on $\prod_{-\infty}^{\infty}\{0,1\}$, where 0 and 1 each have weight $\frac{1}{2}$. Let $\tau_n(\omega) = (-1)^{\omega_n}$. The finite products of the τ_n, together with the constant function 1, form an orthonormal basis for L^2. Partition this basis into equivalence classes by saying that $f \sim g$ if $f\sigma^n = g$ for some $n \in \mathbb{Z}$. There will be countably many equivalence classes, so σ has countable Lebesgue spectrum.

Proposition 5.9 A m.p.t. which has Lebesgue spectrum is strongly mixing.
Proof: The relation

(1) $\quad (U^n f, g) \to (f, 1)(1, g)$

clearly holds if either f or g is 1. For any other elements of the distinguished orthonormal basis, clearly

$$(f_{\lambda, j+n}, f_{\lambda', j'}) = 0$$

for all large enough n. Thus (1) holds for all basis vectors, and hence for all $f \in L^2$.

We say that a m.p.t. T on $(X, \mathscr{B}, \mu)$ is a *K-automorphism* (for Kolmogorov) if there is a sub-σ-algebra $\mathscr{A} \subset \mathscr{B}$ such that

(i) $T^{-1}\mathscr{A} \subset \mathscr{A}$,

(ii) $\bigcup_{n=-\infty}^{\infty} T^n \mathscr{A}$ generates $\mathscr{B}$,

(iii) $\bigcap_{n=-\infty}^{0} T^n \mathscr{A}$ is trivial.

Proposition 5.10 Bernoulli shifts are K-automorphisms.
Proof: If $\mathscr{B}(p_1, \ldots p_N)$ is the shift on the measure space $(\Omega, \mathscr{F}, \mu)$, where $\Omega = \{1, 2, \ldots, N\}^{\mathbb{Z}}$ and μ is the product measure determined by the weights $p_1, \ldots p_N$, let $\mathscr{A}$ be the σ-algebra generated by all cylinder sets of the form

$$\{\omega \in \Omega : \omega_i = j\},$$

where $j = 1, 2, \ldots N$ and $i \geq 0$. Then clearly $\sigma^{-1}\mathscr{A} \subset \mathscr{A}$ and $\bigcup_{-\infty}^{\infty} \sigma^n \mathscr{A}$ generates $\mathscr{B}$. Now if $A \in \sigma^{-n}\mathscr{A}$, then A is in the σ-algebra generated by cylinder sets determined by coordinates at least n, so if B is in the σ-algebra generated by cylinder sets determined by coordinates less than n we will have $\mu(A \cap B) = \mu(A)\mu(B)$. Hence if $A \in \bigcap_{n=0}^{\infty} \sigma^{-n}\mathscr{A}$, A will be independent of every cylinder set B. This implies that $\mu(A) = \mu(A)^2$, and (iii) follows.

2.5. Strong mixing

Proposition 5.11 Every K-automorphism has countable Lebesgue spectrum.

Proof: Let $\mathscr{A}$ be the distinguished sub-σ-algebra of $\mathscr{B}$, and let M denote the closed linear span in L^2 of the characteristic functions of all the sets in $\mathscr{A}$. Then we will have

$$\text{constants} = \bigcap_{n=0}^{\infty} U^n M \subset \ldots \subset U^2 M \subset UM \subset M \subset U^{-1}M \subset \ldots \subset$$

$$\subset \overline{\bigcup_{n=-\infty}^{\infty} U^n M} = L^2.$$

Choose an orthonormal basis $\{e_i\}$ for the orthogonal complement of UM in M. For each i, let

$$M_i = \text{c.l.s.} \{U^n e_i : n \in \mathbb{Z}\}.$$

Then $UM_i = M_i$ for all i, and

$$\sum_i \oplus M_i = (\text{constants})^\perp.$$

If we let $f_{i,j} = U^j e_i$, we will obtain the functions required by the definition of Lebesgue spectrum. Exercise 1 shows that $\dim(M \ominus UM) = \infty$, so the transformation has countable Lebesgue spectrum.

It can be proved that *T is a K-automorphism if and only if every non-trivial factor of T has positive entropy* (see Chapter 5 for the definition of entropy). Ornstein constructed K-automorphisms which are not Bernoulli (i.e. not isomorphic to any Bernoulli shift). On the other hand, there are m.p.t.s with Lebesgue spectrum and zero entropy, for example certain Gaussian systems (Girsanov 1958) and horocycle flows (Parasjuk 1953 and Gurevich 1961). (See also Rokhlin 1960 and Newton and Parry 1966.)

Finally, we cannot end this section without mentioning Rokhlin's old, still-unsolved problem involving *higher degrees of mixing*. A m.p.t. T is said to be *n-mixing* if for any choice of measurable sets $A_1, A_2, \ldots A_n$,

$$\lim_{\substack{\inf_i m_i \to \infty \\ \inf_{i \neq j} |m_i - m_j| \to \infty}} \mu(T^{m_1} A_1 \cap \ldots \cap T^{m_n} A_n) = \mu(A_1) \ldots \mu(A_n).$$

Thus 2-mixing coincides with strong mixing. It is still not known whether there are any 2-mixing transformations that are not 3-mixing.

Exercises

1. Complete the proof of Proposition 2.5.11 by showing that $\dim(M \ominus UM) = \infty$. (Hint: Choose a nonzero $f \in M \ominus UM$. Note that $\mathscr{A}$ must be nonatomic, choose disjoint $\mathscr{A}$-measurable

subsets $E_1, E_2, \ldots$ of the support of f; and consider the functions $f\chi_{E_i}$.)
2. Prove Proposition 2.5.2.
3. Prove Proposition 2.5.3.
4. Prove that a weakly mixing Markov shift is strongly mixing.
5. Prove that T is weakly mixing if and only if

$$\lim_{n\to\infty} \frac{1}{n} \sum_{k=0}^{n-1} |\mu(T^{-k}A \cap A) - \mu(A)^2| = 0$$

for each $A \in \mathscr{B}$.
6. Prove that Bernoulli shifts, mixing Markov shifts, and ergodic automorphisms of compact abelian groups are n-mixing for all n.
7. (Ambrose, Halmos, Kakutani–see Halmos 1949a.) Show that there is no concept of 'uniform mixing' for m.p.t.s: If

$$\mu(T^{-n}A \cap B) \to \mu(A)\mu(B)$$

uniformly for all $A, B \in \mathscr{B}$ with $A \subset B$, then every set in $\mathscr{B}$ has measure 0 or 1 and so $(X, \mathscr{B}, \mu)$ is isomorphic with the space consisting of a single point. (Hint: It is easy if we remove the phrase 'with $A \subset B$' above. But the weaker hypothesis implies the stronger: make

$$|\mu(T^{-n}(A \cap B) \cap B) - \mu(A \cap B)\mu(B)|$$

and

$$|\mu(T^{-n}(A \cap B^c) \cap B^c) - \mu(A \cap B^c)\mu(B^c)|$$

small simultaneously.)

2.6. Weak mixing

Recall that a m.p.t. $T: X \to X$ on a probability space $(X, \mathscr{B}, \mu)$ is said to be *weakly mixing* in case for every pair of measurable sets A and B,

$$\lim_{n\to\infty} \frac{1}{n} \sum_{k=0}^{n-1} |\mu(T^k A \cap B) - \mu(A)\mu(B)| = 0.$$

That the concept of weak mixing is natural and important can be seen from the following theorem, according to which a transformation is weakly mixing if and only if its only measurable eigenfunctions are the constants. Thus weakly mixing transformations lie at the other extreme from those with discrete spectrum, *i.e.*, the group rotations. Most of this theorem seems to go back to Koopman and von Neumann (1932).

2.6. Weak mixing

Theorem 6.1 Let $T: X \to X$ be a m.p.t. on a probability space $(X, \mathcal{B}, \mu)$. Then the following are equivalent:

(1) T is weakly mixing.

(2) $\lim_{n \to \infty} \frac{1}{n} \sum_{k=0}^{n-1} |(U_T^k f, g) - (f, 1)\overline{(g, 1)}| = 0$ for all $f, g \in L^2(X, \mathcal{B}, \mu)$.

(3) Given $A, B \in \mathcal{B}$, there is a set $J \subset \mathbb{Z}^+$ of density 0 such that
$$\lim_{n \to \infty, n \notin J} \mu(T^n A \cap B) = \mu(A)\mu(B).$$

(4) $T \times T$ is weakly mixing.

(5) $T \times S$ is ergodic on $X \times Y$ for each ergodic $(Y, \mathcal{C}, \nu, S)$.

(6) $T \times T$ is ergodic.

(7) T has no measurable eigenfunctions other than the constants.

We will prove this Theorem presently. The implications (1)⇔(2) and (5)⇒(6) are clear. The argument for (1)⇔(3) consists of a real-variables analysis of the type of convergence of sequences involved in the definition of weak mixing and is based on a lemma of Koopman and von Neumann. The proof that (3)⇒(4) turns just on the observation that the union of two sets of density 0 has density 0. The arguments for (4)⇒(5), (6)⇒(7), and (6)⇒(3) (and hence (6)⇒(1)) will be found to be straightforward and easy. The major difficulty of the proof lies in the implication (7)⇒(6). We will first give the usual proof, which makes use of the Spectral Theorem. This will be followed in Section 4.1 by a second, more constructive proof, which does not need the Spectral Theorem.

Notice that (7) implies that T is ergodic; a characterization of weak mixing in terms of eigenvalues, however, would have to mention ergodicity explicity: T is weakly mixing if and only if T is ergodic and has no eigenvalues other than 1.

Let us proceed towards the proof of the Theorem. Recall that a subset $E \subset \mathbb{N}$ of the positive integers is said to have *density zero* in case
$$\lim_{n \to \infty} \frac{1}{n} \sum_{k=0}^{n-1} \chi_E(k) = 0.$$

Lemma 6.2 (*Koopman-von Neumann* 1932) Let $f: \mathbb{N} \to \mathbb{R}$ be nonnegative and bounded. A necessary and sufficient condition that
$$\lim_{n \to \infty} \frac{1}{n} \sum_{k=0}^{n-1} f(k) = 0$$
is that there exist a subset $E \subset \mathbb{N}$ of density zero such that
$$\lim_{n \to \infty, n \notin E} f(n) = 0.$$

Proof: Suppose first that there is such a set E. We may write
$$\frac{1}{n}\sum_{k=0}^{n-1} f(k) = \frac{1}{n}\sum_{k\le n-1, k\in E} f(k) + \frac{1}{n}\sum_{k\le n-1, k\notin E} f(k).$$
The first term is arbitrarily small for large n, since E has density 0 and f is bounded. The second term is arbitrarily small for large n, since $f(k)\chi_{E^c}(k) \to 0$ and ordinary convergence implies Cesaro convergence.

Conversely, suppose that
$$\lim_{n\to\infty} \frac{1}{n}\sum_{k=0}^{n-1} f(k) = 0.$$
For each $m = 1, 2, \ldots$ let
$$E_m = \left\{ k\in\mathbb{N} : f(k) > \frac{1}{m} \right\}.$$
Then $E_1 \subset E_2 \subset \ldots$, and each E_m has density zero, since $(1/m)\chi_{E_m} < f$. Therefore, for each $m = 1, 2, \ldots$ we may choose $i_m > 0$ such that $i_1 < i_2 < \ldots$ and
$$\frac{1}{n}\sum_{k=0}^{n-1} \chi_{E_m}(k) < \frac{1}{m} \text{ for } n \ge i_{m-1} \ (m = 2, 3, \ldots).$$
Now let $i_0 = 0$ and define
$$E = \bigcup_{m=1}^{\infty} E_m \cap (i_{m-1}, i_m);$$
we will show that E has density zero and
$$\lim_{n\to\infty, n\notin E} f(n) = 0.$$
Let us consider $f(k)$ for $k\notin E$. In the interval $(0, i_1]$ we have removed those ks for which $f(k) > 1$; in $(i_1, i_2]$ we have removed those ks for which $f(k) > \frac{1}{2}$; and in $(i_{m-1}, i_m]$ we have removed those ks for which $f(k) > 1/m$. Thus clearly
$$\lim_{n\to\infty, n\notin E} f(n) = 0.$$
However, we have only removed a set of ks of density zero. For if $i_{m-1} < n \le i_m$, then
$$\frac{1}{n}\sum_{k=0}^{n-1} \chi_E(k) = \frac{1}{n}\sum_{k=0}^{i_{m-1}-1} \chi_E(k) + \frac{1}{n}\sum_{k=i_{m-1}}^{n-1} \chi_E(k)$$
$$\le \frac{1}{n}\sum_{k=0}^{i_{m-1}-1} \chi_{E_{m-1}}(k) + \frac{1}{n}\sum_{k=0}^{n-1} \chi_{E_m}(k)$$

2.6. Weak mixing

$$\leq \frac{1}{i_{m-1}} \sum_{k=0}^{i_{m-1}-1} \chi_{E_{m-1}}(k) + \frac{1}{n} \sum_{k=0}^{n-1} \chi_{E_m}(k)$$

$$< \frac{1}{m-1} + \frac{1}{m}.$$

Proof of the Theorem.

(1) $\Rightarrow$ (2): This is clearly true when f and g are characteristic functions of measurable sets. The general statement follows by forming linear combinations and approximating.

(2) $\Rightarrow$ (1): Trivial.

(1) $\Leftrightarrow$ (3): This follows directly from the Lemma, taking

$$f(n) = |\mu(T^n A \cap B) - \mu(A)\mu(B)|.$$

(3) $\Rightarrow$ (4): Let $A, B, C, D \in \mathcal{B}$. By (3), we may choose sets $J_1, J_2 \subset \mathbb{N}$, each of density zero, such that

$$\lim_{n \to \infty, n \notin J_1} |\mu(T^n A \cap C) - \mu(A)\mu(C)| = 0,$$

$$\lim_{n \to \infty, n \notin J_2} |\mu(T^n B \cap D) - \mu(B)\mu(D)| = 0.$$

Let $J = J_1 \cup J_2$. Then J has density zero, and

$$\lim_{n \to \infty, n \notin J} |\mu \times \mu (T \times T)^n ((A \times B) \cap (C \times D))$$
$$\quad - \mu \times \mu(A \times B) \mu \times \mu(C \times D)|$$
$$= \lim_{n \to \infty, n \notin J} |\mu(T^n A \cap C) \mu(T^n B \cap D) - \mu(A)\mu(B)\mu(C)\mu(D)|$$
$$\leq \lim_{n \to \infty, n \notin J} [\mu(T^n A \cap C) |\mu(T^n B \cap D) - \mu(B)\mu(D)|$$
$$\quad + \mu(B)\mu(D) |\mu(T^n A \cap C) - \mu(A)\mu(C)|] = 0.$$

The result then follows from the known equivalence (3) $\Leftrightarrow$ (1) applied to $T \times T$.

(4) $\Rightarrow$ (5): If $T \times T$ is weakly mixing, then so is T itself. Let ergodic $(Y, \mathcal{C}, v, S)$ be given. In order to show that $T \times S$ is ergodic on $X \times Y$, it is enough to show that if $A, B \in \mathcal{B}$ and $C, D \in \mathcal{C}$, then

$$\frac{1}{n} \sum_{k=0}^{n-1} \mu \times v [(T \times S)^k (A \times C) \cap (B \times D)] \to \mu(A)\mu(B)v(C)v(D).$$

However, the left-hand side is

$$\frac{1}{n} \sum_{k=0}^{n-1} \mu(T^k A \cap B) v(S^k C \cap D)$$

$$= \frac{1}{n} \sum_{k=0}^{n-1} \{\mu(A)\mu(B) v(S^k C \cap D)$$
$$\quad + [\mu(T^k A \cap B) - \mu(A)\mu(B)] v(S^k C \cap D)\}.$$

By ergodicity of S, the first term tends to $\mu(A)\mu(B)\nu(C)\nu(D)$. The second term is dominated by

$$\frac{1}{n}\sum_{k=0}^{n-1} |\mu(T^k A \cap B) - \mu(A)\mu(B)|,$$

which tends to 0 as $n \to \infty$ because T is weakly mixing.

(5) $\Rightarrow$ (6): T must itself be ergodic, if it satisfies (5), since $T \times \{1\}(\{1\} =$ the identity transformation on a single point) is ergodic. Therefore, under (5), $T \times T$ is ergodic.

(6) $\Rightarrow$ (3): If $A, B \in \mathscr{B}$, then

$$\frac{1}{n}\sum_{k=0}^{n-1} [\mu(T^k A \cap B) - \mu(A)\mu(B)]^2$$

$$= \frac{1}{n}\sum_{k=0}^{n-1} \mu(T^k A \cap B)^2 - 2\mu(A)\mu(B)\frac{1}{n}\sum_{k=0}^{n-1} \mu(T^k A \cap B) + [\mu(A)\mu(B)]^2$$

$$= \frac{1}{n}\sum_{k=0}^{n-1} \mu \times \mu[(T \times T)^k(A \times A) \cap (B \times B)] - 2\mu(A)\mu(B) \cdot$$

$$\frac{1}{n}\sum_{k=0}^{n-1} \mu \times \mu[(T \times T)^k(A \times X) \cap (B \times X)] + [\mu(A)\mu(B)]^2.$$

Since $T \times T$ is ergodic, as $n \to \infty$ this tends to

$$\mu \times \mu(A \times A)\mu \times \mu(B \times B)$$
$$- 2\mu(A)\mu(B)\mu \times \mu(A \times X)\mu \times \mu(B \times X) + [\mu(A)\mu(B)]^2$$
$$= \mu(A)^2\mu(B)^2 - 2\mu(A)^2\mu(B)^2 + \mu(A)^2\mu(B)^2 = 0.$$

By the Lemma, there is a set J of density zero such that

$$\lim_{n \to \infty, n \notin J} |\mu(T^n A \cap B) - \mu(A)\mu(B)|^2 = 0.$$

Then (3) is immediate.

(6) $\Rightarrow$ (7): Suppose $f \in L^2(X)$ is an eigenfunction, so $Tf = \lambda f$ for some $\lambda \in \mathbb{C}$ with $|\lambda| = 1$. For $(x, y) \in X \times X$, let $g(x, y) = f(x)\overline{f(y)}$. Then

$$(T \times T)g(x, y) = g(Tx, Ty) = f(Tx)\overline{f(Ty)} = \lambda\bar{\lambda}g(x, y) = g(x, y) \text{ a.e.,}$$

and thus g is an invariant function. Assuming that $T \times T$ is ergodic, g, and hence f, must be constant a.e.

(7) $\Rightarrow$ (1): We use the Spectral Theorem. Let V denote the closed linear span in $L^2(X)$ of the eigenfunctions of T. We will show first that if $f \in V^\perp$ and $g \in L^2(X)$, then

$$\lim_{n \to \infty} \frac{1}{n}\sum_{k=0}^{n-1} |(U_T^k f, g)|^2 = 0.$$

2.6. Weak mixing

Fix $f \in V^\perp$ and $g \in L^2(X)$, and let μ be the complex measure on the spectrum $\sigma(U_T)$ of U_T defined by $\mu(A) = (E(A)f, g)$ for all Borel sets $A \subset \sigma(U_T)$ (see p. 19). Note that if $\lambda_0 \in \mathbb{K}$, then

$$U_T E\{\lambda_0\} f = \int_{\sigma(U_T)} \lambda \chi_{\{\lambda_0\}} \, dE(\lambda) f = \lambda_0 \int_{\sigma(U_T)} \chi_{\{\lambda_0\}} \, dE(\lambda) f = \lambda_0 E\{\lambda_0\} f,$$

so $E\{\lambda_0\} f$ is an eigenfunction of U_T. But since $f \in V^\perp$,

$$0 = (E\{\lambda_0\} f, f) = (E\{\lambda_0\}^2 f, f) = (E\{\lambda_0\} f, E\{\lambda_0\} f),$$

so $E\{\lambda_0\} f = 0$. Consequently $\mu\{\lambda_0\} = 0$ for each $\lambda_0 \in \mathbb{K}$.

Now

$$\frac{1}{n} \sum_{k=0}^{n-1} |(U_T^k f, g)|^2 = \frac{1}{n} \sum_{k=0}^{n-1} \left| \int_{\sigma(U_T)} \lambda^k \, d(E(\lambda) f, g) \right|^2$$

$$= \frac{1}{n} \sum_{k=0}^{n-1} \left| \int_{\sigma(U_T)} \lambda^k \, d\mu(\lambda) \right|^2 = \frac{1}{n} \sum_{k=0}^{n-1} \left(\int_{\sigma(U_T)} \lambda^k \, d\mu(\lambda) \right) \left(\int_{\sigma(U_T)} \bar{\zeta}^k \, d\bar{\mu}(\zeta) \right)$$

$$= \frac{1}{n} \sum_{k=0}^{n-1} \iint_{\sigma(U_T) \times \sigma(U_T)} \lambda^k \bar{\zeta}^k \, d\mu(\lambda) \, d\bar{\mu}(\zeta)$$

$$= \iint_{\sigma(U_T) \times \sigma(U_T)} \left(\frac{1}{n} \sum_{k=0}^{n-1} \lambda^k \bar{\zeta}^k \right) d\mu(\lambda) \, d\bar{\mu}(\zeta)$$

$$= \iint_{\sigma(U_T) \times \sigma(U_T)} \frac{1}{n} \frac{1 - (\lambda\bar{\zeta})^n}{1 - \lambda\bar{\zeta}} \, d\mu(\lambda) \, d\bar{\mu}(\zeta).$$

(This is permissible since $1 - \lambda\bar{\zeta} = 0$ if and only if $\lambda = \zeta$. But the diagonal Δ of $\sigma(U_T) \times \sigma(U_T)$ has measure zero by Tonnelli's Theorem, since μ assigns measure zero to individual points:

$$\iint \chi_\Delta \, d(\mu \times \bar{\mu}) = \int_{\sigma(U_T)} \left(\int_{\sigma(U_T)} \chi_\Delta(x, y) \, d\mu(x) \right) d\bar{\mu}(y)$$

$$= \int_{\sigma(U_T)} \mu\{y\} \, d\bar{\mu}(y) = 0.)$$

Now $|(1/n)(1 - (\lambda\bar{\zeta})^n)/(1 - \lambda\bar{\zeta})| \to 0$ a.e., so by the Bounded Convergence Theorem

$$\lim_{n \to \infty} \frac{1}{n} \sum_{k=0}^{n-1} |(U_T^k f, g)|^2 = 0.$$

70   2. The fundamentals of ergodic theory

From Lemma 6.2,

$$\lim_{n\to\infty}\frac{1}{n}\sum_{k=0}^{n-1}|(U_T^k f, g)| = 0 \quad \text{for } f\in V^\perp \text{ and } g\in L^2(X).$$

Now let $f, g \in L^2(X)$ be arbitrary. Then, assuming that the only eigenfunctions of T are the constants, $f - (f, 1) \in V^\perp$. Therefore,

$$0 = \lim_{n\to\infty}\frac{1}{n}\sum_{k=0}^{n-1}|(U_T^k f - (f, 1), g)|$$

$$= \lim_{n\to\infty}\frac{1}{n}\sum_{k=0}^{n-1}|(U_T^k f, g) - (f, 1)\overline{(g, 1)}|,$$

so T is weakly mixing.

Remark 6.3. In statement (3) of Theorem 2.6.1, it is in fact possible to select a single set $J \subset \mathbb{Z}^+$ of density 0 such that

$$\lim_{n\to\infty, n\notin J} |\mu(T^n A \cap B) - \mu(A)\mu(B)| = 0 \text{ for all } A, B\in \mathscr{B}.$$

This has been noted by S. Kakutani and by L. K. Jones (1971). We select a set $J_{i,j}$ (as in (3)) corresponding to each pair A_i, A_j in a countable dense set in $\mathscr{B}$. Renumber these as $J_1, J_2, \ldots$. Let $\{\varepsilon_n\}$ be a sequence decreasing to 0. Choose n_1 so that

$$\frac{1}{n}\sum_{k=0}^{n-1}\chi_{J_1}(k) < \varepsilon, \text{ if } n \geq n_1.$$

Choose $n_2 > n_1$ so that

$$\frac{1}{n}\sum_{k=0}^{n-1}\chi_{J_1}(k) < \tfrac{1}{2}\varepsilon_2 \text{ and } \frac{1}{n}\sum_{k=0}^{n-1}\chi_{J_2}(k) < \tfrac{1}{2}\varepsilon_2 \text{ if } n \geq n_2.$$

Continue in this manner, choosing $n_r > n_{r-1}$ so that

$$\frac{1}{n}\sum_{k=0}^{n-1}\chi_{J_i}(k) < \frac{\varepsilon_r}{r} \quad \text{for } i = 1, \ldots r \text{ if } n \geq n_r.$$

Define

$$J = \bigcup_{i=1}^{\infty}[J_i \cap (n_i, \infty)].$$

Then J has density 0, since if $n_r \leq n \leq n_{r+1}$, we have

$$\frac{1}{n}\sum_{k=0}^{n-1}\chi_J(k) \leq \frac{1}{n}\sum_{k=0}^{n-1}\sum_{i=1}^{r}\chi_{J_i}(k) \leq \varepsilon_r,$$

which tends to 0 as $r \to \infty$. However, since J contains a tail of each J_r,

2.6. Weak mixing

we have
$$\lim_{n\to\infty,\, n\notin J} a_n = \lim_{n\to\infty,\, n\notin J_r} a_n \text{ for each } r,$$
and hence
$$\lim_{n\to\infty,\, n\notin J} |\mu(T^n A_i \cap A_j) - \mu(A_i)\mu(A_j)| = 0$$
for each pair A_i, A_j in the countable dense set in $\mathcal{B}$. Then given $A, B \in \mathcal{B}$ and $\varepsilon > 0$, if we choose A_i, A_j with $\mu(A_i \Delta A) < \varepsilon$ and $\mu(A_j \Delta B) < \varepsilon$, we will have
$$\lim_{n\to\infty,\, n\notin J} |\mu(T^n A \cap B) - \mu(A)\mu(B)|$$
$$\leq \lim_{n\to\infty,\, n\notin J} \{|\mu(T^n A \cap B) - \mu(T^n A_i \cap A_j)|$$
$$+ |\mu(T^n A_i \cap A_j) - \mu(A_i)\mu(A_j)|$$
$$+ |\mu(A_i)\mu(A_j) - \mu(A)\mu(B)|\} < 4\varepsilon$$
(e.g., the third term is no greater than
$$|\mu(A_i)\mu(A_j) - \mu(A_i)\mu(B)| + |\mu(A_i)\mu(B) - \mu(A)\mu(B)| < 2\varepsilon).$$
Since $\varepsilon > 0$ was arbitrary,
$$\lim_{n\to\infty,\, n\notin J} |\mu(T^n A \cap B) - \mu(A)\mu(B)| = 0 \quad \text{for all } A, B \in \mathcal{B}.$$

It is not easy to give examples of m.p.t.s which are weakly mixing but not strongly mixing – see Section 4.5. But there is a sense in which *almost every m.p.t. is weakly mixing but not strong mixing*. Because of the isomorphism theorem and von Neumann's theorem, every m.p.t. on a Lebesgue space is isomorphic to a m.p.t. on the ordinary measure space of the unit interval, $[0, 1)$. We will regard two m.p.t.s on $[0, 1)$ as *equivalent* in case they differ only on a set of measure 0. The resulting set G of equivalence classes is the *group of all automorphisms of* $[0, 1)$.

G is a complete metric topological group with respect to each of the following two topologies:

the *weak topology*, in which $T_n \to T$ if and only if $T_n A \to TA$ (in the sense that $\mu(T_n A \Delta TA) \to 0$) for all $A \in \mathcal{B}$; and

the *strong topology*, determined by either of the equivalent metrics

$$d_1(S, T) = \mu\{x \in X : Sx \neq Tx\} \quad \text{or}$$
$$d_2(S, T) = \sup\{\mu(SA \Delta TA) : A \in \mathcal{B}\}.$$

(The weak topology is the one induced in G by both the strong and weak topologies for operators on L^2, which coincide for unitary operators.)

Halmos (1944) proved that *with respect to the weak topology, the set of weakly mixing m.p.t.s is residual* (i.e. the complement of a first category set); while Rokhlin (1948) showed that *with respect to the weak topology the set of all strongly mixing transformations is of the first category.*

Thus in this particular sense, *the 'generic' m.p.t. is weakly mixing but not strongly mixing.*

The details of the proof, which involves approximation of arbitrary m.p.t.s by periodic m.p.t.s, can be found in Halmos (1956). Katok and Stepin (1967) investigated the possible speeds of such approximations, analogous to the speeds of approximation of irrational numbers by rationals. This refined theory has many applications to the study of spectra, isomorphism, entropy, and genericity of m.p.t.s and is useful for the construction of examples.

There is another sense in which weak mixing is generic. Historically, the first genericity result was the theorem of Oxtoby and Ulam (1941), which stated that *for a finite-dimensional compact manifold with a nonatomic measure which is positive on open sets the set of ergodic measure-preserving homeomorphisms is generic in the strong topology.* (Note that in G, with respect to the strong topology the set of all ergodic m.p.t.s is *nowhere dense*!) Their purpose was to prove the *existence* of ergodic m.p.t.s on manifolds; cf. the realization problem discussed in 4.4. Katok and Stepin (1970) showed that in fact the weakly mixing measure-preserving homeomorphisms of a compact manifold are generic. Recently Alpern (1978) has shown how to obtain these results and the Halmos–Rokhlin theorems simultaneously. See also Oxtoby (1973), White (1974), Prasad (1979), Anosov and Katok (1970) and Fathi and Herman (1977) for recent work on this circle of questions.

Exercises

1. Can a weakly mixing transformation have nonmeasurable (hence nonconstant) eigenfunctions?
2. Prove that if T is weakly mixing, then so are T^n and $\sqrt[n]{T}$ for any $n = 1, 2, \ldots$ (Here $\sqrt[n]{T}$ means any m.p.t. S for which $S^n = T$.)
3. Prove that there is a universal system $(Y, \mathscr{C}, \nu, S)((Y, \mathscr{C}, \nu)$ is a probability space, $S: Y \to Y$ a m.p.t.) such that a m.p.t. $T: X \to X$ on a probability space $(X, \mathscr{B}, \mu)$ is weakly mixing if and only if $T \times S$ is ergodic.
4. There are examples of weakly mixing m.p.t.s that are not strongly mixing (see 4.5). For now, consider some easier counterexamples.
   (a) Find an example of a sequence $\{a_n\}$ for which
   $$\lim_{n \to \infty} \frac{1}{n} \sum_{k=0}^{n-1} |a_k| = 0$$

2.6. Weak mixing

but

$$\lim_{n\to\infty} a_n \neq 0.$$

(b) A sequence $\{A_1, A_2, \ldots\}$ of measurable sets, each having measure α, is called *strongly mixing* if

$$\lim_{n\to\infty} \mu(A_n \cap B) = \alpha\mu(B) \quad \text{for all } B \in \mathcal{B},$$

and it is called *weakly mixing* if

$$\lim_n \frac{1}{n} \sum_{k=0}^{n-1} |\mu(A_k \cap B) - \alpha\mu(B)| = 0.$$

Give an example of a sequence which is weakly mixing but not strongly mixing.

(c) The sequence $\{A_1, A_2, \ldots\}$ as above is called *mixing of order k* ($k = 1, 2, \ldots$) if

$$\lim_{\substack{\inf_i n_i \to \infty \\ \inf_{i\neq j}|n_i - n_j| \to \infty}} \mu(A_{n_1} \cap A_{n_2} \cap \ldots \cap A_{n_k} \cap B) = \alpha^k \mu(B) \text{ for all } B \in \mathcal{B}.$$

Give an example of a sequence that is mixing of order 1 but not of order 2.

5. (England and Martin 1968). Show that T is weakly mixing if and only if for each pair of sets $A, B \in \mathcal{B}$ with positive measure, there is a set $J \subset \mathbb{N}$ of density 0 such that

$$\mu(T^n A \cap B) > 0 \quad \text{for all } n \notin J.$$

Can such a set J be found that works simultaneously for all $A, B \in \mathcal{B}$?

6. (Ornstein 1972). If T^n is ergodic for all n, and if there is c such that
$$\lim\sup_{n\to\infty} \mu(T^n A \cap B) \leq c\mu(A)\mu(B) \quad \text{for all } A, B \in \mathcal{B},$$
then T is strongly mixing. (Hint: First prove that T is weakly mixing. Then let m be a cluster point of the measures v_n on $X \times X$ defined by

$$\int f(x, y) dv_n = \int f(x, T^n x) d\mu.$$

Show that $m = \mu \times \mu$.)

7. Find a metric for the weak topology on G.
8. Which of the following classes of m.p.ts are closed under the formation of Cartesian products (and not just Cartesian squares):
   (a) ergodic;
   (b) weakly mixing;
   (c) strongly mixing?

3
More about almost everywhere convergence

Many of the fundamental convergence theorems of ergodic theory, analysis, and probability can be proved by the technique of maximal inequalities and Banach's Principle. We give proofs of this kind for the Ergodic Theorem, Local Ergodic Theorem (via the Lebesgue Differentiation Theorem), and existence of the ergodic Hilbert transform. For novelty, the Martingale Convergence Theorem and Chacon-Ornstein Theorem are proved without using maximal inequalities. We begin with a careful study of the ergodic maximal function, and find that for one-parameter ergodic flows the maximal inequality is actually an equality. We show that the Ergodic Theorem cannot be improved, in that no statement is possible about the speed with which the convergence of the ergodic averages takes place.

3.1. More about the Maximal Ergodic Theorem

In this section we undertake a deeper study of the ergodic maximal functions

$$f^*(x) = \sup_{n \geq 1} \frac{1}{n} \sum_{k=0}^{n-1} f(T^k x) \quad \text{(for a single m.p.t. } T\text{)}$$

and

$$f^*(x) = \sup_{t > 0} \frac{1}{t} \int_0^t f(T_s x) \, ds \quad \text{(for a flow } \{T_t\}\text{)},$$

including the following topics:

(A) Hopf's extension of the maximal inequality to the case of a positive contraction T on $L^1(X)$.

(B) Proof that the maximal inequality is actually an *equality* for ergodic flows.

(C) The theory of sign changes of

$$F_t(x) = \int_0^t f(T_s x) \, ds \quad \text{and} \quad S_n f(x) = \sum_{k=0}^{n-1} f(T^k x)$$

that follows from (B).

3.1. More about the maximal ergodic theorem

(D) A characterization of when the maximal function f^* of an L^1 function is itself integrable: Wiener's Dominated Ergodic Theorem together with the converse that follows from (B).

A. Positive contractions.

Suppose that $T: L^1(X) \to L^1(X)$ is *positive contraction*. This means that if $f \in L^1$, then $f \geq 0$ a.e. implies $Tf \geq 0$ a.e., and $\|T\| \leq 1$, i.e. $\|Tg\|_1 \leq \|g\|_1$ for all $g \in L^1$. Ergodic theorems can be proved for certain such operators T also, whether or not they arise from m.p.ts on X (see Section 3.8). Usually the essential first step is the proof of the corresponding maximal inequality.

Theorem 1.1 Maximal Ergodic Theorem for Operators (Hopf 1954) If T is a positive contraction on $L^1(X)$ and $f \in L^1$, then

$$\int_{\{f^*>0\}} f \, d\mu \geq 0.$$

(Here $f^*(x) = \sup_{n \geq 1} (1/n) \sum_{k=0}^{n-1} T^k f(x)$, as might be guessed.)

Proof (Garsia 1965): For each $n = 1, 2, \ldots$ let

$$f_n = \sup_{1 \leq j \leq n} \sum_{k=0}^{j-1} T^k f.$$

Then

$$f = f_1 \leq f_2 \leq \ldots$$

and

$$\{f^* > 0\} = \bigcup_{n=1}^{\infty} \{f_n > 0\}.$$

Also,

$$f_n \leq f + Tf_n^+ \text{ for all } n \geq 1;$$

for clearly

$$f_1 = f \leq f + Tf_1^+,$$

and for $1 < j \leq n$

$$\sum_{k=0}^{j-1} T^k f = f + T\left(\sum_{k=0}^{j-2} T^k f\right) \leq f + Tf_{j-1} \leq f + Tf_n \leq f + Tf_n^+.$$

Therefore

3. More about almost everywhere convergence

$$\int_{\{f_n>0\}} f\,d\mu \geq \int_{\{f_n>0\}} f_n\,d\mu - \int_{\{f_n>0\}} Tf_n^+\,d\mu$$

$$= \int f_n^+\,d\mu - \int_{\{f_n>0\}} Tf_n^+\,d\mu \geq \int f_n^+\,d\mu - \int Tf_n^+\,d\mu \geq 0,$$

since $\|T\| \leq 1$ implies $\int Tg\,d\mu \leq \int g\,d\mu$ if $0 \leq g \in L^1$. The conclusion follows from the Monotone Convergence Theorem, letting $n \to \infty$.

In 3.8 we will examine the relationship of this maximal inequality with the filling scheme and will obtain a proof of the Chacon–Ornstein Theorem that actually avoids the maximal inequality.

B. The maximal equality

Let us consider now a one-parameter flow $\{T_t : -\infty < t < \infty\}$ of measure-preserving transformations on a probability space $(X, \mathscr{B}, \mu)$. It is assumed that $T_t(x)$ is jointly measurable in x and t and $T_{s+t} = T_s T_t$ for $s, t \in \mathbb{R}$. For $f \in L^1(X)$, define

$$F_t(x) = \int_0^t f(T_s x)\,ds$$

and

$$f^*(x) = \sup_{t>0} \frac{1}{t} F_t(x) = \sup_{t>0} \frac{1}{t} \int_0^t f(T_s x)\,ds.$$

Recall the Maximal Ergodic Theorem.

Theorem 1.2 *Maximal Ergodic Theorem (Wiener 1939, Yosida and Kakutani 1939)* If $f \in L^1$ and $\alpha \in \mathbb{R}$, then

$$\int_{\{f^*>\alpha\}} f\,d\mu \geq \alpha\mu\{f^* > \alpha\}.$$

Actually this inequality is often an equality.

Theorem 1.3 *Maximal Ergodic Equality (Marcus and Petersen 1979; see also Engel and Kakutani 1981).* If $\{T_t\}$ is *ergodic*, in that every measurable subset A of X which is invariant under the flow ($T_t A = A$ for $t \in \mathbb{R}$) has measure 0 or 1, if $f \in L^1(X)$, and if $\alpha \geq \int f\,d\mu$, then

$$\int_{\{f^*>\alpha\}} f\,d\mu = \alpha\mu\{f^* > \alpha\}.$$

(If $\alpha < \int f\,d\mu$, then $\mu\{f^* > \alpha\} = 1$.)

3.1. More about the maximal ergodic theorem

We will prove these results simultaneously, using Hartman's approach (1947), which is based on the following simple lemma.

Lemma 1.4 Rising Sun Lemma (F. Riesz 1931, 1932) Let h be a real-valued continuous function on an interval $[a, b] \subset \mathbb{R}$ and let

$$S = \{t \in (a, b) : \text{there is } t' > t \text{ with } h(t') > h(t)\}.$$

(If the graph of h represented a mountain range, S would be the set of points that were in the shade as the sun rose at the positive end of the x axis.) Then
  (i) S is open;
  (ii) if $S = \bigcup (a_k, b_k)$ is the decomposition of S as the union of pairwise disjoint open intervals, then $h(a_k) \leq h(b_k)$ for all k; and
  (iii) equality holds in (ii) except possibly when $a_k = a$.

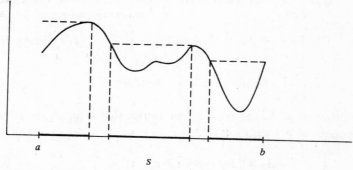

The proof of the Lemma is easy and indeed is already clear from the picture.

Proof of the Theorems: We let

$$\underline{A} = \{x : F_t(x) \geq 0 \text{ for all } t > 0\}$$

and notice that we need to prove that

$$\int_{\underline{A}^c} f \, d\mu \leq 0,$$

and that equality holds in case $\int f \, d\mu \geq 0$ and $\{T_t\}$ is ergodic – for, as before, the results will then follow after replacing f by $\alpha - f$. (If $f = \alpha - g$, then $\underline{A}^c(f) = \{x : \text{there is } t > 0 \text{ with } F_t(x) < 0\} = \{x : \text{there is } t > 0 \text{ with } \alpha t - \int_0^t g(T_s x) \, ds < 0\} = \{g^* > \alpha\}$, and $\int_{\underline{A}^c} f \, d\mu \leq 0$ says $\int_{\{g^* > \alpha\}} (\alpha - g) \, d\mu \leq 0$, or $\int_{\{g^* > \alpha\}} g \, d\mu \geq \alpha \mu \{g^* > \alpha\}$.)

For each $x \in X$, let

$$W_x = \{t \in \mathbb{R} : T_t x \in \underline{A}^c\} = \{t \in \mathbb{R} : \text{there is } t' > t \text{ with } F_{t'}(x) < F_t(x)\},$$

and let
$$W_x^1 = W_x \cap (0, 1).$$
By the Rising Sun Lemma (with h replaced by $-h$),
$$W_x^1 = \bigcup (a_k, b_k)$$
with
$$F_{a_k}(x) \geq F_{b_k}(x) \text{ for all } k.$$
Thus
$$\int_{W_x^1} f(T_s x) ds = \sum_k \int_{a_k}^{b_k} f(T_s x) ds = \sum_k [F_{b_k}(x) - F_{a_k}(x)] \leq 0,$$
so that
$$0 \geq \int_X \int_{W_x^1} f(T_s x) ds d\mu(x) = \int\int_{\{(x,s): 0 < s < 1, T_s x \in \underline{A}^c\}} f(T_s x) ds d\mu(x)$$
$$= \int_0^1 \int_{\{x: T_s x \in \underline{A}^c\}} f(T_s x) d\mu(x) ds = \int_0^1 \int_{T_s^{-1}\underline{A}^c} f(T_s x) d\mu(x) ds$$
$$= \int_0^1 \int_{\underline{A}^c} f(x) d\mu(x) ds = \int_{\underline{A}^c} f(x) d\mu(x).$$

In order to prove the *equality*, we notice that if (a_k, b_k) is a bounded component of W_x, then $a_k \notin W_x$, so
$$\int_{a_k}^t f(T_s x) ds \geq 0 \text{ for every } t \geq a_k,$$
and in particular
$$\int_{a_k}^{b_k} f(T_s x) ds \geq 0, \text{ hence } \int_{a_k}^{b_k} f(T_s x) ds = 0.$$
If $\mu(\underline{A}) > 0$, then (for almost all x) all the components of W_x must be bounded, since $\{T_t\}$ is ergodic.

But if $\mu(\underline{A}) = 0$, then by the inequality just proved,
$$0 \leq \int f d\mu = \int_{\underline{A}} f d\mu + \int_{\underline{A}^c} f d\mu = \int_{\underline{A}^c} f d\mu \leq 0,$$
so that the equality
$$\int_{\underline{A}^c} f d\mu = 0$$
holds in this case. We suppose then that the components of W_x are bounded.

3.1. More about the maximal ergodic theorem

For each $x \in X$, let

$$J_x = \begin{cases} \text{the component of } W_x \text{ containing } 0, & \text{if } x \in \underline{A}^c \\ \varnothing, & \text{if } x \in \underline{A} \end{cases}$$

and

$$\lambda(x) = \begin{cases} \dfrac{1}{l(J_x)} & \text{if } x \in \underline{A}^c \\ 0 & \text{if } x \in \underline{A} \end{cases}$$

Then if $x \in \underline{A}^c$, we see that $-s \in J_x$ if and only if $s \in J_{T_{-s}x}$. Thus $\chi_{\underline{A}^c}(x) = \int_{\{s:\, -s \in J_x\}} \lambda(T_{-s}x)\,ds$, and applying the measure-preserving change of variables $(x, s) \to (T_s x, s)$ on $X \times \mathbb{R}$ allows us to compute that

$$\int_{\underline{A}^c} f(x)\,d\mu(x) = \int_X f(x)\chi_{\underline{A}^c}(x)\,d\mu(x)$$

$$= \int_X f(x)\int_{\{s:\, -s \in J_x\}} \lambda(T_{-s}x)\,ds\,d\mu(x)$$

$$= \int_X f(x)\int_{\{s:\, s \in J_{T_{-s}x}\}} \lambda(T_{-s}x)\,ds\,d\mu(x)$$

$$= \int_X \int_{J_x} \lambda(x)f(T_s x)\,ds\,d\mu(x) = \int_X \lambda(x)\left(\int_{J_x} f(T_s x)\,ds\right)d\mu(x) = 0$$

(since we have seen that the integrals of f over the components of W_x are all 0).

Remark 1.5 If $A = \{x \in X : F_t(x) > 0 \text{ for all } t > 0\}$ and $\int f\,d\mu \geq 0$, then

$$\int_{A^c} f\,d\mu = \int_{\underline{A}^c} f\,d\mu = 0.$$

Proof: Because of the Local Ergodic Theorem (see 3.3),

$$\lim_{t \to 0^+} \frac{1}{t}\int_0^t f(T_s x)\,ds = f(x) \text{ a.e.};$$

thus if $x \in \underline{A}, f(x) \geq 0$ almost surely.

It is enough to prove, then, that

$$\mu\{x \in \underline{A}\setminus A : f(x) > 0\} = 0.$$

If $f(x) > 0$, then almost surely $F_t(x) > 0$ for all small $t > 0$. Let

$$E_n = \{x \in \underline{A}\setminus A : f(x) > 0 \text{ and } F_t(x) > 0 \text{ for } 0 < t < 1/n\};$$

we will show that $\mu(E_n) = 0$ for each $n = 1, 2, \ldots$.

If $\mu(E_n) > 0$, by a theorem of von Neumann (1932 c) it is possible to choose t with $0 < t < 1/n$ and $\mu(E_n \cap T_{-t} E_n) > 0$. If $x \in E_n \cap T_{-t} E_n$, since $x \in \underline{A} \setminus A$ we can find $t_0 > 0$ with $F_{t_0}(x) = 0$. Then $t_0 \geq 1/n > t$, and

$$0 = F_{t_0}(x) = F_t(x) + F_{t_0-t}(T_t x),$$

which is impossible since $F_t(x) > 0$ and $F_{t_0-t}(T_t x) \geq 0$.

C. *Sign changes of the partial sums. The one-parameter flow case*

Suppose that $\{T_t\}$ is ergodic and $\int f d\mu > 0$. Then $\underline{A}^c$ coincides with the *oscillation set*

$$\mathcal{O} = \{x \in X : \text{there are } t, t' > 0 \text{ with } F_t(x) > 0 \text{ and } F_{t'}(x) < 0\},$$

while A^c coincides with the *crossing set*

$$\mathscr{C} = \{x \in X : \text{there is } t > 0 \text{ with } F_t(x) = 0\}.$$

(This observation uses the Ergodic Theorem.) The Maximal Equality asserts, then, that

$$\int_{\mathcal{O}} f d\mu = \int_{\mathscr{C}} f d\mu = 0.$$

We claim that this formula is true whether or not $\{T_t\}$ is ergodic and irrespective of the sign of $\int f d\mu$.

First, if $\int f d\mu < 0$, then by considering $-f$, for which $\int -f d\mu > 0$, we see that

$$\mathcal{O}(-f) = \mathcal{O}(f) \text{ and } \mathscr{C}(-f) = \mathscr{C}(f),$$

so that still

$$\int_{\mathcal{O}} f d\mu = \int_{\mathscr{C}} f d\mu = 0.$$

If $\int f d\mu = 0$, then we have

$$\int_{A(f)} f d\mu = \int_{A(-f)} f d\mu = \int f d\mu = 0,$$

so that

$$0 = \int_{A(f) \cup A(-f)} f d\mu = \int f d\mu$$

and

$$\int_{\mathscr{C}} f d\mu = \int_{[A(f) \cup A(-f)]^c} f d\mu = 0.$$

3.1. More about the maximal ergodic theorem

Also, by the preceding Remark,

$$\int_{\mathscr{C}\setminus\mathscr{O}} f\,d\mu = \int_{A(-f)\setminus A(-f)} f\,d\mu + \int_{A(f)\setminus A(f)} f\,d\mu = 0,$$

so that also

$$\int_{\mathscr{O}} f\,d\mu = 0$$

in this case. We have proved the formula, then, in the ergodic case, no matter what the sign of $\int f\,d\mu$ may be.

In order to deal with the general, nonergodic case, we appeal to the theorem on ergodic decompositions. We may assume that X is the disjoint union of sets X_ω, each equipped with a σ-algebra $\mathscr{B}_\omega$ and a probability measure μ_ω, and that T acts ergodically on each $(X_\omega, \mathscr{B}_\omega, \mu_\omega)$. The indexing set is another probability space $(\Omega, \mathscr{F}, P)$, and we have the formula

$$\int_X f\,d\mu = \int_\Omega \int_{X_\omega} f\,d\mu_\omega\,dP(\omega) \quad \text{for } f \in L^1(X).$$

Now for each ω,

$$\mathscr{O}_\omega = \mathscr{O} \cap X_\omega \text{ and } \mathscr{C}_\omega = \mathscr{C} \cap X_\omega$$

(where $\mathscr{O}_\omega$, for example, is the oscillation set in X_ω of f restricted to X_ω, with respect to T restricted to X_ω); therefore

$$\int_\mathscr{O} f\,d\mu = \int_\Omega \int_{\mathscr{O}_\omega} f\,d\mu_\omega\,dP(\omega) = 0$$

and

$$\int_\mathscr{C} f\,d\mu = \int_\Omega \int_{\mathscr{C}_\omega} f\,d\mu_\omega\,dP(\omega) = 0.$$

We have proved the following comprehensive result.

Theorem 1.6 For any measure-preserving flow $\{T_t\}$ and any $f \in L^1(X)$,

$$\int_\mathscr{O} f\,d\mu = \int_\mathscr{C} f\,d\mu = 0.$$

David Engel has pointed out that one can prove this result without making use of the full force of the theorem on ergodic decompositions. First, we extend the Maximal Equality to the non-ergodic case as follows.

Theorem 1.7 Let $\{T_t\}$ be a measure-preserving flow on $(X, \mathscr{B}, \mu)$, let $f \in L^1(X)$, and denote by $\mathscr{I}$ the σ-algebra of all T_t-invariant measurable sets. If $E(f|\mathscr{I}) \geq 0$, then

$$\int_{I \cap A^c} f \, d\mu = 0 \quad \text{for each } I \in \mathscr{I}.$$

Proof: In case $E(f|\mathscr{I}) > 0$, we use the Ergodic Theorem to show that the components of W_x are bounded. On an invariant set where $E(f|\mathscr{I}) = 0$, replace f by $f_\varepsilon = f + \varepsilon$. Then

$$\int_{I \cap A(f_\varepsilon)^c} f \, d\mu = 0$$

and $\bigcap_{\varepsilon > 0} A(f_\varepsilon) \doteq A(f)$, so the conclusion follows by approximation. The rest of the proof carries through as before.

Corollary 1.8 (Engel and Kakutani 1981) If $\alpha(x)$ is measurable with respect to $\mathscr{I}$ and $\alpha(x) \geq E(f|\mathscr{I})$ a.e., then

$$\int_{I \cap \{x : f^*(x) > \alpha(x)\}} f \, d\mu = \int_{I \cap \{x : f^*(x) > \alpha(x)\}} \alpha \, d\mu \quad \text{for each } I \in \mathscr{I}.$$

Now the proof that $\int_\mathscr{O} f \, d\mu = \int_\mathscr{C} f \, d\mu = 0$ proceeds as before. We divide $\mathscr{O}$ into invariant sets according to the value of sgn $E(f|\mathscr{I})$ and observe that the integral of f over each of these sets is 0.

We may also consider crossings of, and oscillations about, levels other than 0. For any $\alpha \in \mathbb{R}$, let

$$\mathscr{O}_\alpha = \left\{ x \in X : \text{there are } t, t' > 0 \text{ with } \frac{1}{t} F_t(x) > \alpha \text{ and } \frac{1}{t'} F_{t'}(x) < \alpha \right\}$$

$$\mathscr{C}_\alpha = \left\{ x \in X : \text{there is } t > 0 \text{ with } \frac{1}{t} F_t(x) = \alpha \right\}.$$

Theorem 1.9 For any measure-preserving flow $\{T_t\}$, $f \in L^1(X)$, and $\alpha \in \mathbb{R}$,

$$\int_{\mathscr{O}_\alpha} f \, d\mu = \alpha \mu(\mathscr{O}_\alpha) \quad \text{and} \quad \int_{\mathscr{C}_\alpha} f \, d\mu = \alpha \mu(\mathscr{C}_\alpha).$$

With a slight auxiliary argument for the converse, the following result is also immediate. (Here $\doteq$ means equals up to a set of measure 0.)

Corollary 1.10 If $\mathscr{C}_\alpha \doteq X$, then $\alpha = \int f \, d\mu$. The converse holds if $\{T_t\}$ is ergodic.

3.1. More about the maximal ergodic theorem

Thus $\alpha = \int f \, d\mu$ is the only level which can be crossed by almost all the graphs $\{(1/t)F_t(x): -\infty < t < \infty\}$, even if $\{T_t\}$ is not ergodic.

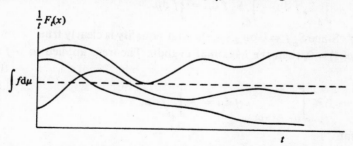

If $\alpha \neq \int f \, d\mu$, there is a positive-measure set of trajectories which stay entirely below the level α, or such a set of trajectories which stay entirely above the level α.

Remark 1.11 The following are equivalent:
(1) $C_\alpha \doteq X$.
(2) $\int_{X_\omega} f \, d\mu_\omega = \alpha$ for almost all ω.
(3) $\bar{f} = \lim\limits_{t \to \infty} \dfrac{1}{t} F_t(x)$ equals α a.e.

The discrete case

If $T: X \to X$ is a single measure-preserving transformation, it is not hard to see that equality cannot hold in the Maximal Ergodic Theorem the same way as it does for a flow. (If we deal, however, with a function that takes values only in $\{-1, 0, 1\}$, then again there is a sort of continuity and the equality can be recovered.) For $f \in L^1(X)$, we consider the partial sums

$$S_n f(x) = \sum_{k=0}^{n-1} f(T^k x),$$

the extrema

$$S_* f(x) = \inf_{n \geq 1} S_n f(x) \quad \text{and} \quad S^* f(x) = \sup_{n \geq 1} S_n f(x),$$

and the sets of constant sign

$$A = \{x \in X : S_n f(x) > 0 \text{ for all } n \geq 1\},$$
$$\underline{A} = \{x \in X : S_n f(x) \geq 0 \text{ for all } n \geq 1\},$$
$$E = \{x \in X : S_n f(x) < 0 \text{ for all } n \geq 1\},$$
$$\underline{E} = \{x \in X : S_n f(x) \leq 0 \text{ for all } n \geq 1\}.$$

3. More about almost everywhere convergence

Theorem 1.12 If T is ergodic and $\int f\,d\mu \geq 0$, then

$$\int_A S_* f\,d\mu = \int_{\underline{A}} S_* f\,d\mu = \int f\,d\mu.$$

Proof: Since $S_* f = 0$ on $\underline{A}\setminus A$, the first equality is clearly true.

If $\mu(\underline{A}) = 0$, then the Maximal Ergodic Theorem applied to $-f$ shows that

$$0 \leq \int_{\{(-f)^* > 0\}} -f\,d\mu = -\int_{\underline{A}^c} f\,d\mu,$$

so that

$$\int_{\underline{A}^c} f\,d\mu \leq 0$$

and we have

$$0 \leq \int f\,d\mu = \int_{\underline{A}^c} f\,d\mu \leq 0,$$

whence $\int f\,d\mu = 0 = \int_A S_* f\,d\mu$, which proves the Theorem in this case.

Suppose now that $\mu(\underline{A}) > 0$. Then for almost every $x \in X$ there is a smallest $n(x) \geq 1$ such that

$$T^{n(x)} \in \underline{A}.$$

I claim that

$$S_* f(x) = \sum_{k=0}^{n(x)-1} f(T^k x).$$

For $T^{n(x)}x \in \underline{A}$, so that

$$\sum_{k=n(x)}^{m} f(T^k x) \geq 0 \quad \text{for all } m \geq n(x).$$

This implies that the infimum $S_* f(x)$ is achieved by some i with $1 \leq i \leq n(x)$:

$$S_* f(x) = \sum_{k=0}^{i-1} f(T^k x).$$

Then for $m \geq i$,

$$\sum_{k=i}^{m} f(T^k x) = \sum_{k=0}^{m} f(T^k x) - \sum_{k=0}^{i-1} f(T^k x) \geq 0,$$

so that $T^i x \in \underline{A}$ and hence $i = n(x)$.

3.1. More about the maximal ergodic theorem

Now we may compute that

$$\int f\, d\mu = \int_{\underline{A}} \sum_{k=0}^{n(x)-1} f(T^k x)\, d\mu(x) = \int_{\underline{A}} S_* f\, d\mu.$$

Remark 1.13 The preceding argument involves a sort of discrete Rising Sun Lemma. The Theorem is also, via suspensions (i.e. flows built under constant functions), a corollary of the Maximal Equality for flows.

Corollary 1.14 If T is ergodic, $f \in L^1(X)$ with $\int f\, d\mu \geq 0$, and f takes values only in $\{-1, 0, 1\}$, then

$$\mu(A) = \int_A f\, d\mu = \int f\, d\mu.$$

This last result is familiar in the theory of random walks. Suppose we flip a coin with probability of heads equal to $p > 1/2$ and that of tails equal to $q = 1 - p$, and we move up one unit for each head obtained and down one for each tail. (i.e., T is the Bernoulli shift $\mathscr{B}(p, q)$ on the symbols ± 1.) Then $S_n f(x)$ gives the walker's height after n (independent) flips. The Corollary states that the probability of always remaining above 0 is $p - q$. Our result applies to all stationary processes, not just the independent, identically-distributed ones; on the other hand, the conclusion of the Corollary remains true for some non-stationary processes too, as for example in the Ballot Problem (see Feller 1950, p. 69).

Let us turn now to the unrestricted discrete case: we no longer assume that T is ergodic or $\int f\, d\mu \geq 0$.

Theorem 1.15 If $T: X \to X$ is a measure-preserving transformation and $f \in L^1(X)$, then

$$\int_A S_* f\, d\mu + \int_E S^* f\, d\mu = \int f\, d\mu.$$

Proof: We again use an ergodic decomposition of X into $\{(X_\omega, \mathscr{B}_\omega, \mu_\omega): \omega \in \Omega\}$. (As before, arguments avoiding ergodic decompositions are also available.) If

$$\int_{X_\omega} f\, d\mu_\omega \geq 0, \text{ then } \int_{X_\omega} f\, d\mu_\omega = \int_{A \cap X_\omega} S_* f\, d\mu_\omega;$$

while if

$$\int_{X_\omega} f\, d\mu_\omega < 0, \text{ then } \int_{X_\omega} f\, d\mu_\omega = \int_{E \cap X_\omega} S^* f\, d\mu_\omega$$

(as may be seen by considering $-f$ in place of f). Thus

$$\int f\,d\mu = \int_\Omega \int_{X_\omega} f\,d\mu_\omega\,dP(\omega) = \int_{\{\omega:\int_{X_\omega} f\,d\mu_\omega \geq 0\}} \int_{A \cap X_\omega} S_* f\,d\mu_\omega\,dP(\omega)$$
$$+ \int_{\{\omega:\int_{X_\omega} f\,d\omega < 0\}} \int_{E \cap X_\omega} S^* f\,d\mu_\omega\,dP(\omega)$$
$$= \int_A S_* f\,d\mu + \int_E S^* f\,d\mu,$$

since, for example, $\omega \in A \cap X_\omega$ implies that $\int_{X_\omega} f\,d\mu_\omega \geq 0$.

If f takes only the values $-1, 0, 1$, as in a simple random walk, then the Theorem states that *the probability that the walker first returns to 0 from above equals the probability that he first returns to 0 from below*, no matter what the probabilities of moving up and down are (and even in the case of dependent increments)! For $S_* f = 1$ on A and $S^* f = -1$ on E, so that according to the Theorem,

$$\mu(A) - \mu(E) = \mu\{f > 0\} - \mu\{f < 0\},$$

or

$$\mu\{f > 0\} - \mu(A) = \mu\{f < 0\} - \mu(E)\},$$

that is,

$$\mu\{x: f(x) > 0, S_n f(x) \leq 0 \text{ for some } n \geq 1\}$$
$$= \mu\{x: f(x) < 0, S_n f(x) \geq 0 \text{ for some } n \geq 1\}$$

(the probability of first returning to 0 from above equals that of first returning to 0 from below). By considering $\underline{A}$ and $\underline{E}$ in place of A and E, it follows similarly that

$$\mu\{x: f(x) > 0, S_n f(x) < 0 \quad \text{for some } n \geq 1\}$$
$$= \mu\{x: f(x) < 0, S_n f(x) > 0 \quad \text{for some } n \geq 1\}.$$

In the I.I.D. case of a true random walk (i.e., T a Bernoulli shift), such results can be easily proved by symmetry considerations:

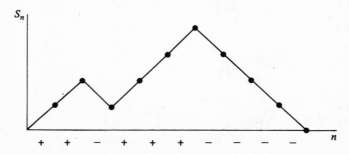

each path which returns to 0 corresponds (under a measure-preserving transformation obtained by considering the increments in reverse order, starting at the first impact on 0) to a path on the other side of 0:

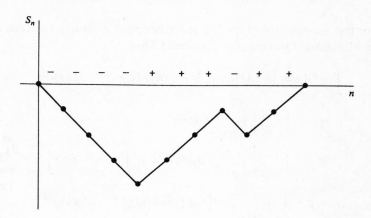

Our theorem shows that such symmetry is present in every stationary stochastic process. For further discussion of random walks with stationary (dependent) increments, see Derriennic (1980).

D. *The Dominated Ergodic Theorem and its converse*

Wiener's Dominated Ergodic Theorem gives a sufficient condition that $f^* \in L^1$: that $f \in L \log L$, i.e.

$$\int_X |f(x)| \log^+ |f(x)| \, d\mu(x) < \infty.$$

(For f^* to be in L^1 means that the a.e. convergence of $(1/n)S_n f$ is of the kind in the Dominated Convergence Theorem, hence the name.) Because of the Maximal Equality, we know the distribution of f^* well enough to see that for a nonnegative function this condition is also necessary.

Theorem 1.16 (*Wiener* 1939, *Petersen* 1979) Let $\{T_t\}$ be an ergodic measure-preserving flow on $(X, \mathcal{B}, \mu)$ and $0 \leq f \in L^1(X)$. Then $f^* \in L^1$ if and only if $f \in L \log L$.

Proof: For each $\alpha \geq 0$, let

$$G(\alpha) = \mu\{x \in X : f^*(x) > \alpha\}.$$

From (B), for $\alpha \geq \int f \, d\mu$ we have

$$G(\alpha) = \frac{1}{\alpha} \int_{\{f^* > \alpha\}} f \, d\mu.$$

3. More about almost everywhere convergence

Now for any continuous increasing function h on $[0, \infty)$ with $h(0) = 0$, we have

$$\int_X h(f^*(x))\,d\mu(x) = -\int_0^\infty h(y)\,dG(y) = \int_0^\infty G(y)\,dh(y).$$

(For the last equality, check it first for bounded functions and then apply the Monotone Convergence Theorem.) Thus

$$\int_X f^*\,d\mu = \int_0^\infty G(y)\,dy = \int_0^{\|f\|_1} G(y)\,dy + \int_{\|f\|_1}^\infty G(y)\,dy$$

$$= c + \int_{\|f\|_1}^\infty \frac{1}{y}\int_{\{f^*>y\}} f\,d\mu\,dy$$

$$\geq c + \int_{\|f\|_1}^\infty \frac{1}{y}\int_{\{f>y\}} f\,d\mu\,dy = c + \int_{\{f>\|f\|_1\}} f(x)\int_{\|f\|_1}^{f(x)} \frac{dy}{y}\,d\mu(x)$$

$$= c + \int_{\{f>\|f\|_1\}} f(x)[\log f(x) - \log\|f\|_1]\,d\mu(x),$$

so that $f^* \in L^1$ implies $f \in L\log L$.

For the converse, we use a neat trick of Wiener's: if $g = f \cdot \chi_{\{f > (\alpha/2)\}}$, then $f^* \leq g^* + \alpha/2$, and the Maximal Ergodic Theorem applied to g says that

$$\mu\{f^* > \alpha\} \leq \mu\{g^* > (\alpha/2)\} \leq \frac{2}{\alpha}\int_{\{g^*>(\alpha/2)\}} g\,d\mu \leq \frac{2}{\alpha}\int_X g\,d\mu$$

$$= \frac{2}{\alpha}\int_{\{f>(\alpha/2)\}} f\,d\mu.$$

Thus

$$\int_X f^*\,d\mu = c + \int_{2\|f\|_1}^\infty \mu\{f^* > y\}\,dy$$

$$\leq c + \int_{2\|f\|_1}^\infty \frac{2}{y}\int_{\{f>(y/2)\}} f\,d\mu\,dy$$

$$= c + \int_{\{f>\|f\|_1\}} f(x)\int_{\|f\|_1}^{f(x)} \frac{dy}{y}\,d\mu(x)$$

$$\leq c + \int_{\{f>\|f\|_1\}} f(x)\log^+ f(x)\,d\mu(x),$$

so that $f \in L\log L$ implies $f^* \in L^1$.

Reverse maximal inequalities and converse dominated theorems are found in other contexts as well: see Burkholder (1962), Stein (1969),

3.1. More about the maximal ergodic theorem

Gundy (1969), Ornstein (1971b), Derriennic (1973), Jones (1977). Part of the interest in the question of exactly when f^* is integrable arises from the general theory of *Hardy spaces*: frequently one says that $f \in H^p$ if an appropriate maximal function of f is in L^p. See Hardy and Littlewood (1930), Burkholder, Gundy and Silverstein (1971), and Fefferman and Stein (1972) for the original results in this direction and Petersen (1977) and Garsia (1973) for an introduction to this deep and fascinating subject.

Exercises

1. Show that if T is a positive contraction on L^1, $0 \leq f, g \in L^1$, and
$$E = \left\{ \sum_{k=0}^{n-1} T^k f > \sum_{k=0}^{n-1} T^k g \text{ for some } n \geq 1 \right\},$$
then
$$\int_E g \, d\mu \leq \int_E f \, d\mu.$$

2. Prove that the sets A, $\underline{A}$ are measurable.
3. Give examples for which $\underline{A} \neq A$, in both the flow and discrete cases.
4. Prove that if $\{T_t\}$ is ergodic, then $\mathscr{C}_\alpha \doteq X$ if and only if $\alpha = \int f \, d\mu$. (Hint: First consider the discrete case. If $\int f \, d\mu = 0$, use the tower decomposition of X with respect to A to prove that $\mu(A) = 0$. Then consider the time 1 map of $\{T_t\}$ and $f_1(x) = \int_0^1 f(T_s x) \, ds$.)
5. Prove Remark 1.11.
6. Fill in the details of the proof of Theorem 1.7 and prove Corollary 1.8.
7. Show that the Maximal Ergodic Theorem (see Section 2.2) still holds if we only require that $f^+ \in L^1$, rather than $f \in L^1$.
8. If $f \in L^1$ and $\mathscr{I}$ is the family of invariant sets, then, without assuming ergodicity,
    (a) in the discrete case $E(f|\mathscr{I}) = E(S_* f \cdot \chi_A + S^* f \cdot \chi_E | \mathscr{I})$ a.e.
    (b) in the flow case, $E(f|\mathscr{I}) = E(f \cdot (\chi_{\underline{A}} + \chi_{\underline{E}}) | \mathscr{I})$ a.e. (where $\underline{E} = \{x : F_t(x) \leq 0 \text{ for all } t \geq 0\}$).
9. Work Exercise 8 on the basis of 1.12 and 1.3, avoiding ergodic decompositions. (Hint: (a) Consider first an integer-valued f, and work separately on the sets where $E(f|\mathscr{I}) > 0$, $E(f|\mathscr{I}) = 0$, and $E(f|\mathscr{I}) < 0$. Mimic the proof of 1.12. (b) Where $E(f|\mathscr{I}) = 0$, replace f by $f + \varepsilon$.)

3.2. More about the Pointwise Ergodic Theorem

This section includes two main topics. First, we will give a proof of Birkhoff's theorem which shows more clearly exactly how the maximal inequality produces the convergence theorem. The same type of argument will also appear in the next few sections, when we prove the Fundamental Theorem of Calculus and the existence of the real-variable and ergodic Hilbert transforms. Second, we will show that in general no statement can be made about the *speed* with which the convergence takes place in the Ergodic Theorem.

A. Maximal inequalities and convergence theorems

Already in 1925, Kolmogorov (1925) proved a maximal (or *weak-type* (1,1)) inequality for the conjugate of a harmonic function, and soon thereafter (Kolmogorov 1928, 1929–30) he established a weak-type (2, 2) inequality for sums of independent random variables. Hardy and Littlewood (1930) further demonstrated the usefulness of maximal functions in analysis. They considered both the *Hardy–Littlewood maximal function*

$$Mf(\theta) = \sup_{-\pi \leq t \leq \pi} \frac{1}{t} \int_0^t |f(\theta + x)| \, dx \quad (f \in L^1[-\pi, \pi])$$

and the *nontangential maximal function*

$$N_\sigma F(\theta) = \sup_{z \in \Omega_\sigma(\theta)} |F(z)| \quad (F \text{ analytic on } D = \text{the open unit disk}),$$

where $\Omega_\sigma(\theta)$ is as shown,

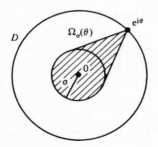

and used these functions to study the nontangential convergence of analytic functions to their boundary values and what we now call Hardy spaces. F. Riesz (1931) based a direct proof of the Hardy–Littlewood Maximal Theorem (originally proved from its discrete version) on his Rising Sun Lemma and the (consequent) maximal inequality

$$m\{Mf > \alpha\} \leq \frac{1}{\alpha} \int |f| \, dm$$

3.2. More about the pointwise ergodic theorem

for the Hardy–Littlewood maximal function. Riesz was primarily interested in applying these ideas to differentiation theory, and indeed he showed that the Fundamental Theorem of Calculus for the Lebesgue integral,

$$\frac{1}{\varepsilon}\int_0^\varepsilon [f(x-t) - f(x)]\, dt \to 0 \quad \text{a.e.} \quad \text{as } \varepsilon \to 0,$$

was easily proved via this approach. Later, F. Riesz contributed important simplifications to the proof of the Ergodic Theorem; as we have seen, however, it was because of Wiener (1939) and Yosida and Kakutani (1939) that the Maximal Ergodic Theorem achieved its central position in the argument. For certain stochastic processes $\{X_k(\omega): k = 1, 2, \ldots\}$ (submartingales), maximal inequalities like

$$P\{\sup_{k \leq n} X_k(\omega) \geq \alpha\} \leq \frac{1}{\alpha}\int_{\{\sup_k X_k \geq \alpha\}} X_n\, dP$$

were used by Paul Lévy (1937) and Jean Ville (1939); these led to the martingale convergence theorems of Doob (1940, 1953). Finally, we should mention that the discoveries of recent workers such as Fefferman and Stein (1972) and Calderón (1950, 1977), have led to a very deep understanding of maximal functions and, of course, have also produced a set of new problems of correspondingly greater depth.

The basic reason why the maximal function and the maximal inequalities are so useful in analysis was already formulated by Banach (1926).

Theorem 3.1 *Banach's Principle* Let $1 \leq p < \infty$ and let $\{T_n\}$ be a sequence of bounded linear operators on L^p. If

$$T^*f = \sup_n |T_n f| < \infty \quad \text{a.e.} \quad \text{for each } f \in L^p,$$

then the set of f for which $\{T_n f\}$ converges a.e. is closed in L^p.

For the proof, see the Exercises.

Frequently it is easy to exhibit a dense set $\mathscr{D}$ in L^p such that $\{T_n f\}$ converges a.e. for each $f \in \mathscr{D}$. If one can prove that $T^*f < \infty$ a.e. for each $f \in L^p$, then the a.e. convergence of $\{T_n f\}$ for all $f \in L^p$ will follow from Banach's Principle.

In most cases, that $T^*f < \infty$ a.e. for each $f \in L^p$ is established by proving that T^*f satisfies a *weak-type* (p, q) *inequality*:

$$\mu\{T^*f > \lambda\} \leq \left(\frac{c}{\lambda}\|f\|_p\right)^q.$$

(Indeed, Stein 1961, Sawyer 1966, and Garsia 1970 have shown that in many situations T^*f satisfies such an inequality if and only if

$T^*f<\infty$ a.e.) Of course the Maximal Ergodic Theorem is an example of such a result.

We will follow this scheme of proof several times in the following pages and will thus have a chance to see it operating in various contexts. First, let us apply it to the Ergodic Theorem.

Approximation proof of the Ergodic Theorem 2.2 We will prove the essential part, namely that the ergodic averages
$$\frac{1}{n}\sum_{k=0}^{n-1} fT^k \quad \text{converge a.e.}$$
First, if $g\in L^\infty$ and $f = g - Tg$ (so that f is a 'coboundary'), then
$$\frac{1}{n}\sum_{k=0}^{n-1} fT^k = \frac{1}{n}\sum_{k=0}^{n-1}(gT^k - gT^{k+1}) = \frac{1}{n}(g - gT^n) \to 0 \text{ a.e.}$$
Also, if $f\in L^1$ is invariant ($fT = f$ a.e.), then
$$\frac{1}{n}\sum_{k=0}^{n-1} fT^k \to f \text{ a.e.}$$
Thus if
$$\mathscr{D} = \{f + (g - Tg) : f\in L^1 \text{ is invariant}, g\in L^\infty\}$$
and
$$\mathscr{C} = \left\{f\in L^1 : \frac{1}{n}\sum_{k=0}^{n-1} fT^k \text{ converges a.e.}\right\},$$
then $\mathscr{D}\subset\mathscr{C}$ and $\mathscr{C}$ is a linear subspace of L^1. We only need to show, then, that (1) $\mathscr{D}$ is dense in L^1 and (2) $\mathscr{C}$ is closed in L^1.

In order to prove (1), let
$$\mathscr{D}_2 = \{f + (g - Tg) : f\in L^2 \text{ is invariant}, g\in L^\infty\}.$$
We will show that $\mathscr{D}_2 \subset \mathscr{D}$ is dense in L^2, and hence in L^1 (recall that on a finite measure space $\|\ \|_1 \leq c\|\ \|_2$ by Hölder's Inequality).

If $h\in L^2$ and $h \perp (g - Tg)$ for all $g\in L^\infty$, then
$$\int_A h\, d\mu = \int_A hT^{-1}\, d\mu \quad \text{for all } A\in\mathscr{B},$$
and hence $h = hT^{-1}$ a.e. This implies that L^2 is even the orthogonal direct sum of the invariant functions and the closure of $\{g - Tg : g\in L^\infty\}$. At any rate, $\mathscr{D}_2$ is dense in L^2 and hence in L^1.

To prove (2), let
$$A_n f = \frac{1}{n}\sum_{k=0}^{n-1} fT^k$$

3.2. More about the pointwise ergodic theorem

and note that by the Maximal Ergodic Theorem

$$\mu\{\sup_n |A_n f| > \alpha\} \leq \mu\{\sup_n A_n |f| > \alpha\} \leq \frac{1}{\alpha}\int_{\{|f|^* > \alpha\}} |f|\, d\mu \leq \frac{\|f\|_1}{\alpha}.$$

Suppose then that $f_k \in \mathscr{C}$ and $f_k \to f$ in L^1; we will prove that $\{A_n f\}$ is fundamental (i.e., Cauchy) a.e., so that also $f \in \mathscr{C}$. Now

$$|A_m f - A_n f| \leq |A_m f_k - A_n f_k| + |A_m (f - f_k)| + |A_n (f_k - f)|,$$

and the first term tends to 0 a.e. as $m, n \to \infty$, since $\{A_n f_k\}$ is fundamental a.e.; therefore, for each $\alpha > 0$

$$\mu\{\limsup_{m,n \to \infty} |A_m f - A_n f| > \alpha\}$$

$$\leq \mu\{2\sup_j |A_j(f - f_k)| > \alpha\} \leq \frac{2}{\alpha}\|f - f_k\|_1,$$

which can be made arbitrarily small by a suitable choice of k. Thus

$$\mu\{\limsup_{m,n \to \infty} |A_m f - A_n f| > \alpha\} = 0 \quad \text{for each } \alpha > 0.$$

and hence, letting $\alpha \to 0$, we see that $\{A_n f\}$ is fundamental a.e.

B. The speed of convergence in the Ergodic Theorem

Suppose that $T: X \to X$ is an *ergodic* m.p.t., so that

$$\frac{1}{n}\sum_{k=0}^{n-1} f(T^k x) \to \int f\, d\mu \quad \text{a.e. for each } f \in L^1.$$

For each $n \geq 1$, let

$$S_n f(x) = f(x) + f(Tx) + \ldots + f(T^{n-1} x).$$

Let us consider only functions $f \in L^1$ for which $\int f\, d\mu = 0$. For such an f,

$$\frac{S_n f}{n} \to 0 \quad \text{a.e.}$$

It is not unreasonable to suppose that for some T a series such as

$$\sum_{n=1}^{\infty} \frac{S_n f}{n^2}$$

might converge a.e. After all, not only does

$$\frac{S_n f(x)}{n} \to 0 \quad \text{a.e.,}$$

but it usually changes sign frequently enough so that the convergence of this series, even if not absolute, might be enhanced by a great deal of cancellation, as in the case of an alternating series. Indeed, the series does converge a.e. in many cases: if the fT^k are independent, as when T is a Bernoulli shift and f is a coordinate function (by the Law of the Iterated Logarithm); if $Tx = x + \alpha$ (mod 1), f is the characteristic function of an interval, and α is not a Liouville number, i.e., for almost all α (by number-theoretic estimates on the 'discrepancy'); if $f = g - gT$ for some $g \in L^\infty$; or if f is a nonconstant eigenfunction of T. (For sums of independent random variables X_k, the Law of Random Signs asserts that $\sum a_k X_k$ converges a.e. if $\sum a_k^2 < \infty$. The definitive result along these lines is Kolmogorov's Three Series Theorem: If $\{X_n\}$ are independent, then $\sum X_n$ converges a.e. if and only if each of the following three series does, for some $\lambda \in \mathbb{R}$:

$$\sum P\{|X_n| \geq \lambda\}, \quad \sum E(X_n^{(\lambda)}), \quad \sum \mathrm{var}(X_n^{(\lambda)}).$$

Here $X_n^{(\lambda)}$ is X_n truncated at $\pm \lambda$:

$$X_n^{(\lambda)}(\omega) = \begin{cases} X_n(\omega) & \text{if } |X_n(\omega)| \leq \lambda \\ 0 & \text{otherwise.} \end{cases}$$

Statements about series $\sum a_k f(T^k x)$ and $\sum b_n (S_n f(x))/n$ are related because of partial summation.)

We will see, however, that the series

$$\sum_{n=1}^{\infty} \frac{S_n f(x)}{n^2}$$

does not in general converge a.e., and that in fact $(S_n f(x))/n$ does not tend to 0 fast enough to make any divergent series convergent: given any sequence $b_n \geq 0$ with $\sum_{n=1}^{\infty} b_n = \infty$ and any ergodic T, there is an $f \in L^\infty$ such that

$$\sum_{n=1}^{\infty} b_n \frac{S_n f(x)}{n} \quad \text{diverges a.e.}$$

Thus, no general statement can be made about the speed with which convergence takes place in the Ergodic Theorem.

We will actually construct such a function f for each given sequence $\{b_n\}$ and each ergodic T, using one of the basic tools of ergodic theory, the Kakutani–Rokhlin tower (see Lemma 2.4.7). (A category argument for the existence of such functions has been given by del Junco and Rosenblatt (1979).)

Theorem 2.3 (Kakutani and Petersen 1981) Let $T: X \to X$ be an ergodic m.p.t. on a nonatomic probability space $(X, \mathcal{B}, \mu)$, and let $b_k \geq 0$ be any sequence for which

$$\sum_{k=1}^{\infty} b_k = \infty.$$

3.2. More about the pointwise ergodic theorem

Then there is an $f \in L^\infty(X)$ with $\int f \, d\mu = 0$ for which

$$\sup_l \left| \sum_{k=1}^{l} b_k \frac{S_k f(x)}{k} \right| = \infty \quad \text{a.e.;}$$

in particular, the series

$$\sum_{n=1}^{\infty} b_n \frac{S_n f(x)}{n}$$

diverges a.e.

Proof: The function f will be defined as an infinite series,

$$f = \sum_{n=1}^{\infty} \frac{f_n}{\sqrt{K_n}}.$$

The constants $K_n \nearrow \infty$ and a sequence $l_n \nearrow \infty$ will be defined inductively, and each f_n will be defined as a simple function with respect to (i.e., constant on each set of) a Kakutani–Rokhlin tower in X. Let

$$G_l f(x) = \sum_{k=1}^{l} b_k \frac{S_k f(x)}{k};$$

we will show that for large N, G_{l_N} is very large on a set of measure very close to 1: there are sequences $M_n \nearrow \infty$ and $\varepsilon_n \searrow 0$ such that

$$\mu\{x : |G_{l_N}(x)| \geq M_N\} \geq 1 - \varepsilon_N.$$

This estimate will be carried out by splitting the series for

$$G_l f(x) = \sum_{k=1}^{l} \frac{b_k}{k} \sum_{n=1}^{\infty} S_k \frac{f_n(x)}{\sqrt{K_n}} = \sum_{n=1}^{\infty} \sum_{k=1}^{l} \frac{b_k}{k} S_k \frac{f_n(x)}{\sqrt{K_n}}$$

into three terms – those for which $n < N$, $n > N$, and $n = N$ – and showing that the term for which $n = N$ dominates the other two. To begin, let $K_1 = 1, l_1 = 1$, and find a tower of height 2 in X: That is, find a measurable set $A_1 \subset X$ such that $A_1 \cap TA_1 = \phi$. Define f_1 by $f_1 \equiv 1$ on A_1, $f_1 \equiv -1$ on TA_1, and $f_1 \equiv 0$ on $X - (A_1 \cup TA_1)$. Notice that $S_k f_1(x)$ takes only the values 1, 0, and -1, for all k and x.

Suppose that K_j, l_j, and f_j have already been defined for all $j < n$, and that

$$c_n = \sup_{j<n} \sup_{k \geq 1} \|S_k f_j\|_\infty < \infty.$$

(1) Choose K_n large enough that
  (a) $K_n \geq K_{n-1}$,
  (b) $K_n \geq 2^{4n}$,
  (c) $\sqrt[4]{K_n} \geq \sum_{k=1}^{l_j} b_k$ for all $j < n$.

(2) Choose l_n large enough that

(a) $l_n \geq l_{n-1}$

(b) $\sum_{k=1}^{l_n} b_k \geq K_n c_n \sum_{k=1}^{l_n} \frac{b_k}{k}$.

This is possible since $\sum_{k=1}^{\infty} b_k = \infty$ implies that

$$\frac{\sum_{k=1}^{l} b_k}{\sum_{k=1}^{l} b_k/k} \to \infty.$$

(3) f_n is defined by using a tower of height $2nl_n$ in X, with 'remainder' having measure no greater than $1/n$. That is, find a measurable set $A_n \subset X$ such that $A_n, TA_n, \ldots, T^{2nl_n-1} A_n$ are pairwise disjoint and

$$\mu\left(\bigcup_{i=0}^{2nl_n-1} T^i A_n\right) \geq 1 - \frac{1}{n}.$$

We think of the tower as consisting of a lower half, n blocks each of height l_n, and an upper half, again n blocks each of height l_n.

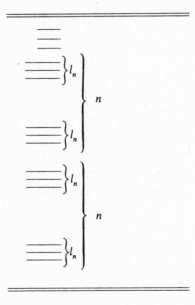

3.2. More about the pointwise ergodic theorem

The function f_n is defined to be 1 on the entire lower half of the tower, -1 on the entire upper half, and 0 on the remainder:

$$f_n \equiv 1 \text{ on } T^j A_n \quad \text{for } 0 \leq j \leq nl_n - 1,$$
$$f_n \equiv -1 \text{ on } T^j A_n \quad \text{for } nl_n \leq j \leq 2nl_n - 1,$$
$$f_n \equiv 0 \text{ on } X \setminus \bigcup_{j=0}^{2nl_n - 1} T^j A_n.$$

Notice that $f_n \in L^\infty$, $\int f_n d\mu = 0$, and $S_k f_n(x)$ takes values only in $[-nl_n, nl_n]$, for all k and x. Thus $c_{n+1} < \infty$.

The K_n, l_n, and f_n are therefore defined by induction for all $n = 1, 2, \ldots$. We let

$$f(x) = \sum_{k=1}^{\infty} \frac{f_n(x)}{\sqrt{K_n}}.$$

This series converges in L^∞, since $|f_n| \leq 1$ a.e. and $\sum (1/\sqrt{K_n}) < \infty$. We have $f \in L^\infty$ and $\int f d\mu = 0$.

We will now show that, for large N, $G_{l_N} f(x)$ is very large on a set of measure very close to 1. We may write

$$G_{l_N} f(x) = \sum_{k=1}^{l_N} b_k \frac{S_k f(x)}{k} = \sum_{k=1}^{l_N} \frac{b_k}{k} \sum_{n=1}^{\infty} \frac{S_k f_n(x)}{\sqrt{K_n}}$$

$$= \sum_{n=1}^{\infty} \frac{1}{\sqrt{K_n}} \sum_{k=1}^{l_N} \frac{b_k}{k} S_k f_n(x) = \sum_{n<N} \frac{1}{\sqrt{K_n}} \sum_{k=1}^{l_N} \frac{b_k}{k} S_k f_n(x)$$

$$+ \sum_{n>N} \frac{1}{\sqrt{K_n}} \sum_{k=1}^{l_N} \frac{b_k}{k} S_k f_n(x)$$

$$+ \frac{1}{\sqrt{K_N}} \sum_{k=1}^{l_N} \frac{b_k}{k} S_k f_N(x) = \text{I} + \text{II} + \text{III}.$$

Now

$$|\text{I}| \leq \sum_{n<N} \frac{1}{\sqrt{K_n}} \sum_{k=1}^{l_N} \frac{b_k}{k} |S_k f_n(x)| \leq \sum_{n<N} \frac{1}{\sqrt{K_n}} \sum_{k=1}^{l_N} \frac{b_k}{k} c_N$$

$$= c_N \left(\sum_{k=1}^{l_N} \frac{b_k}{k} \right) \left(\sum_{n<N} \frac{1}{\sqrt{K_n}} \right) \leq c_N \left(\sum_{k=1}^{l_N} \frac{b_k}{k} \right) \left(\sum_{n=1}^{\infty} \frac{1}{\sqrt{K_n}} \right)$$

$$= c c_N \sum_{k=1}^{l_N} \frac{b_k}{k} \text{ (by 1(b))}, \text{ where } c \text{ is a finite constant}.$$

Further, since $|S_k f_n(x)| \le k$ for all k and x,

$$|\mathrm{II}| \le \sum_{n>N} \frac{1}{\sqrt{K_n}} \sum_{k=1}^{l_N} b_k \le \sum_{n>N} \frac{1}{\sqrt{K_n}} \sqrt[4]{K_n} \text{ (by 1(c))}$$

$$\le \sum_{n=1}^{\infty} \frac{1}{\sqrt[4]{K_n}} = c', \text{ another finite constant.}$$

However, $S_k f_N(x) = k$ for $k = 1, \ldots l_N$ if x is in the bottom $n-1$ blocks of the lower half of the N'th tower, and $S_k f_N(x) = -k$ for $k = 1, \ldots l_N$ if x is in the bottom $n-1$ blocks of the upper half of the N'th tower. This set has measure at least $((N-1)/N)^2$. More precisely,

$S_k f_N(x) = k$ for $k = 1, \ldots l_N$ if $x \in T^j A_N$ and $0 \le j \le (N-1)l_N - 1$;

$S_k f_N(x) = -k$ for $k = 1, \ldots l_N$

if $x \in T^j A_N$ and $Nl_N \le j \le (2N-1)l_N - 1$;

$$\mu\left(\bigcup_{\substack{0 \le j \le (N-1)l_N - 1 \\ Nl_N \le j \le (2N-1)l_N - 1}} T^j A_N\right) \ge \left(\frac{N-1}{N}\right)^2.$$

Therefore, on this set

$$|\mathrm{III}| = \frac{1}{\sqrt{K_N}} \sum_{k=1}^{l_N} b_k \ge \sqrt{K_N} c_N \sum_{k=1}^{l_N} \frac{b_k}{k}.$$

Thus $\dfrac{|\mathrm{III}|}{|\mathrm{I}| + |\mathrm{II}|} \to \infty$ as $N \to \infty$, so also $|\mathrm{I} + \mathrm{II} + \mathrm{III}| \to \infty$ as $N \to \infty$. We have proved that

$$\mu\left\{x : \limsup_{l \to \infty} \left|\sum_{k=1}^{l} b_k \frac{S_k f(x)}{k}\right| = +\infty\right\} = 1.$$

While for $\{fT^k\}$ independent we have the Law of the Iterated Logarithm (Khintchine 1923, Kolmogorov 1929) – according to which

$$|S_n f(x)| = 0(\sqrt{n \log \log n}) \text{ a.e.} -$$

because of this Theorem no analogous statement can be made for stationary stochastic processes in general, for any choice of the comparison speed. Some applications of this result to questions about the convergence of series of the form

$$\sum_{k=1}^{\infty} a_k f(T^k x)$$

are mentioned in the Exercises and in section 3.6.

3.2. More about the pointwise ergodic theorem

Exercises

1. Prove Banach's Principle. (Hint: First use the Baire Category Theorem to show that if $T^*f < \infty$ a.e., then there is a function $\delta(\alpha)$ such that $\delta(\alpha) \to 0$ as $\alpha \to \infty$ and $\mu\{T^*f > \alpha \|f\|_p\} \leq \delta(\alpha)$ for all $f \in L^p$ and all $\alpha > 0$. For this purpose, consider the set
$$\{f \in L^p : \mu\{T^*f > m\} \leq \varepsilon\} = \bigcap_n \{f : \mu\{\sup_{k \leq n}|T_k f| > m\} \leq \varepsilon\}.$$
Then proceed as in 2.2.)

2. Extend the approximation proof of the Ergodic Theorem to the infinite measure case. (Hint: $\{f + g - gT : f, g \in L^2 \text{ and } f = fT \text{ a.e.}\}$ is dense in L^2, and weak $(1, 1)$ implies (using the integration technique on p. 88) that $\|T^*f\|_2 \leq c\|f\|_2$ for $f \in L^2$.)

3. Prove that no statement concerning the speed of convergence in the Ergodic Theorem is possible, in the following sense: Given any ergodic transformation T on a nonatomic measure space and any sequence $a_n \to 0$, there is a bounded measurable function f on X such that
$$\limsup_{n \to \infty} \frac{|[S_n f(x)/n] - \int f d\mu|}{a_n} = \infty \text{ a.e.}$$
(See (Krengel 1978)), where it is shown that if $X = [0, 1)$, then f can in fact be chosen to be continuous or the characteristic function of a measurable set.)

4. Suppose $c_k \geq c_{k+1} \geq 0$ for all $k = 1, 2, \ldots$, $\{kc_k\}$ is bounded, and $\sum_{k=1}^{\infty} c_k = \infty$. Prove that for any ergodic transformation T on a nonatomic measure space X, there is a function $f \in L^\infty(X)$ with $\int f d\mu = 0$ such that
$$\sum_{k=1}^{\infty} c_k f(T^k x)$$
diverges a.e. Thus series like
$$\sum_{k=1}^{\infty} \frac{f(T^k x)}{k}, \sum_{k=3}^{\infty} \frac{f(T^k x)}{k \log k \log \log k}, \text{ etc.}$$
can be made to diverge a.e. (But see 3.6.) (Hint: Let $b_k = k(c_k - c_{k+1})$ and use partial summation.)

5. Prove the *Ulam-von Neumann Random Ergodic Theorem* (Ulam and von Neumann 1945) Suppose that $(X, \mathscr{B}, \mu)$ and $(\Gamma, \mathscr{C}, \nu)$ are probability spaces, and that for each $\gamma \in \Gamma$ there is a m.p.t. $T_\gamma : X \to X$. Assume that for each measurable f on X, the function

100   3. More about almost everywhere convergence

$f(T_\gamma x)$ is jointly measurable in x and γ. Let $\Omega = \prod_{-\infty}^{\infty}(\Gamma, \mathscr{C}, \nu)$ be the product measure space; each $\omega \in \Omega$ is a sequence $\omega = \{\gamma_n(\omega)\}$ of elements of Γ. If $f \in L^1(X)$, then for almost all $\omega \in \Omega$ there is $\bar{f}_\omega \in L^1(X)$ such that

$$\lim_{n \to \infty} \frac{1}{n} \sum_{k=1}^{n} f(T_{\gamma_{k-1}(\omega)} \cdots T_{\gamma_1(\omega)} T_{\gamma_0(\omega)} x) = \bar{f}_\omega(x) \quad \text{a.e. } d\mu.$$

(Hint: Apply the Ergodic Theorem to a suitable skew product.)

6. (R.L. Jones) Let $T: X \to X$ be a m.p.t. and $f \in L^1(X)$.
   (a) Show that $f(T^k x)/k \to 0$ a.e.
   (b) Show that if $p > 1$, then

$$\sum_{k=1}^{\infty} \left(\frac{f(T^k x)}{k}\right)^p$$

converges a.e. (Hint: Assume $f \geq 0$ and show that

$$\mu\left\{x : \left[\sum_{k=1}^{\infty}\left(\frac{f(T^k x)}{k}\right)^p\right]^{1/p} > \lambda\right\} \leq \frac{c\|f\|_1}{\lambda}$$

by writing

$$g_k(x) = f(T^k x)\chi_{\{fT^k \leq \lambda k\}}(x)$$

and establishing such inequalities for g_k and $h_k = fT^k - g_k$ separately. Recall that

$$\int_X [g_k(x)]^p d\mu = \int_0^\infty p\alpha^{p-1} \mu\{g_k > \alpha\} d\alpha.)$$

   (c) If $p > 1$, then

$$\sum_{k=1}^{\infty} \frac{f(T^k x)}{k^p}$$

converges a.e. (Hint: Use partial summation.)

3.3. Differentiation of integrals and the Local Ergodic Theorem

For a real-valued function $g \in L^1(\mathbb{R})$, we define the *Hardy–Littlewood maximal function* Mg of g by

$$Mg(t) = \sup_{\varepsilon > 0} \frac{1}{2\varepsilon} \int_{-\varepsilon}^{\varepsilon} |g(t+s)| ds.$$

We first prove the original maximal inequality of F. Riesz.

Theorem 3.1 (F. Riesz 1931) If $g \in L^1(\mathbb{R})$ and $\lambda > 0$, then

$$m\{x \in \mathbb{R} : Mg(x) > \lambda\} \leq \frac{2}{\lambda}\|g\|_1.$$

3.3. Differentiation of integrals and the local ergodic theorem

Proof: Let $A \subset \{x: Mg(x) > \lambda\}$ be closed and bounded. For each $x \in A$ there is an open interval I_x centered at x with

$$\frac{1}{m(I_x)} \int_{I_x} |g| \, dm > \lambda.$$

A finite subcollection $I_1, I_2, \ldots, I_n$ of these still covers A. We may assume that no point of $\mathbb{R}$ belongs to more than two of the I_k. (For if t is in three I_k's the interval with left endpoint farthest from t and the one with right endpoint farthest from t cover the union of the three intervals.)

Thus

$$\sum_k \chi_{I_k} \leq 2\chi_{\bigcup_k I_k}$$

so that

$$m(A) \leq \sum_k m(I_k) \leq \frac{1}{\lambda} \sum_k \int_{I_k} |g| \, dm$$

$$= \frac{1}{\lambda} \int \sum_k |g| \chi_{I_k} \leq \frac{2}{\lambda} \int |g| \chi_{\bigcup_k I_k} \leq \frac{2}{\lambda} \|g\|_1.$$

Theorem 3.2 **Lebesgue Differentiation Theorem** (Lebesgue 1904)
If g is locally integrable on $\mathbb{R}$ (i.e. $f \in L^1(K)$ for each compact $K \subset \mathbb{R}$), then

$$\lim_{\varepsilon \to 0^+} \frac{1}{2\varepsilon} \int_{-\varepsilon}^{\varepsilon} g(t+s) \, ds = g(t) \text{ a.e.}$$

Proof: Fix an open interval I and suppose that $g \in L^1(I)$; we will show that the formula holds a.e. on I.

If g is continuous, then as in elementary calculus

$$\frac{1}{2\varepsilon} \int_{-\varepsilon}^{\varepsilon} |g(t+s) - g(t)| \, ds$$

can easily be made arbitrarily small by taking ε small enough.

Let

$$A_\varepsilon g(t) = \frac{1}{2\varepsilon} \int_{-\varepsilon}^{\varepsilon} g(t+s) \, ds$$

and define

$$\mathscr{C} = \{g \in L^1(I): A_\varepsilon g \to g \text{ a.e. on } I\}.$$

We have just seen that $\mathscr{C}$ contains the continuous functions on I, a dense set in $L^1(I)$, so that it is enough to prove that $\mathscr{C}$ is closed in $L^1(I)$.

Suppose then that $g_k \in \mathscr{C}$ and $g_k \to g$ in $L^1(I)$. Since
$$|A_\varepsilon g(t) - g(t)| \leq |A_\varepsilon(g - g_k)(t)| + |A_\varepsilon g_k(t) - g_k(t)| + |g_k(t) - g(t)|$$
and
$$\limsup_{\varepsilon \to 0^+} |A_\varepsilon g_k(t) - g_k(t)| = 0 \quad \text{on } I,$$
we see that for each $\lambda > 0$
$$m\{t \in I : \limsup_{\varepsilon \to 0^+} |A_\varepsilon g(t) - g(t)| > \lambda\}$$
$$\leq m\left\{t \in I : \sup_{\varepsilon \to 0} |A_\varepsilon(g - g_k)(t)| > \frac{\lambda}{2}\right\} + m\left\{t \in I : |g_k(t) - g(t)| > \frac{\lambda}{2}\right\}$$
$$\leq \frac{2}{\lambda} \|g - g_k\|_1 + m\left\{t \in I : |g_k(t) - g(t)| > \frac{\lambda}{2}\right\},$$
which tends to 0 as $k \to \infty$. (Recall that L^1 convergence implies convergence in measure). Thus
$$m\left\{t \in I : \limsup_{\varepsilon \to 0^+} |A_\varepsilon g(t) - g(t)| > \lambda\right\} = 0 \quad \text{for each } \lambda > 0,$$
and letting $\lambda \to 0$ shows that
$$\limsup_{\varepsilon \to 0^+} |A_\varepsilon g(t) - g(t)| = 0 \quad \text{a.e.}$$

Remark 3.3 This result can be extended to show that
$$\frac{1}{2\varepsilon} \int_{-\varepsilon}^{\varepsilon} |g(t + s) - g(t)| ds \to 0 \quad \text{for a.e. } t.$$
The set of t for which this happens is called the *Lebesgue set* of g.

Theorem 3.4 Local Ergodic Theorem (Wiener 1939) Let $\{T_t\}$ be a measure-preserving flow on a probability space $(X, \mathscr{B}, \mu)$, and suppose that $f \in L^1(X)$. Then
$$\lim_{\varepsilon \to 0^+} \frac{1}{2\varepsilon} \int_{-\varepsilon}^{\varepsilon} f(T_s x) ds = f(x) \quad \text{a.e. } d\mu.$$

Proof: Direct proofs are also possible, and of course the result would follow immediately from the representation of $\{T_t\}$ as a flow under a function if this could be established independently, but we will give Wiener's original argument, which uses the Lebesgue Differentiation Theorem and Fubini's Theorem.

There is a set of x of full measure for which $f(T_s x)$, as a function of

$s \in \mathbb{R}$, is locally integrable on $\mathbb{R}$. For any fixed x in this set, by the preceding Theorem

$$\left\{ t \in \mathbb{R} : \lim_{\varepsilon \to 0^+} \frac{1}{2\varepsilon} \int_{-\varepsilon}^{\varepsilon} f(T_s T_t x) ds \neq f(T_t x) \right\}$$

has measure 0. Hence

$$\{(t, x) : \text{convergence fails at } T_t x\}$$

has product measure 0, and therefore for almost all t

$$\mu\{x : \text{convergence fails at } T_t x\} = 0.$$

If we choose and fix a single such t, then, letting

$$F = \{x : \text{convergence fails at } x\},$$

we have that

$$0 = \mu\{x : T_t x \in F\} = \mu(T_{-t} F) = \mu(F),$$

which proves the Theorem.

3.4. The martingale convergence theorems

Let $(\Omega, \mathscr{F}, P)$ be a probability space and $\mathscr{F}_1 \subset \mathscr{F}_2 \subset \ldots$ an increasing sequence of sub-σ-algebras of $\mathscr{F}$. A sequence $X_1, X_2, \ldots$ of functions in $L^1(\Omega)$ such that X_n is measurable with respect to $\mathscr{F}_n$ for $n = 1, 2, \ldots$ is called

a *submartingale* if $E(X_{n+1} | \mathscr{F}_n) \geq X_n$ a.e.,

a *martingale* if $E(X_{n+1} | \mathscr{F}_n) = X_n$ a.e., and

a *supermartingale* if $E(X_{n+1} | \mathscr{F}_n) \leq X_n$ a.e.

If X_n is the fortune of a gambler after n plays, in the case of a submartingale the game is favorable to the player. The word 'martingale' was formerly used for a certain betting system, possibly appropriated for that use from its usual meaning as part of a bridle. The word 'supermartingale' is unfortunate – it is in no way connected with grocery stores. Here $\{X_n\}$ is properly called a martingale (or submartingale or supermartingale) *with respect to* $\{\mathscr{F}_n\}$; the simple word 'martingale' may be reserved for the case when $\mathscr{F}_n$ is the σ-algebra generated by $X_1, X_2, \ldots, X_n$.

The fundamental result about martingales is the Martingale Convergence Theorem of Doob, which is comparable in importance as well as content to the Ergodic Theorem. The following extremely elementary proof was shown to me by J. Horowitz. In contrast to the usual proofs and the other arguments in this chapter it relies on no maximal inequalities or upcrossing lemmas. It is no harder to prove the a.e. convergence of L^1-bounded submartingales, and we will do so. First a simple decomposition will reduce the statement to showing that every non-negative supermartingale converges, and the proof will be completed with the help of an elementary optional sampling (i.e., stopping time) theorem.

Theorem 4.1 Submartingale Convergence Theorem Every L^1-bounded submartingale (i.e., $\sup_n E|X_n| < \infty$) converges a.e.

Lemma 4.2 (*Krickeberg Decomposition* – see Neveu 1975) Every L^1-bounded submartingale $\{X_n\}$ is the difference of a non-negative martingale $\{M_n\}$ and a non-negative supermartingale $\{S_n\}$: $X_n = M_n - S_n$ for all n.

Proof: Notice that $\{X_n^+\}$ is a non-negative submartingale, since
$$E(X_{n+1}^+|\mathscr{F}_n) \geq E(X_{n+1}|\mathscr{F}_n) \geq X_n,$$
and hence, since $E(X_{n+1}^+|\mathscr{F}_n) \geq 0$ and X_n is either X_n^+ or $-X_n^-$,
$$E(X_{n+1}^+|\mathscr{F}_n) \geq X_n^+ \quad \text{a.e.} \quad \text{for all } n.$$
For fixed $m \leq n$,
$$E(X_{n+1}^+|\mathscr{F}_m) = E^{\mathscr{F}_m} E^{\mathscr{F}_n} X_{n+1}^+ \geq E^{\mathscr{F}_m} X_n^+ = E(X_n^+|\mathscr{F}_m),$$
so that $E(X_n^+|\mathscr{F}_m)$ increases in n for fixed m and hence has an increasing limit
$$M_m = \lim_n \uparrow E(X_n^+|\mathscr{F}_m).$$
I claim that $\{M_m\}$ is a martingale. For
$$E(M_{n+1}|\mathscr{F}_n) = E(\lim_k \uparrow E(X_k^+|\mathscr{F}_{n+1})|\mathscr{F}_n)$$
$$= \lim_k \uparrow E^{\mathscr{F}_n} E^{\mathscr{F}_{n+1}} X_k^+ = \lim_k E^{\mathscr{F}_n} X_k^+ = M_n.$$
Also, each $M_n \in L^1$, since
$$E(M_n) = \lim_k \uparrow E(E(X_k^+|\mathscr{F}_n)) = \lim_k \uparrow E(X_k^+)$$
$$\leq \sup_k E|X_k| < \infty.$$

Then let $S_n = M_n - X_n$ for $n = 1, 2, \ldots$. Clearly $S_n \geq 0$ a.e. (see the second line of the proof) and each S_n is in $L^1(\Omega)$. Also,
$$E(S_{n+1}|\mathscr{F}_n) = E(M_{n+1}|\mathscr{F}_n) - E(X_{n+1}|\mathscr{F}_n) \leq M_n - X_n = S_n \text{ a.e.},$$
so that $\{S_n\}$ is a non-negative supermartingale.

3.4. The martingale convergence theorems

Lemma 4.3 (*Optional Sampling*) Let $\{X_n\}$ be a non-negative supermartingale with respect to $\{\mathcal{F}_n\}$ and $\sigma, \tau: \Omega \to \{1, 2, \ldots, \infty\}$ *stopping times* (i.e., $\{\omega: \sigma(\omega) \leq n\} \in \mathcal{F}_n$ for all $n = 1, 2, \ldots$, and similarly for τ). Define

$$X_\tau(\omega) = \begin{cases} X_{\tau(\omega)}(\omega) & \text{if } \tau(\omega) < \infty \\ 0 & \text{if } \tau(\omega) = \infty, \end{cases}$$

and similarly for $X_\sigma(\omega)$. If $\sigma(\omega) \leq \tau(\omega)$ a.e., then

$$E(X_\sigma) \geq E(X_\tau).$$

(In the case of a supermartingale, a gambler's fortune tends to decrease with time. According to this Lemma, the sooner he stops, the greater his expected fortune.)

Proof: Fix $m \geq 1$. If $n \geq m$, then

$$\int_{\{\sigma=m\}} X_{\tau \wedge n} \, dP = \int_{\{\sigma=m, \tau \leq n\}} X_\tau \, dP + \int_{\{\sigma=m, \tau > n\}} X_n \, dP$$

$$\geq \int_{\{\sigma=m, \tau \leq n\}} X_\tau \, dP + \int_{\{\sigma=m, \tau > n\}} X_{n+1} \, dP = \int_{\{\sigma=m\}} X_{\tau \wedge (n+1)} \, dP$$

(since $\{\sigma = m, \tau > n\} \in \mathcal{F}_n$ and the supermartingale condition says that

$$\int_A X_{n+1} \, dP \leq \int_A X_n \, dP \text{ if } A \in \mathcal{F}_n).$$

Also, for $n \geq m$, on $\{\sigma = m\}$ we have $m = \sigma \leq \tau$, so that

$$\int_{\{\sigma=m\}} X_\sigma \, dP = \int_{\{\sigma=m\}} X_m \, dP = \int_{\{\sigma=m\}} X_{\tau \wedge m} \, dP,$$

and by the first part of the proof this is greater than, or equal to,

$$\int_{\{\sigma=m\}} X_{\tau \wedge n} \, dP \geq \int_{\{\sigma=m, \tau < \infty\}} X_{\tau \wedge n} \, dP$$

(since each $X_i \geq 0$).

Thus

$$\int_{\{\sigma=m\}} X_\sigma \, dP \geq \int_{\{\sigma=m, \tau < \infty\}} X_{\tau \wedge n} \, dP$$

for all $n \geq m$.

We want to apply Fatou's Lemma; note that

$$X_{\tau \wedge n} \to X_\tau \text{ as } n \to \infty \text{ on } \{\sigma = m, \tau < \infty\}$$

and

$$X_\tau = 0 \text{ on } \{\tau = \infty\}.$$

3. More about almost everywhere convergence

Thus

$$\int_{\{\sigma=m\}} X_\sigma \, dP \geq \liminf_n \int_{\{\sigma=m,\,\tau<\infty\}} X_{\tau \wedge n} \, dP$$

$$\geq \int_{\{\sigma=m,\,\tau<\infty\}} \liminf_n X_{\tau \wedge n} \, dP = \int_{\{\sigma=m,\,\tau<\infty\}} X_\tau \, dP$$

$$= \int_{\{\sigma=m\}} X_\tau \, dP.$$

Summing on m gives

$$\int X_\sigma \, dP \geq \int X_\tau \, dP.$$

Corollary 4.4 Under the hypotheses of Lemma 4.3, $X_\tau \in L^1$

Proof: Take $\sigma \equiv 1$.

Proof 4.5 (*of the Submartingale Convergence Theorem*) Because of Lemma 4.2 it is enough to show that *every non-negative L^1-bounded supermartingale converges a.e.* Let $\{X_n\}$ be a non-negative supermartingale which does not converge a.e. Then there are α, β with $0 \leq \alpha < \beta < \infty$ such that

$$E = \{\omega : \liminf_n X_n(\omega) < \alpha < \beta < \limsup_n X_n(\omega)\}$$

has positive measure.

Define a sequence of stopping times $\{\tau_i\}$ by

$$\tau_0 \equiv 1,$$
$$\tau_{2i+1}(\omega) = \inf\{n > \tau_{2i}(\omega) : X_n(\omega) > \beta\} \quad \text{for } i \geq 0,$$

and

$$\tau_{2i}(\omega) = \inf\{n > \tau_{2i-1}(\omega) : X_n(\omega) < \alpha\} \quad \text{for } i \geq 1,$$

where as usual $\inf \phi = +\infty$. Then

$$\tau_0 \leq \tau_1 \leq \tau_2 \leq \ldots,$$
$$X_{\tau_{2i}} < \alpha \text{ if } \tau_{2i} \leq \infty, \quad \text{and}$$
$$X_{\tau_{2i+1}} > \beta \text{ if } \tau_{2i+1} < \infty.$$

Let $p_j = P\{\omega : \tau_j(\omega) < \infty\}$. We have $p_{2i} \leq p_{2i-1}$ and

$$\beta p_{2i+1} \leq \int X_{\tau_{2i+1}} \, dP \leq \int X_{\tau_{2i}} \, dP \leq \alpha p_{2i},$$

by Lemma 4.3. Therefore,
$$p_{2i+1} \leq \frac{\alpha}{\beta} p_{2i} \leq \frac{\alpha}{\beta} p_{2i-1} \leq \cdots \leq \left(\frac{\alpha}{\beta}\right)^i p_1 \to 0,$$
but this is impossible since
$$p_{2i+1} \geq P(E) > 0.$$

3.5. The maximal inequality for the Hilbert transform

For a function $f \in L^1(\mathbb{R})$, the *Hilbert transform* Hf of f is defined by
$$Hf(t) = -\frac{1}{\pi} \lim_{\varepsilon \to 0^+} \int_{|s| \geq \varepsilon} \frac{f(t-s)}{s} \, ds.$$
Considerable argument is required to show that this limit exists. We will show here that the *maximal Hilbert transform*
$$H^*f(t) = \sup_{\varepsilon > 0} \left| \frac{1}{\pi} \int_{|s| \geq \varepsilon} \frac{f(t-s)}{s} ds \right|$$
satisfies a weak-type (1, 1) inequality,
$$m\{t \in \mathbb{R} : H^*f(t) > \lambda\} \leq c \frac{\|f\|_1}{\lambda} \quad \text{for each } \lambda > 0.$$

While the convergence would, as usual, follow readily from this inequality, we will not stop to prove it, but will move on to the ergodic case.

We give the ingenious proof due to Loomis (1946), who rediscovered an idea that actually goes back to Boole (1857). This argument, however, works only in one dimension, and for the higher-dimensional versions one must resort to covering lemmas and Calderón–Zygmund decompositions (see Stein 1970, Stein and Weiss 1971).

The Hilbert transform Hf is the analogue on $\mathbb{R}$ of the Fourier conjugate $\tilde{f}$ of an L^1 function on the unit circle Γ. Recall that a harmonic function u on the unit disk D has a unique *conjugate function* $\tilde{u}$ which is harmonic on D with $\tilde{u}(0) = 0$ and such that $u + i\tilde{u}$ is analytic on D. If u has boundary values $f \in L^p$ for some $p > 1$, then $\tilde{u}$ has boundary values $\tilde{f}$. The conjugate function $\tilde{f}$ is also given by a certain singular integral:
$$\tilde{f}(x) = -\frac{1}{\pi} \lim_{\varepsilon \to 0^+} \int_{\varepsilon \leq |t| \leq \pi} f(x+t) \frac{1}{2} \cot \frac{t}{2} dt$$
(see Zygmund 1959). Moreover, there is a simple relationship between the Fourier coefficients $\{a_n\}$ of f and $\{\tilde{a}_n\}$ of $\tilde{f}$: $\tilde{a}_n = -i \operatorname{sgn} n \cdot a_n$. Similarly, if

$f \in L^p(\mathbb{R})$ then f determines by a Poisson integral a harmonic function u on the upper half-plane. If $\tilde{u}$ is the imaginary part of a suitable analytic completion of u, then $\tilde{u}$ has boundary values Hf on $\mathbb{R}$. The situation in case $p \leq 1$ is too complicated and too interesting to be discussed here; it is already hard to prove that $\tilde{f} \in L^p$ when $f \in L^p$ for $p > 1$.

We are studying the Hilbert transform here because we wish to establish in the next section the existence of the *ergodic Hilbert transforms*

$$\tilde{f}(x) = \lim_{\varepsilon \to 0} \int_{\varepsilon < |s| < 1/\varepsilon} \frac{f(T_s x)}{s} \, ds,$$

for a flow $\{T_s\}$, and

$$\tilde{f}(x) = \sum_{k=1}^{\infty} \frac{f(T^k x) - f(T^{-k} x)}{k},$$

for a single transformation T. The existence of these limits was first proved by Cotlar (1955). We will use Calderón's approach (1968), which transfers the real-variable theorems to the ergodic-theoretic context.

The first step toward the maximal inequality for the Hilbert transform is a simple covering lemma, which seems to be due to Sierpinski and appears also in the paper of Wiener (1939). A version of this Lemma also holds in $\mathbb{R}^n$.

Lemma 5.1 Let A be a bounded subset of and $\mathbb{R}$ and $\{I_1, I_2, \ldots I_n\}$ a family of intervals which covers A. Then there is a subfamily $\{J_1, \ldots J_r\}$ of intervals with disjoint interiors such that

$$m(A) \leq 3 \sum_{k=1}^{r} m(J_k).$$

Proof: Again, by discarding some of the I_k's we can produce a covering of A with the property that no more than two I_k's intersect at any one point. Choose J_1 to be an interval of maximal length among the remaining $\{I_k\}$. We include J_1 in our subfamily, and we delete from A the interval J_1 and any of the I_k that meet J_1. There remains a set A_1 of measure

$$m(A_1) \geq m(A) - 3m(J_1).$$

Now repeat the process with A_1, choosing an interval J_2 of maximal length from among the remaining I_k's and deleting a set of measure at most $3m(J_2)$. There remains a set A_2 of measure

$$m(A_2) \geq m(A) - 3[m(J_1) + m(J_2)].$$

At the r'th stage, we remove the last of the I_k's. Then we have

$$0 \geq m(A) - 3 \sum_{k=1}^{r} m(J_k),$$

so that
$$m(A) \leq 3 \sum_{k=1}^{r} m(J_k).$$

A deep study of the function appearing in the following Lemma can be found in a paper of Boole (1857).

Lemma 5.2 If $a_1, a_2, \ldots a_n > 0$ and $g(t) = \sum_{i=1}^{n} a_i/(s_i - t)$, then for each $\lambda > 0$
$$m\{t : g(t) > \lambda\} = \frac{1}{\lambda} \sum_{i=1}^{n} a_i = m\{t : g(t) < -\lambda\}.$$

Proof: Notice that $g'(t) > 0$ if $t \notin \{s_1, \ldots s_n\}$, and
$$\lim_{t \to s_i^+} g(t) = -\infty, \quad \lim_{t \to s_i^-} g(t) = \infty,$$
so that g has the following type of graph:

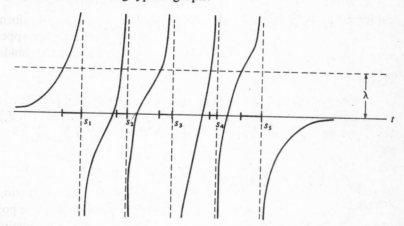

The set where $g > \lambda$ consists of exactly n intervals, as shown. There are exactly n points, $m_1, m_2, \ldots m_n$, with $g(t) = \lambda$, and
$$m\{g > \lambda\} = \sum_{i=1}^{n} (s_i - m_i).$$
The m_i are the roots of the equation
$$\sum_{i=1}^{n} \frac{a_i}{s_i - t} = \lambda,$$
which may be rewritten as
$$\sum_{i=1}^{n} \left[a_i \prod_{j \neq i} (s_j - t) \right] = \lambda \prod_{j=1}^{n} (s_j - t),$$

or
$$(-1)^{n-1}(\sum a_i)t^{n-1} + \ldots = \lambda\left[(-1)^n t^n + (-1)^{n-1}\sum_{j=1}^n s_j t^{n-1} + \ldots\right],$$
or
$$t^n - \left(\sum s_j - \frac{1}{\lambda}\sum a_i\right)t^{n-1} + \ldots = 0.$$

The coefficient of t^{n-1} is the negative of the sum of the roots; therefore
$$\sum m_i = \sum s_j - \frac{1}{\lambda}\sum a_i$$
and
$$\sum(s_i - m_i) = \frac{1}{\lambda}\sum a_i.$$

If $p_1, \ldots p_n$ are the roots of $g(t) = -\lambda$, then
$$m\{g < -\lambda\} = \sum(p_i - s_i),$$
and for $t = p_j$ we have
$$\sum \frac{a_i}{s_i - t} = -\lambda,$$
or
$$\sum_{i=1}^n a_i \prod_{j \neq i}(s_j - t) = -\lambda \prod_{j=1}^n (s_j - t),$$
$$(-1)^{n-1}(\sum a_i)t^{n-1} + \ldots = -\lambda[(-1)^n t^n + (-1)^{n-1}(\sum s_i)t^{n-1} + \ldots],$$
or
$$t^n - \left(\sum s_i + \frac{1}{\lambda}\sum a_i\right)t^{n-1} + \ldots = 0,$$
so that
$$\sum p_i = \sum s_i + \frac{1}{\lambda}\sum a_i$$
and
$$\sum(p_i - s_i) = \frac{1}{\lambda}\sum a_i.$$

Theorem 5.3 There is a constant c such that if $f \in L^1(\mathbb{R})$ and $\lambda > 0$, then
$$m\left\{s \in \mathbb{R}: \sup_{\varepsilon > 0}\left|\int_{|t| \geq \varepsilon} \frac{f(s-t)}{t}dt\right| > \lambda\right\} \leq c\frac{\|f\|_1}{\lambda}.$$

3.5. The maximal inequality for the Hilbert transform

Proof: We suppose first that $f \geq 0$. Let A be a closed and bounded subset of the set where

$$\sup_{\varepsilon > 0}|H_\varepsilon f(s)| = \sup_{\varepsilon > 0}\left|\int_{|t|\geq\varepsilon}\frac{f(s-t)}{t}dt\right| = \sup_{\varepsilon > 0}\left|\int_{|t-s|\geq\varepsilon}\frac{f(t)}{s-t}dt\right| > \lambda.$$

For each $s \in A$ there is an interval I_s, say of radius ε_s, centered at s such that

$$\left|\int_{|t-s|\geq\varepsilon_s}\frac{f(t)}{t-s}dt\right| > \lambda.$$

A finite number $I_1, I_2, \ldots I_n$ of these intervals cover A. Let $\{J_k\}$, with centers s_k and radii ε_k, be as in Lemma 5.1. Then the J_k are pairwise disjoint,

$$m(A) \leq 3\sum m(J_k),$$

and

$$\left|\int_{t\notin J_k}\frac{f(t)}{s_k - t}dt\right| > \lambda \quad \text{for each } k.$$

Since

$$F(s) = \int_{-\infty}^{s} f(t)dt$$

is a continuous function of bounded variation and

$$\int_{t\notin J_k}\frac{f(t)}{s_k - t}dt = \int_{t\notin J_k}\frac{dF(t)}{s_k - t}$$

exists as a Riemann–Stieltjes integral, this integral can be approximated by its Riemann sums (Rudin 1953 p. 112):

$$\sum_i \frac{1}{s_k - t_i}\Delta F_i = \sum_i \frac{1}{s_k - t_i}\int_{t_i}^{t_{i+1}} f(t)dt.$$

Choose a partition $t_1, \ldots t_n$ including the endpoints and centers s_k of the J_k such that

$$\left|\sum_{(t_i,t_{i+1})\not\subset J_k}\frac{\Delta F_i}{s_k - t_i}\right| > \lambda \quad \text{for each } k.$$

Let us deal with the inequality

$$\sum_{(t_i,t_{i+1})\not\subset J_k}\frac{\Delta F_i}{s - t_i} > \lambda \quad \text{for } s = s_k;$$

a similar argument will apply in case the sum is less than $-\lambda$. The left side is a decreasing function of s, so the inequality holds for $s_k - \varepsilon_k < s \le s_k$. For each such s, then, either

$$\sum_i \frac{\Delta F_i}{s - t_i} > \frac{\lambda}{2} \quad \text{or} \quad \sum_{(t_i, t_{i+1}) \subset J_k} \frac{\Delta F_i}{s - t_i} < -\frac{\lambda}{2}.$$

In the first case, by Lemma 5.2 s falls into a set of measure

$$\frac{2}{\lambda} \sum_i \Delta F_i \le 2 \frac{\|f\|_1}{\lambda},$$

and in the second case into a set of measure

$$\frac{2}{\lambda} \sum_{(t_i, t_{i+1}) \subset J_k} \Delta F_i.$$

Taking the union on k, we find that a set of measure $\sum \varepsilon_k$ is covered by a set of measure $2\|f\|_1/\lambda$ (the set of the first case does not change with k) and a set of measure

$$\sum_k \frac{2}{\lambda} \sum_{(t_i, t_{i+1}) \subset J_k} \Delta F_i \le \frac{2}{\lambda} \|f\|_1.$$

Thus (taking also into account the case when the Riemann sum is less than $-\lambda$), we find that

$$\sum \varepsilon_k \le 2 \frac{4}{\lambda} \|f\|_1$$

and

$$m(A) \le 3 \sum m(J_k) = 6 \sum \varepsilon_k \le \frac{48}{\lambda} \|f\|_1.$$

Of course in case f were negative a similar argument would apply. For an arbitrary $f \in L^1$,

$$|H_\varepsilon f(s)| \le |H_\varepsilon f^+(s)| + |H_\varepsilon f^-(s)|,$$

so that,

$$\{H^* f > \lambda\} \subset \left\{ H^* f^+ > \frac{\lambda}{2} \right\} \cup \left\{ H^* f^- > \frac{\lambda}{2} \right\},$$

and hence

$$m\{H^* f > \lambda\} \le \frac{96}{\lambda} (\|f^+\|_1 + \|f^-\|_1) = \frac{96}{\lambda} \|f\|_1.$$

3.6. The ergodic Hilbert transform

We will prove that if $\{T_t\}$ is a measure-preserving flow on a probability space $(X, \mathcal{B}, \mu)$ and $f \in L^1(X)$, then

$$\tilde{f}(x) = \lim_{\varepsilon \to 0^+} \int_{\varepsilon \leq |t| \leq 1/\varepsilon} \frac{f(T_t x)}{t} dt$$

exists a.e. A similar argument shows that for a single measure-preserving transformation $T: X \to X$, the series

$$\sum_{k=1}^{\infty} \frac{f(T^k x) - f(T^{-k} x)}{k}$$

converges a.e. (It might be conjectured that if $f \in L^1$ with $\int f \, d\mu = 0$, then in fact

$$\sum_{k=1}^{\infty} \frac{f(T^k x)}{k}$$

converges a.e. However, this is not true; partial summation shows that if

$$S_n f = \sum_{k=1}^{n} f T^k,$$

then

$$\sum_{k=1}^{\infty} \frac{f T^k}{k} = \sum_{n=1}^{\infty} \frac{S_n f}{n(n+1)} - f,$$

and we have seen (in 3.2. B) that the series on the right does not converge a.e. in general, even for $f \in L^\infty$.)

The method of proof involves first establishing a weak-type (1, 1) inequality for the maximal function

$$f^h(x) = \sup_{\varepsilon > 0} \left| \int_{\varepsilon \leq |t| \leq 1/\varepsilon} \frac{f(T_t x)}{t} dt \right|.$$

Theorem 6.1 There is a constant c such that if $f \in L^1(X)$ and $\lambda > 0$, then

$$\mu\{x \in X : f^h(x) > \lambda\} \leq \frac{c}{\lambda} \|f\|_1.$$

Proof: Fix N and let

$$f_N^h(X) = \sup_{\varepsilon \geq 1/N} \left| \int_{\varepsilon \leq |t| \leq 1/\varepsilon} \frac{f(T_t x)}{t} dt \right| = \sup_{\varepsilon \geq 1/N} \left| \int_{\varepsilon \leq |t| \leq 1/\varepsilon} \frac{f(T_{-t} x)}{t} dt \right|.$$

We will find a constant c independent of N such that

$$\mu\{x : f_N^h(x) > \lambda\} \leq \frac{c}{\lambda} \|f\|_1 \quad \text{for all } f \in L^1;$$

the conclusion will follow upon letting $N \to \infty$. The technique, due to Calderón, involves transferring the real-variable result of the previous section to the ergodic context by considering one orbit at a time and then applying Fubini's Theorem. This is similar to the idea behind Wiener's proof of the Local Ergodic Theorem, but this time the argument is a bit more complicated; in particular, certain truncations must be made, since in the ergodic case we cannot integrate over an entire orbit.

Now

$$H_N^* g(t) = \sup_{\varepsilon \geq 1/N} \left| \int_{\varepsilon \leq |s| \leq 1/\varepsilon} \frac{g(t-s)}{s} \, ds \right|, \quad \text{for } g \in L^1(\mathbb{R}),$$

is a *semilocal* operator, in that the support of $H_N^* g$ is contained in a neighborhood of the support of g:

$$H_N^* g(t) = \sup_{\varepsilon \geq 1/N} \left| \int_{\varepsilon \leq |t-s| \leq 1/\varepsilon} \frac{g(s)}{t-s} \, ds \right|$$

is 0 at points t which are farther than N from supp g.

Let

$$F(t, x) = f(T_t x), \quad \text{and}$$

$$G(t, x) = H_N^* F(t, x) = \sup_{\varepsilon \geq 1/N} \left| \int_{\varepsilon \leq |s| \leq 1/\varepsilon} \frac{F(t-s, x)}{s} \, ds \right|$$

$$= \sup_{\varepsilon \geq 1/N} \left| \int_{\varepsilon \leq |s| \leq 1/\varepsilon} \frac{f(T_{t-s} x)}{s} \, ds \right| = f_N^h(T_t x).$$

Then

$$F(t, T_s x) = F(s + t, x)$$
$$G(t, T_s x) = G(s + t, x),$$

so that for fixed t_1 and t_2, $F(t_1, x)$ and $F(t_2, x)$ are equimeasurable functions of x, as are $G(t_1, x)$ and $G(t_2, x)$.

Fix $a > 0$ and define

$$F_a(t, x) = \begin{cases} F(t, x) & \text{if } |t| \leq a \\ 0 & \text{otherwise} \end{cases}$$

and

$$G_a(t, x) = H_N^* F_a(\cdot, x)(t).$$

Now

$$G(t, x) = H_N^* F = H_N^*(F_{a+N} + (F - F_{a+N}))$$
$$\leq H_N^* F_{a+N} + H_N^*(F - F_{a+N})$$

and $F - F_{a+N}$ has support in $|t| > a + N$, so $H_N^*(F - F_{a+N})$ is 0 for $|t| \leq a$,

3.6. The ergodic Hilbert transform

and hence
$$G \leq G_{a+N} \quad \text{for } |t| \leq a.$$
Fix $\lambda > 0$, and let
$$E = \{x \in X : G(0, x) > \lambda\} = \{x \in X : f_N^h(x) > \lambda\},$$
$$\bar{E} = \{(t, x) : G_{a+N}(t, x) > \lambda\}, \quad \text{and}$$
$$\bar{E}_y = \{t : (t, y) \in \bar{E}\}, \quad \text{for } y \in X.$$
For any fixed $t \in \mathbb{R}$, let
$$E^t = \{x : G(t, x) > \lambda\};$$
by the equimeasurability of the $G(t, \cdot)$, $\mu(E^t) = \mu(E)$ for all t. Thus
$$2a\mu(E) = \int_{-a}^{a} \mu(E^t)\, dt,$$
and, since $G \leq G_{a+N}$ for $|t| \leq a$ implies that $(t, x) \in \bar{E}$ if $-a \leq t \leq a$ and $x \in E^t$, this is no greater than
$$\int_{-a}^{a} \mu\{x : (t, x) \in \bar{E}\}\, dt \leq \int_{-\infty}^{\infty} \mu\{x : (t, x) \in \bar{E}\}\, dt$$
$$= m \times \mu(\bar{E}) = \int_X m(\bar{E}_x)\, d\mu(x)$$
$$= \int_X m\{t : G_{a+N}(t, x) > \lambda\}\, d\mu(x)$$
$$= \int_X m\{t : H_N^* F_{a+N}(t, x) > \lambda\}\, d\mu(x).$$

Thus, by the weak-type $(1, 1)$ inequality for the real-variables Hilbert transform,
$$2a\mu(E) \leq \int_X \frac{96}{\lambda} \|F_{a+N}(t, x)\|_{L^1(dt)}\, d\mu(x)$$
$$= \frac{96}{\lambda} \int_X \int_{-\infty}^{\infty} |F_{a+N}(t, x)|\, dt\, d\mu(x)$$
$$= \frac{96}{\lambda} \int_X \int_{-(a+N)}^{a+N} |F(t, x)|\, dt\, d\mu(x)$$
$$= \frac{96}{\lambda} \int_{-(a+N)}^{a+N} \int_X |F(t, x)|\, d\mu(x)\, dt = \frac{96}{\lambda} \int_{-(a+N)}^{a+N} \int_X |F(0, x)|\, d\mu(x)\, dt$$
$$= \frac{96}{\lambda} \int_{-(a+N)}^{a+N} \|f\|_1\, dt = 2(a+N) \frac{96}{\lambda} \|f\|_1.$$

Therefore
$$\mu(E) \leq \frac{a+N}{a}\frac{96}{\lambda}\|f\|_1 \quad \text{for each } a > 0,$$
and letting $a \to \infty$ gives
$$\mu(E) \leq \frac{96}{\lambda}\|f\|_1.$$

By an analogous argument, but convolving with a step-function kernel rather than $1/t$, one can also obtain:

Corollary 6.2 There is a constant c such that if $f \in L^1(X)$ and $\lambda > 0$, then
$$\mu\left\{x: \sup_{n \geq 1}\left|\sum_{k=1}^{n}\frac{f(T^k x) - f(T^{-k}x)}{k}\right| > \lambda\right\} \leq \frac{c}{\lambda}\|f\|_1.$$

We will now prove the existence a.e. of the ergodic Hilbert transform
$$\tilde{f}(x) = \lim_{\varepsilon \to 0^+}\int_{\varepsilon \leq |t| \leq 1/\varepsilon}\frac{f(T_t x)}{t}dt$$
of a function $f \in L^1(X)$. The pattern of proof is familiar: convergence is established relatively easily for a dense subset of L^1, and then the maximal inequality is used to show that the set of functions for which convergence takes place is closed in L^1.

Theorem 6.3 If $\{T_t: -\infty < t < \infty\}$ is a measure-preserving flow on a probability space $(X, \mathcal{B}, \mu)$ and $f \in L^1(X)$, then the limit
$$\tilde{f}(x) = \lim_{\varepsilon \to 0^+}\int_{\varepsilon \leq |t| \leq 1/\varepsilon}\frac{f(T_t x)}{t}dt$$
exists a.e. on X.

Proof: With
$$V_\varepsilon f(x) = \int_{\varepsilon \leq |t| \leq 1/\varepsilon}\frac{f(T_t x)}{t}dt$$
and
$$f^h(x) = \sup_{\varepsilon > 0}|V_\varepsilon f(x)|,$$
we know that
$$\mu\{x: f^h(x) > \lambda\} \leq \frac{c}{\lambda}\|f\|_1.$$

Let $\mathcal{D}$ be the subspace of $L^1(X)$ generated by all invariant functions and all functions of the form
$$f(x) = \int_{-\infty}^{\infty}f_0(T_s x)\phi(s)ds,$$

3.6. The ergodic Hilbert transform

where $f_0 \in L^\infty(X)$ and $\phi \in \mathscr{C}^1(\mathbb{R})$ has compact support and 0 integral. We will show that $\tilde{f}(x)$ exists a.e. if $f \in \mathscr{D}$ and that $\mathscr{D}$ is dense in L^1.

If f is a function of this form, then, with

$$k_\varepsilon(t) = \begin{cases} \dfrac{1}{t} & \text{if } \varepsilon \leq |t| \leq \dfrac{1}{\varepsilon} \\ 0 & \text{otherwise} \end{cases},$$

we have that

$$(1) \quad V_\varepsilon f(x) = \int_{-\infty}^\infty k_\varepsilon(t) f(T_t x)\,dt = \int_{-\infty}^\infty k_\varepsilon(t) \int_{-\infty}^\infty f_0(T_{s+t}x)\phi(s)\,ds\,dt$$

$$= \int_{-\infty}^\infty f_0(T_t x) \int_{-\infty}^\infty k_\varepsilon(t-s)\phi(s)\,ds\,dt = \int_{-\infty}^\infty f_0(T_t x)[k_\varepsilon * \phi(t)]\,dt.$$

Let $K_\varepsilon(s) = 1/s$ for $|s| \geq \varepsilon$ and 0 otherwise. We will show that (i) $K_\varepsilon * \varphi(t)$ converges for a.e. t as $\varepsilon \to 0$, (ii) $|K_\varepsilon * \phi(t)| \leq g(t) \in L^1(\mathbb{R})$, and (iii) $\|k_\varepsilon * \phi - K_\varepsilon * \phi\|_1 \to 0$ as $\varepsilon \to 0$. This will imply that $k_\varepsilon * \phi$ converges in $L^1(\mathbb{R})$ as $\varepsilon \to 0$, hence $\lim_{\varepsilon \to 0} V_\varepsilon f(x)$ exists a.e. for our particular f.

To prove (i), write

$$K_\varepsilon * \phi(t) = \int_{|s-t| \geq \varepsilon} \frac{\phi(s)}{t-s}\,ds = \int_{|s-t| \geq 1} \frac{\phi(s)}{t-s}\,ds + \int_{1 \geq |s-t| \geq \varepsilon} \frac{\phi(s)}{t-s}\,ds.$$

The first term is of course constant with respect to ε, while the second is

$$\int_{1 \geq |s-t| \geq \varepsilon} \frac{\phi(s) - \phi(t)}{t-s}\,ds = \int \frac{\phi(s) - \phi(t)}{t-s} \chi_{\{\sigma: 1 \geq |\sigma-t| \geq \varepsilon\}}(s)\,ds.$$

The integrand is bounded by an absolute constant and converges a.e. ds to $[(\phi(s) - \phi(t))/(t-s)]\chi_{(t-1, t+1)}(s)$, so the integrals converge as $\varepsilon \to 0$.

For (ii), suppose that $\operatorname{supp} \phi \subset [-K, K]$. If $|t| \geq 2K$, then

$$|K_\varepsilon * \phi(t)| = \left|\int_{-\infty}^\infty K_\varepsilon(t-s)\phi(s)\,ds\right| = \left|\int_{-\infty}^\infty [K_\varepsilon(t-s) - K_\varepsilon(t)]\phi(s)\,ds\right| \left(\text{because } \int_{-\infty}^\infty \phi(s)\,ds = 0\right)$$

$$= \left|\int_{-K}^K [K_\varepsilon(t-s) - K_\varepsilon(t)]\phi(s)\,ds\right|$$

$$\leq \int_{-K}^K \left|\frac{1}{t-s} - \frac{1}{t}\right| |\phi(s)|\,ds = \int_{-K}^K \frac{|s|}{|t||t-s|} |\phi(s)|\,ds$$

$$\leq \frac{c}{t^2} \text{ for some constant } c \text{ depending on } \phi.$$

If $|t| \leq 2K$, then

$$K_\varepsilon * \phi(t) = \int_{-\infty}^{\infty} K_\varepsilon(s)\phi(t-s)\,ds = \int_{-\infty}^{\infty} K_\varepsilon(s)[\phi(t-s) - \phi(t)]\,ds$$

$$= \int_{\varepsilon \leq |s|} \frac{\phi(t-s) - \phi(t)}{s}\,ds = \int_{\varepsilon \leq |s| \leq 3K} \frac{\phi(t-s) - \phi(t)}{s}\,ds,$$

which is again bounded by a constant. Thus $|K_\varepsilon * \phi(t)| \leq (c_1/t^2) + c_2 \chi_{[-2K, 2K]}(t) \in L^1(\mathbb{R})$.

For (iii), notice that $k_\varepsilon * \phi(t)$ and $K_\varepsilon * \phi(t)$ only differ because $k_\varepsilon(t-s) = 0$ for $|t-s| \geq 1/\varepsilon$ while $K_\varepsilon(t-s) = 1/(t-s)$ for all large $|t-s|$. Thus

$$K_\varepsilon * \phi(t) - k_\varepsilon * \phi(t) = \int_{|s-t| \geq 1/\varepsilon} \frac{\phi(s)}{t-s}\,ds$$

$$= \int_{|s| \geq 1/\varepsilon} \frac{\phi(t-s)}{s}\,ds \leq \frac{c}{t^2}\chi_{(-1/\varepsilon + K, 1/\varepsilon - K)^c}(t)$$

as in (ii), and this tends to 0 in L^1.

To show that the invariant functions together with all the functions $f(x) = \int_{-\infty}^{\infty} f_0(T_t x)\phi(t)\,dt$ are dense in L^1, it is enough to show that if $h \in L^\infty(X)$ and $\int_X hf\,d\mu = 0$ for all such f, then h is invariant. (By the Hahn–Banach Theorem, the closed subspace generated by $\mathcal{D}$ can be separated by a linear functional from any function not in it.) But

$$0 = \int fh\,d\mu = \int_X f_0(x) \int_{-\infty}^{\infty} h(T_{-t}x)\phi(t)\,dt\,d\mu(x) \quad \text{for all } f_0 \in L^\infty(X)$$

implies that

$$\int_{-\infty}^{\infty} h(T_{-t}x)\phi(t)\,dt = 0 \quad \text{for all such } \phi.$$

Considering a $\mathscr{C}_c^1$ (smooth) version ϕ of $(1/m(I))\chi_I - (1/m(J))\chi_J$ where I and J are intervals and letting I and J shrink to arbitrary points shows that h is a.e. constant along almost every orbit.

Finally, let us show that the set $\mathscr{E}$ of all $f \in L^1(X)$ for which $V_\varepsilon f(x)$ converges a.e. is closed in L^1. Let $f_k \in \mathscr{E}, k = 1, 2, \ldots$ and suppose that $f_k \to f$ in L^1. Since

$$|V_\varepsilon f(x) - V_{\varepsilon'} f(x)| = |V_\varepsilon[f_k + (f - f_k)](x) - V_{\varepsilon'}[f_k + (f - f_k)](x)|$$

$$\leq |V_\varepsilon f_k(x) - V_{\varepsilon'} f_k(x)| + 2\sup_{\eta > 0} |V_\eta(f - f_k)(x)|$$

3.7. The filling scheme

and for fixed k the first term approaches 0 as $\varepsilon, \varepsilon' \to 0$, we see that for any $\lambda > 0$,

$$\mu\{x: \limsup_{\varepsilon,\varepsilon' \to 0} |V_\varepsilon f(x) - V_{\varepsilon'} f(x)| > \lambda\}$$

$$\leq \mu\left\{x: \sup_{\eta > 0} |V_\eta(f - f_k)(x)| > \frac{\lambda}{2}\right\} \leq \frac{c}{\lambda}\|f - f_k\|_1,$$

which tends to 0 as $k \to \infty$. Therefore $\{V_\varepsilon f(x)\}$ is fundamental a.e., and hence converges a.e.

Remark 6.4 In most of the arguments of this type, we do not need the weak-type (1, 1) inequality for the full maximal function – such an inequality for an 'eventual' maximal function, in which the sup is replaced by a lim sup, would be sufficient. (This was Birkhoff's method in his proof of the Ergodic Theorem.) Also, for the purposes of the theorem on differentiation of integrals, the absolute value signs could have been moved outside the integral. However, there is a theorem of Stein (1961) according to which the existence of a convergence theorem frequently already forces the corresponding full maximal function to satisfy a weak-type (1, 1) inequality, so such apparent gains in power may be only illusory.

Corollary 6.5 If $T: X \to X$ is a m.p.t. and $f \in L^1(X)$, then the series

$$\tilde{f}(x) = \sum_{k=1}^\infty \frac{f(T^k x) - f(T^{-k} x)}{k}$$

converges a.e.

Proof: Use Corollary 6.2 and mimic the proof of Theorem 6.3.

3.7. The filling scheme

In the next few sections we undertake the proof of the Chacon–Ornstein Ergodic Theorem (1960), an extremely general pointwise convergence result which includes many other ergodic theorems as special cases. This theorem deals with an operator $T: L^1(X) \to L^1(X)$, and thus it extends the case when T arises from a m.p.t. $X \to X$ to situations that are found in interesting applications, for example in the theory of Markov processes. We will give the proof of the conservative case with the simplifications due to Neveu (1979), who saw how to base the argument on certain key features of the filling scheme. (The filling scheme appeared already in Chacon and Ornstein 1960).

3. More about almost everywhere convergence

We deal with a σ-finite measure space $(X, \mathscr{B}, \mu)$, and assume that
$$T: L^1(X) \to L^1(X)$$
is a *positive contraction*:

if $0 \leq f \in L^1$, then $0 \leq Tf \in L^1$, and

$\|Tf\|_1 \leq \|f\|_1$ for all $f \in L^1$.

Such a T is sometimes called a *sub-Markov* operator; if $T^*1 = 1$, where T^* is the adjoint of T on L^∞ (defined by $\int \phi \cdot Tf \, d\mu = \int T^*\phi \cdot f \, d\mu$ for all $\phi \in L^\infty$ and all $f \in L^1$) then T is called a *Markov* operator.

The *filling scheme* is defined as follows. We begin with $0 \leq f, g \in L^1$. Think of f as giving the amount of sand piled above each point of X and g as giving the depth of a hole below some points of X:

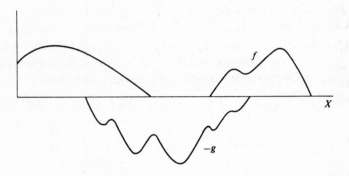

When we allow whatever sand will do so to fall into the hole, and consequently decrease the depth of the hole, we obtain a new sand function
$$h^+ = (f-g)^+$$
and a new hole-depth function
$$h^- = (f-g)^-.$$
(Thus $h = h^+ - h^- = f - g$.)

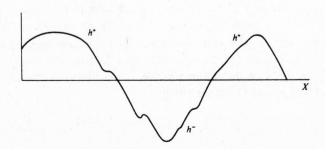

3.7. The filling scheme

Now we *use the transformation T to transport the sand*: at the next stage the sand has distribution Th^+, while the hole still has depth h^-. When the sand and hole interact, the distribution of the outcome is given by $h_1 = Th^+ - h^-$. Continuing in this manner, we build a sequence $h_0, h_1, h_2, \ldots$ of L^1 functions. The inductive definition is

$$h_0 = h = f - g$$
$$h_{n+1} = Th_n^+ - h_n^- \quad \text{for } n = 0, 1, 2, \ldots$$

A short proof of the Chacon–Ornstein Theorem, for the conservative ergodic case, will emerge (in the next section) from the observation that if $\int h \, d\mu < 0$ then $\sum_{n=0}^{\infty} h_n^+ < \infty$ a.e.

The filling scheme has the following three basic properties.

(1) $h_{n+1}^- \leq h_n^-$ for all $n \geq 0$.

This just says that the depth of the hole continues to decrease at each point. Let $H^- = \lim \downarrow h_n^-$ (the decreasing limit of the h_n^-) and $A = \{H^- > 0\}$.

(2) $h_{n+1}^+ \leq Th_n^+$ for all $n \geq 0$

This says that after each time the sand is moved, some of it may fall into the hole. In fact this statement can be slightly strengthened. Notice that $h_n^+ = 0$ on A, since

$$h_n^- \geq H^- > 0 \quad \text{on } A.$$

Thus

$$h_{n+1}^+ \leq Th_n^+$$

implies that in fact

$$h_{n+1}^+ \leq T(\chi_{A^c} h_n^+) = (TM_{A^c})h_n^+$$

(M_{A^c} denotes multiplication by the characteristic function of A^c), and hence, by induction,

$$h_{n+1}^+ \leq (TM_{A^c})^{n+1} h^+ \quad \text{for } n = 0, 1, 2, \ldots.$$

(3) If T is a Markov operator, so that $\int Tf \, d\mu = \int f \, d\mu$ for all $f \in L^1$, then

$$\int h_{n+1} \, d\mu = \int h_n \, d\mu \quad \text{for } n = 0, 1, 2, \ldots.$$

This says that the difference between the amount of sand and the size of the hole remains constant under the hypothesis that $T^*1 = 1$. This automatically holds, as we shall see, in case T is *conservative*.

Then for all n,

(*) $$\int h_n \, d\mu = \int h \, d\mu = \int h_n^+ \, d\mu - \int h_n^- \, d\mu \to \lim \downarrow \int h_n^+ \, d\mu - \int H^- \, d\mu.$$

The filling scheme is useful in ergodic theory because the sequence $\{h_n\}$ is related to the sequence of partial sums of $\{T^k f\}$, as the following Proposition shows.

Proposition 7.1 Let $0 \leq f, g \in L^1(X)$ and $h = f - g$. Then there are $u_n \in L^1$ with $u_n \geq 0$ a.e. and

$$\sum_{k=0}^{n-1} T^k f = \sum_{k=0}^{n-1} h_k^+ + u_n \quad \text{and}$$

$$\sum_{k=0}^{n-1} T^k g = \sum_{k=0}^{n-1} T^{n-k-1} h_k^- + u_n \quad \text{for } n = 1, 2, \ldots.$$

Proof: We define

$$u_0 = 0,$$
$$u_n = (g - h_{n-1}^-) + T u_{n-1} \quad \text{for } n \geq 1.$$

We will prove by induction that these u_n satisfy the required equations.

For $n = 1$, the formulas say that

$$f = h_0^+ + u_1 = h^+ + u_1 = h^+ + (g - h^-) + T0 = f$$

and

$$g = h_0^- + u_1 = h^- + u_1 = h^- + (g - h^-) + T0 = g.$$

Note also that

$$u_1 = g - h^- \geq 0$$

since $g \geq h^-$.

Assume now that the formulas hold for n, and let us prove them for $n + 1$. Notice that

$$u_{n+1} \geq 0$$

since

$$g \geq h^- \geq h_n^-$$

by (1) above. Further,

$$T h_k^+ = h_{k+1} + h_k^- = h_{k+1}^+ + (h_k^- - h_{k+1}^-),$$

so by induction

$$T \sum_{k=0}^{n-1} T^k f = T \sum_{k=0}^{n-1} h_k^+ + T u_n = \sum_{k=0}^{n-1} h_{k+1}^+ + (h^- - h_n^-) + T u_n$$

3.7. The filling scheme

and hence

$$\sum_{k=0}^{n} T^k f = f + T \sum_{k=0}^{n-1} T^k f = f + \sum_{k=0}^{n-1} h_{k+1}^+ + (h^- - g)$$

$$+ \underbrace{(g - h_n^-) + Tu_n}_{u_{n+1}} = \sum_{k=0}^{n} h_k^+ + u_{n+1}$$

(of course $f - g + h^- = h^+$).

For the other formula, by induction

$$T \sum_{k=0}^{n-1} T^k g = \sum_{k=0}^{n-1} T^{n-k} h_k^- + Tu_n;$$

therefore

$$\sum_{k=0}^{n} T^k g = g + \sum_{k=0}^{n-1} T^{n-k} h_k^- + Tu_n = g + \sum_{k=0}^{n} T^{n-k} h_k^- - h_n^- + Tu_n$$

$$= \sum_{k=0}^{n} T^{n-k} h_k^- + u_{n+1}.$$

Next we will use the filling scheme to establish the *Hopf decomposition* of X into a conservative part and a dissipative part with respect to T. Usually this decomposition is a corollary of the Hopf maximal ergodic theorem, but here we deduce both results from the filling scheme. (Actually the Hopf decomposition can be proved from scratch, as Exercise 2.3.2 does for a partial case (Foguel 1980).) The heart of the Chacon–Ornstein Theorem lies in the case when T is conservative and ergodic.

If $T: L^1(X) \to L^1(X)$ is a positive contraction, then we say that T is *conservative* if

$$0 \leqslant u \in L^1(X) \text{ implies } \sum_{k=0}^{\infty} T^k u \text{ takes only the two values 0 and } \infty \text{ a.e.}$$

T is called *conservative ergodic* if

$$0 \leqslant u \in L^1(X), u \not\equiv 0 \quad \text{implies} \quad \sum_{k=0}^{\infty} T^k u = \infty \text{ a.e.}$$

Lemma 7.2 Let $0 \leqslant f_i \in L^1$ with $f_i \downarrow 0$ a.e. and $0 \leqslant g \in L^1$. Then the H^- functions ${}^i H^-$ associated with the ${}^i h = f_i - g$ increase to g a.e. as $i \to \infty$.

Proof: Since ${}^i h$ decreases with i and the mapping $h \to Th^+ - h^-$ is increasing (since T is positive), the $({}^i h)_k$ decrease in i for fixed k. Therefore $({}^i h)_k^-$ increases in i for fixed k and hence

$${}^i H^- = \lim \downarrow ({}^i h)_k^-$$

increases with i. Also,

$${}^i H^- \leqslant {}^i h^- \leqslant g$$

and (from (1) and (2)),

$$\int h \, d\mu = \int h_0^+ \, d\mu - \int h_0^- \, d\mu \geqslant \int T h_0^+ \, d\mu - \int h_0^- \, d\mu = \int h_1 \, d\mu$$

(since $\|T\| \leqslant 1$ implies $\int T h_0^+ \, d\mu \leqslant \int h_0^+ \, d\mu$), so that

$$\int h \, d\mu \geqslant \int h_1 \, d\mu \geqslant \ldots \to \lim \downarrow \int h_n^+ \, d\mu - \int H^- \, d\mu \geqslant -\int H^- \, d\mu.$$

Therefore, replacing h by ${}^i h$,

$$\int f_i \, d\mu = \int ({}^i h + g) \, d\mu \geqslant \int (g - {}^i H^-) \, d\mu \geqslant 0.$$

Now taking the limit as $i \to \infty$ gives

$${}^i H^- \nearrow g \text{ a.e.}$$

Proposition 7.3 If $0 \leqslant u, v \in L^1(X)$, then $\sum_{n=1}^\infty T^n v = 0$ a.e. on

$$\left\{ x : \sum_{n=1}^\infty T^n u(x) = \infty, \sum_{n=1}^\infty T^n v(x) < \infty \right\}.$$

Proof: Let us show that $v = 0$ a.e. on the set in question; the stronger statement will follow upon replacing v by $T^n v$ for each $n = 1, 2, \ldots$. We apply the Lemma to $f_i = u/i$ and $g = v$ to find

$$v = \lim \uparrow {}^i H^-.$$

But we claim that ${}^i H^- = 0$, for each i, on $\{\sum_{n=0}^\infty T^n u = \infty, \sum_{n=0}^\infty T^n v < \infty\}$. This is so because for each point x in this set there is an $n > 0$ with

$$\sum_{k=0}^{n-1} T^k f_i(x) > \sum_{k=0}^{n-1} T^k v(x),$$

so, by Proposition 7.1, there is an n with ${}^i h_n^+(x) > 0$. For this latter n, ${}^i h_n^-(x) = 0$, and hence also ${}^i h_m^-(x) = 0$ for $m \geqslant n$. Thus on this set

$${}^i H^- = \lim_n \downarrow {}^i h_n^- = 0 \text{ a.e.}$$

3.7. The filling scheme

Theorem 7.4 (*Hopf Decomposition* 1954) Choose any $u \in L^1(X)$ with $u(x) > 0$ a.e. The set

$$C = \left\{ x : \sum_{k=0}^{\infty} T^k u(x) = \infty \right\},$$

called the *conservative part* of X, is independent of u. $D = X \backslash C$ is called the *dissipative part* of X. If $0 \leq u \in L^1$, then $\sum_{k=0}^{\infty} T^k u < \infty$ a.e. on D.

Proof: Let $0 < v \in L^1(X)$. By Proposition 7.3, v must be 0 a.e. on

$$C(u) \backslash C(v) \subset \{\sum T^k u = \infty, \sum T^k v < \infty\},$$

so this set must have measure 0.

If $0 \leq u \in L^1$, there is $u' > 0$ with $u' \in L^1$ and $u' \geq u$ a.e. Then $\sum_{k=0}^{\infty} T^k u' < \infty$ a.e. on D, so $\sum_{k=0}^{\infty} T^k u < \infty$ a.e. on D.

The adjoint operator T^* on the dual $L^\infty(X)$ of $L^1(X)$ is also positive: $\phi \geq 0$ implies $T^*\phi \geq 0$. Sets E for which $T^*\chi_E = \chi_E$ a.e. are called *invariant*.

Proposition 7.5 If T is conservative (i.e., $X = C$ up to a set of measure 0) and $\phi \in L^\infty(X)$ with $T^*\phi \leq \phi$ a.e., then $T^*\phi = \phi$ a.e. In particular, $T^*1 = 1$ when T is conservative (so that (3) above holds).

Proof: We may suppose that $\phi \geq 0$, by adding a constant if necessary. Choose $0 < u \in L^1$ with $\langle \phi, u \rangle = \int \phi u \, d\mu < \infty$. (This is possible because X is σ-finite.) Then for each $n \geq 1$,

$$0 \leq \langle \phi - T^*\phi, \sum_{k=0}^{n} T^k u \rangle = \sum_{k=0}^{n} \langle T^{*k}\phi, u \rangle - \sum_{k=1}^{n+1} \langle T^{*k}\phi, u \rangle$$

$$= \langle \phi, u \rangle - \langle T^{*n+1}\phi, u \rangle \leq \langle \phi, u \rangle < \infty$$

independently of n, so by Fatou's Lemma

$$0 \leq \langle \phi - T^*\phi, \sum_{k=0}^{\infty} T^k u \rangle < \infty,$$

which is impossible, since $\sum_{k=0}^{\infty} T^k u = \infty$ a.e., unless $T^*\phi = \phi$ a.e.

Of course $T^*1 \leq 1$ since $\int 1 \cdot Tf \, d\mu \leq \int 1 \cdot f \, d\mu$ for each $0 \leq f \in L^1$. Thus $T^*1 = 1$:

$$\int Tf \, d\mu = \int f \, d\mu \quad \text{for } f \in L^1.$$

Exercises

1. Use measurable point transformations to construct examples of contractions of each of the following types:
   (*a*) conservative,
   (*b*) dissipative (i.e., $D = X$),

(c) with both C and D nontrivial,
(d) conservative ergodic,
(e) nonconservative ergodic,
(f) conservative nonergodic,
(g) nonconservative nonergodic.

2. Show that when T is induced by a point transformation, D is generated by a *wandering set* (W, $T^{-1}W$, $T^{-2}W$,... pairwise disjoint) and the restriction of T to C is *incompressible* ($T^{-1}B \subset B$ implies $\mu(B \setminus T^{-1}B) = 0$).

3. Let $\mathscr{I}$ denote the family of all invariant subsets of C.
   (a) $\mathscr{I}$ is a σ-algebra.
   (b) $\mathscr{I}$ coincides with the family of all sets of the form $\{\sum fT^k = \infty\}$, for $0 \leqslant f \in L^1$.
   (c) $T\chi_I = \chi_I$ on I implies $I \in \mathscr{I}$; $I \in \mathscr{I}$ implies $T\chi_I = \chi_I$ on C.
   (d) These are equivalent for each non-negative measurable h on C: (i) $Th \leqslant h$ on C, (ii) $Th = h$ on C, (iii) h is $\mathscr{I}$-measurable on C.

4. Use the Filling Scheme to prove the formula
$$\int_A S_* f \, d\mu = \int f \, d\mu$$
of 3.1.

3.8. The Chacon–Ornstein Theorem

Theorem 8.1 (Chacon and Ornstein 1960) If $(X, \mathscr{B}, \mu)$ is σ-finite, $0 \leqslant f, g \in L^1(X)$, and $T: L^1(X) \to L^1(X)$ is a positive contraction, then

$$\frac{\sum_{k=0}^{n-1} T^k f}{\sum_{k=0}^{n-1} T^k g}$$

converges to a finite limit a.e. on $\{x: \sum_{k=0}^{\infty} T^k g(x) > 0\}$.

Remark: On C the limit can be identified as a quotient

$$\frac{E_C(\tilde{f})}{E_C(\tilde{g})}$$

of conditional expectations of modifications of f and g with respect to the σ-algebra of invariant subsets of C (see Foguel 1980, 1.3) while on D it is

3.8. The Chacon–Ornstein theorem

the quotient

$$\frac{\sum_{k=0}^{\infty} T^k f}{\sum_{k=0}^{\infty} T^k g}$$

of two almost surely finite numbers.

The main interest of the Chacon–Ornstein Theorem lies in the case when T is conservative. We will henceforth assume that T is *conservative* and *ergodic*; the general statement follows from this special case with some extra effort, which we will not undertake here. (See Foguel 1980 for a complete discussion.) Also, notice that the hypothesis $f \geq 0$ is not necessary, since the general statement could be proved by applying the theorem to f^+ and f^- separately.

This Theorem does not yet imply that if T is a positive contraction on L^1, then

$$\frac{1}{n} \sum_{k=0}^{n-1} T^k f$$

converges a.e. for each $f \in L^1$. In fact, such a statement is *false* (Chacon 1964). However, M. A. Akcoglu (1975) has shown that if $1 < p < \infty$ and T is a positive contraction on L^p, then $1/n \sum_{k=0}^{n-1} T^k f$ does converge a.e. for each $f \in L^p$.

Suppose then that $T: L^1(X) \to L^1(X)$ is a conservative, ergodic, positive contraction, $0 \leq f, g \in L^1$, $h = f - g$, and $h_1, h_2, \ldots$ are associated to f and g by the filling scheme. Let

$$A = \{x : H^-(x) > 0\}$$

and, for $k = 0, 1, 2, \ldots,$

$$v_k = (M_{A^c} T^*)^k \chi_A.$$

(If T were induced by a point transformation $T: X \to X$ on a probability space, then we would have $v_k(x) = \chi_{\{n_A = k\}}(x)$, where $n_A(y)$ is the first entry time of y to A.)

Lemma 8.2 If $\mu(A) > 0$, then $\sum_{k=0}^{\infty} v_k = 1$ a.e.

Proof: First we will prove by induction that

$$\sum_{k=0}^{n} v_k + (M_{A^c} T^*)^n \chi_{A^c} = 1 \text{ a.e.}$$

For $n = 0$, the formula says that
$$v_0 = \chi_A,$$
which is true. If the formula holds for n, then
$$\sum_{k=0}^{n+1} v_k + (M_{A^c}T^*)^{n+1}\chi_{A^c}$$
$$= \chi_A + (M_{A^c}T^*)\left\{\sum_{k=0}^{n} v_k + (M_{A^c}T^*)^n \chi_{A^c}\right\}$$
$$= \chi_A + (M_{A^c}T^*)(1) = \chi_A + \chi_{A^c} = 1 \text{ a.e.,}$$
since $T^*1 = 1$ when T is conservative.

Since the v_k are non-negative, the $(M_{A^c}T^*)^n \chi_{A^c}$ decrease a.e. with n to $w \geq 0$. We have
$$\sum_{k=0}^{\infty} v_k + w = 1, \quad w = 0 \quad \text{on } A, \, M_{A^c}T^*w = w.$$
The last equation implies that $T^*w \geq w$ a.e., so by a slight extension of Proposition 7.5, $T^*w = w$ a.e. (If $Z = \|w\|_{\infty} - w$, then $T^*Z \leq Z$, so $T^*Z = Z$.)

We need to show, then, that $w = 0$ a.e. Choose $0 \leq f \in L^1$, $f \not\equiv 0$, with $\operatorname{supp} f \subset A$. Then
$$\int f \cdot w \, d\mu = 0,$$
so that
$$\int \sum_{k=0}^{n-1} T^k f \cdot w \, d\mu = \int f \cdot \sum_{k=0}^{n-1} T^* w \, d\mu = \int f \cdot nw \, d\mu = 0;$$
since $\sum_{k=0}^{n-1} T^k f \to \infty$ a.e., this is impossible unless $w = 0$ a.e.

Proposition 8.3 If T is conservative ergodic and $h \in L^1$ with $H^- \not\equiv 0$, then
$$\lim_{n \to \infty} \downarrow \int h_n^+ \, d\mu = 0 \quad \text{and} \quad \sum_{n=0}^{\infty} h_n^+ < \infty \text{ a.e.}$$

($H^- \not\equiv 0$ means $\int f \, d\mu < \int g \, d\mu$, so that at the outset the size of the hole exceeds the amount of sand. The Proposition says that under this condition, in the conservative ergodic case, all the sand disappears eventually, and during the entire process only a finite amount of sand passes over each point of X.)

3.8. The Chacon–Ornstein theorem

Proof: We know from (2) in the preceding section that

$$h_n^+ \leq (TM_{A^c})h_{n-1}^+ \leq \ldots \leq (TM_{A^c})^n h^+ \quad \text{for } n = 0, 1, 2, \ldots.$$

Therefore, for each $n, k = 0, 1, 2, \ldots$,

$$\int h_n^+ v_k \, d\mu \leq \int (TM_{A^c})^n h^+ \cdot v_k \, d\mu = \int h^+ (M_{A^c} T^*)^n v_k \, d\mu$$

$$= \int h^+ v_{n+k} \, d\mu.$$

Now sum over k and use the Dominated Convergence Theorem to find that

$$\int h_n^+ \, d\mu \leq \int h^+ \sum_{k=0}^{\infty} v_{n+k} \, d\mu \searrow 0 \quad \text{as } n \to \infty,$$

since $\sum_{k=0}^{\infty} v_{n+k} \searrow 0$ a.e. as $n \to \infty$.

On the other hand, if we summed over n we would have

$$\int \sum_n h_n^+ v_k \, d\mu \leq \int h^+ \sum_n v_{n+k} \, d\mu \leq \int h^+ \, d\mu < \infty,$$

so that $\sum_{n=0}^{\infty} h_n^+ < \infty$ a.e. on each $E_k = \{v_k > 0\}$, hence a.e. on X, since $\sum_k v_k > 0$ a.e. on X.

Theorem 8.4 Let $T: L^1(X) \to L^1(X)$ be a conservative, ergodic, positive contraction and $0 \leq f, g \in L^1(X)$. Then

$$\lim_{n \to \infty} \frac{\sum_{k=0}^{n-1} T^k f}{\sum_{k=0}^{n-1} T^k g} = \frac{\int f \, d\mu}{\int g \, d\mu}$$

a.e. on $\{x : \sum_{k=0}^{\infty} T^k g(x) > 0\}$.

Proof: We suppose that $g \not\equiv 0$ and consider first the case when $\int f \, d\mu < \int g \, d\mu$. Let $h = f - g$ and apply the filling scheme. Then $\int h \, d\mu < 0$, so $H^- \not\equiv 0$. By Proposition 8.3, $\sum h_n^+ < \infty$ a.e. By Proposition 7.1.

$$\limsup_{n \to \infty} \frac{\sum_{k=0}^{n-1} T^k f}{\sum_{k=0}^{n-1} T^k g} \leq \limsup_{n \to \infty} \frac{\sum_{k=0}^{n-1} h_k^+}{\sum_{k=0}^{n-1} T^k g} + \limsup_{n \to \infty} \frac{u_n}{\sum_{k=0}^{n-1} T^k g} \leq 0 + 1 = 1.$$

3. *More about almost everywhere convergence*

Let

$$Q_n(f,g) = \frac{\sum_{k=0}^{n-1} T^k f}{\sum_{k=0}^{n-1} T^k g}.$$

If $\rho > 0$, then $Q_n(\rho f, g) = \rho Q_n(f, g)$. Thus given arbitrary $0 \leq f, g \in L^1$ with $g \not\equiv 0$, we can choose $\rho > 0$ so that $\int \rho f \, d\mu < \int g \, d\mu$ and apply the above estimate to see that

$$\limsup_{n \to \infty} Q_n(f,g) \leq \frac{1}{\rho}.$$

Letting $\rho \nearrow \dfrac{\int g \, d\mu}{\int f \, d\mu}$, we find that $\displaystyle\limsup_n Q_n(f,g) \leq \frac{\int f \, d\mu}{\int g \, d\mu}$.

If $f \equiv 0$, the result is obvious. Otherwise, we interchange f and g and use the information already gained to conclude that

$$\liminf_{n \to \infty} Q_n(f,g) = \frac{1}{\limsup_{n \to \infty} Q_n(g,f)} \geq \frac{\int f \, d\mu}{\int g \, d\mu}.$$

Finally, we show how the Hopf Maximal Ergodic Theorem, the usual starting point for the proof of operator ergodic theorems, follows from the filling scheme. (By Exercise 2, the implication can also be reversed.)

Theorem 8.5 *Hopf Maximal Ergodic Theorem* (1954) Let $h \in L^1(X)$ and

$$E = \left\{ x : \sup_{n \geq 1} \sum_{k=0}^{n-1} T^k h(x) > 0 \right\}.$$

Then

$$\int_E h \, d\mu \geq 0.$$

3.8. The Chacon–Ornstein theorem

Proof: Note that
$$\{h > 0\} \subset E \subset \{H^- = 0\},$$
since (as in the proof of Proposition 7.3) for each point of E we have (with $f - g = h = h^+ - h^-$)
$$\sum_{k=0}^{n-1} T^k f > \sum_{k=0}^{n-1} T^k g,$$
so by Proposition 7.1 $h_n^+ > 0$ for some n at each point of E. Again we find that
$$H^- = \lim \downarrow h_n^- = 0 \text{ a.e. on } E.$$
Thus
$$\int h \, d\mu \geqslant -\int H^- \, d\mu \quad \text{(see (*) in the preceding section, p. 121)},$$
and hence
$$\int_E h \, d\mu + \int_{E^c} h \, d\mu \geqslant -\int_E H^- \, d\mu - \int_{E^c} H^- \, d\mu = -\int_{E^c} H^- \, d\mu;$$
however, since on E^c we have $h \leqslant 0$ so $h = -h^-$, this also equals
$$\int_E h \, d\mu - \int_{E^c} h^- \, d\mu.$$
We find that
$$\int_E h \, d\mu \geqslant \int_{E^c} h^- \, d\mu - \int_{E^c} H^- \, d\mu;$$
but the latter expression is nonnegative, since $h^- \geqslant H^-$ a.e.

Exercises

1. (Rost) If T is conservative ergodic, then
   (a) $\lim_{n \to \infty} \downarrow \int h_n^+ \, d\mu = 0$ if and only if $\int h \, d\mu \leqslant 0$
   (b) $\lim_{n \to \infty} \downarrow \int h_n^- \, d\mu = 0$ if and only if $\int h \, d\mu \geqslant 0$;
   Interpret in terms of sand and holes.
2. Prove Proposition 8.3 from the Hopf Maximal Ergodic Theorem.
3. Prove *Hurewicz's Ergodic Theorem* (1944): If T is *nonsingular* on $(X, \mathscr{B}, \mu)$ ($\mu T \sim \mu T^{-1} \sim \mu$) and $w_n = d\mu T^n/d\mu$ for $n = 0, 1, 2, \ldots$, then for any $f \in L^1$ and any positive measurable g with
$$\sum_{k=0}^{n-1} g(T^k x) w_k(x) \to \infty \text{ a.e.},$$

3. More about almost everywhere convergence

the limit

$$\lim_{n\to\infty} \frac{\sum_{k=0}^{n-1} f(T^k x) w_k(x)}{\sum_{k=0}^{n-1} g(T^k x) w_k(x)}$$

exists a.e.

4. Extend the Chacon–Ornstein Theorem to the case of a one-parameter group $\{T_t : -\infty < t < \infty\}$ of positive contractions on $L^1(X)$.

5. Show that the Chacon–Ornstein Theorem for general T follows from the case when T is conservative.
   (Hint: Consider the restriction of T to $L^1(C)$.)

4
More about recurrence

This chapter presents several further topics concerning recurrence and mixing. First is a direct construction (based on the theory of almost periodic functions) of eigenfunctions for m.p.t.s which are not weakly mixing. The second section presents the purely topological analogues of recurrence, ergodicity, and weak and strong mixing, and the third introduces Furstenberg's theory of multiple recurrence and his proof of the Szemerédi Theorem. In 4.4 we give the Jewett–Bellow–Furstenberg proof of the existence of uniquely ergodic topological representations of ergodic m.p.t.s. The final section examines two examples of weakly mixing m.p.t.s which are not strongly mixing: an example of Kakutani (which we show is not even topologically strongly mixing), and one of Chacon, which is also prime (i.e. without proper invariant sub-σ-algebras), as was discovered by del Junco.

4.1. Construction of eigenfunctions

We will give now a more direct proof of the existence of eigenfunctions for m.p.t.s which are not weakly mixing. The idea, apparently due to Varadhan, Furstenberg, and Katznelson, is to reduce the statement to the existence of nontrivial characters on compact abelian groups. However, since this is itself usually proved with the help of the Spectral Theorem for compact operators, we will complete the argument by means of an elementary proof, which also introduces some important ideas from topological dynamics and the theory of almost periodic functions.

A. Existence of rigid factors

Theorem 1.1 Let T be a m.p.t. on $(X, \mathscr{B}, \mu)$. If T is ergodic but $T \times T$ is not ergodic, then T has a factor which is an isometry on a compact metric space, hence a rotation on a compact abelian (even monothetic) group.

4. More about recurrence

Proof: Suppose that $T \times T$ is not ergodic, and let $G: X \times X \to \mathbb{C}$ be a nonconstant invariant function in $L^2(X \times X)$. For $x, y \in X$, define

$$d(x, y) = \|G(x, \cdot) - G(y, \cdot)\|_2 = \sqrt{\int_X |G(x, z) - G(y, z)|^2 \, d\mu(z)}.$$

That is, we have a map $\iota: X \to L^2(X)$ defined by $\iota(x) = G(x, \cdot)$, and d is the pull-back to X of the metric on $L^2(X): d(x, y) = \|\iota(x) - \iota(y)\|_2$.

Clearly d is a pseudo-metric on X (i.e., d is symmetric and satisfies the triangle inequality). Since G is T-invariant, T is an *isometry* of X with respect to d.

We form the quotient metric space $\tilde{X}$ of X obtained by identifying points of X which are at zero distance from one another. The quotient map $\pi: X \to \tilde{X}$ is continuous, and if $\tilde{d}(\pi x, \pi y) = d(x, y)$, then $\tilde{d}$ is a metric on $\tilde{X}$. We also obtain a map $\tilde{T}: \tilde{X} \to \tilde{X}$ defined by $\tilde{T}(\pi x) = \pi(Tx)$ (the definition is allowable since $x \sim y$ if and only if $Tx \sim Ty$, where $x \sim y$ means $d(x, y) = 0$). The measure μ is carried to a Borel measure $\tilde{\mu}$ on $\tilde{X}$. Notice that π is measurable since π^{-1} of an ε-ball in $(\tilde{X}, \tilde{d})$ is an ε-ball in (X, d), which is measurable because, looking at the definition of d, d is a measurable function of x and y. Notice that $\tilde{T}$ *is ergodic*, since it is a factor of T.

I claim now that $\tilde{X}$ *has more than one point*. For if $\tilde{X}$ is a singleton, then $G(x, \cdot) = G(y, \cdot)$ in $L^2(X)$ for each $x, y \in X$, so that $g(u) = G(x, u)$ is invariant, and hence constant, for a.e. x.

Next we will show that $\tilde{X}$ *is totally bounded* (i.e., for each $\varepsilon > 0$, $\tilde{X}$ can be covered by finitely many ε-balls).

Notice first that if $B_\varepsilon(x)$ is an ε-ball in X, with center $x \in X$, then $\mu(B_\varepsilon(x))$ is a.e. a constant independent of x. For

$$\mu(B_\varepsilon(x)) = \mu(TB_\varepsilon(x)) = \mu(B_\varepsilon(Tx)),$$

since μ is T-invariant and T is an isometry with respect to d, so $\mu(B_\varepsilon(x))$ is an invariant function and hence, by ergodicity of T, must be a constant a.e. Call this constant c_ε.

We next show that $c_\varepsilon > 0$. First observe that $(\tilde{X}, \tilde{d})$ is separable. This is so because $(\tilde{X}, \tilde{d})$ is a (topological) subspace of L^2, and $L^2(X)$ is separable and hence second countable, so $(\tilde{X}, \tilde{d})$ is second countable and hence separable. Thus $\tilde{X}$ can be covered by countably many ε-balls, and hence one of these must have positive measure. Since they all have measure c_ε, however, it follows that $c_\varepsilon > 0$.

Thus $\tilde{\mu}(B_\varepsilon(\tilde{x})) = c_\varepsilon > 0$ a.e. for each $\varepsilon > 0$, with c_ε independent of $\tilde{x} \in \tilde{X}$. By taking a sequence $\{\varepsilon_n\}$ decreasing to 0 and discarding a set of measure 0

4.1. Construction of eigenfunctions

from $\tilde{X}$ for each n, we may assume that
$$\tilde{\mu}(B_\varepsilon(\tilde{x})) = c_\varepsilon > 0 \quad \text{for all } \tilde{x} \in \tilde{X}$$
for each $\varepsilon = \varepsilon_n$.

Now given any $\delta > 0$, choose n so that $2\varepsilon_n < \delta$. We will show that finitely many $2\varepsilon_n$-balls, and hence finitely many δ-balls, cover $\tilde{X}$. Since $\tilde{T}$ is ergodic,
$$\tilde{\mu}\bigcup_{k=0}^{\infty} \tilde{T}^k(B_{\varepsilon_n}(\tilde{x})) = 1 \quad \text{for each } \tilde{x} \in \tilde{X}.$$
Thus, fixing $\tilde{x} \in \tilde{X}$, we may choose N so that
$$\tilde{\mu}\bigcup_{k=0}^{N} \tilde{T}^k(B_{\varepsilon_n}(\tilde{x})) > 1 - c_{\varepsilon_n}.$$
Then
$$\bigcup_{k=0}^{N} \tilde{T}^k B_{\varepsilon_n}(\tilde{x})$$
must hit every ε_n-ball in $\tilde{X}$, and therefore
$$\bigcup_{k=0}^{N} \tilde{T}^k B_{2\varepsilon_n}(\tilde{x})$$
covers $\tilde{X}$.

We have shown, then, that $\tilde{X}$ is totally bounded. It follows that the completion $\bar{X}$ of $\tilde{X}$ is compact. $\tilde{T}$ extends to an isometry $\bar{T}$ on $\bar{X}$, and $\tilde{\mu}$ on $\tilde{X}$ extends to $\bar{\mu}$ on $\bar{X}$ with
$$\bar{\mu}|_{\tilde{X}} = \tilde{\mu}$$
$$\bar{\mu}(\bar{X}\setminus\tilde{X}) = 0.$$
$(\bar{X}, \bar{T}, \bar{\mu})$ is a factor of (X, T, μ); thus $\bar{T}$ is an *ergodic isometry* on the compact metric space $\bar{X}$. It is an easy exercise to show that $\bar{X}$ is a compact abelian group and $\bar{T}$ is multiplication by a certain member of $\bar{X}$ (Exercise 1). Then any nontrivial character of $\bar{X}$ will produce an eigenfunction of $\bar{T}$ and hence of T.

B. Almost periodicity

Suppose that $T: X \to X$ is an ergodic isometry on a compact metric space X with respect to a Borel measure μ which assigns positive measure to each nonempty open subset of X. (The system $(\bar{X}, \bar{T}, \bar{\mu})$ constructed in the preceding section has these properties.) We will show that if f is a continuous function on X and $x_0 \in X$, then $\{f(T^n x_0): n \in \mathbb{Z}\}$ is an *almost periodic sequence*. A sequence $\{a_n : n \in \mathbb{Z}\}$ is called *almost periodic* (we write $a \in AP$) in case for each $\varepsilon > 0$ the set of all $p \in \mathbb{Z}$ for which

$$\sup_n |a_{n+p} - a_n| < \varepsilon$$

(such p are called ε-*periods of* f) is *relatively dense*: there is a K such that every interval of length K contains at least one such p. (Thus a relatively dense set is one with *bounded gaps*.) Almost periodic functions on other groups than $\mathbb{Z}$ may be defined also, and are also of great interest, although we will not need to consider them here.

To show that $\{f(T^n x_0): n \in \mathbb{Z}\}$ is an almost periodic sequence, we will prove that T itself is *uniformly almost periodic*: given $\varepsilon > 0$, there is a relatively dense set of n for which

$$d(x, T^n x) < \varepsilon \quad \text{for all } x \in X.$$

For a moment, let us consider an arbitrary homeomorphism T on a compact metric space X. For each $x \in X$, let $\mathcal{O}(x) = \{T^n x : n \in \mathbb{Z}\}$. A non-empty closed T-invariant set M is called *minimal* if it contains no proper closed T-invariant subsets. M is minimal if and only if the orbit of each point of M is dense in M. The phrase 'almost periodic' is used in a slightly different way in topological dynamics: a point $x \in X$ is called *almost periodic* if for each $\varepsilon > 0$ there is a relatively dense set of n such that

$$d(x, T^n x) < \varepsilon.$$

Theorem 1.2 (Gottschalk 1944). $\overline{\mathcal{O}(x)}$ is minimal if and only if x is almost periodic.

Proof: Suppose that x is almost periodic. Let $y \in \overline{\mathcal{O}(x)}$ and let U be a compact neighborhood of x: if we can show that $\mathcal{O}(y) \cap U \neq \varnothing$, it will follow that $x \in \overline{\mathcal{O}(y)}$, and hence $\overline{\mathcal{O}(y)} = \overline{\mathcal{O}(x)}$. If $R = \{n : T^n x \in U\}$, then R is relatively dense, so there is a finite interval of integers $I \subset \mathbb{Z}$ such that $\mathbb{Z} = R + I$. Then, writing nx for $T^n x$,

$$\mathcal{O}(x) \subset (R + I)x \subset I(Rx) \subset IU,$$

which is compact, so that

$$\overline{\mathcal{O}(x)} \subset IU.$$

Thus, since $y \in \overline{\mathcal{O}(x)}$, we have $y \in IU$ and hence $\mathcal{O}(y) \cap U \neq \varnothing$.

Conversely, suppose that $\overline{\mathcal{O}(x)}$ is minimal and let U be an open neighborhood of x; we need to show that $R = \{n : T^n x \in U\}$ is relatively dense. Each point of $\overline{\mathcal{O}(x)}$ has an orbit which meets U; therefore,

$$\overline{\mathcal{O}(x)} \subset \bigcup_{n \in \mathbb{Z}} T^{-n} U,$$

4.1. Construction of eigenfunctions

and by compactness there are $n_1, n_2, \ldots n_r$ such that
$$\overline{\mathcal{O}(x)} \subset T^{-n_1}U \cup \ldots \cup T^{-n_r}U.$$
If $n \in \mathbb{Z}$, we can find $i = 1, \ldots r$ such that
$$T^n x \in T^{-n_i}U;$$
then
$$T^{n+n_i}x \in U,$$
so that $n + n_i \in R$, i.e., $n \in R - n_i$. This shows that
$$\mathbb{Z} = R - \{n_1, n_2, \ldots n_r\} = \bigcup_{i=1}^{r}(R - n_i),$$
so that R is relatively dense (finitely many translates of R cover $\mathbb{Z}$).

(This proof is written in language which allows it to extend easily to the case of actions of groups other than $\mathbb{Z}$.)

Remark 1.3 *Uniform almost periodicity* Since T is an ergodic isometry, this Theorem readily implies that T is in fact *uniformly almost periodic*. For each $n = 0, 1, 2, \ldots$ the continuous function
$$f_n(x) = d(x, T^n x)$$
must be constant on X, since $f_n(Tx) = f_n(x)$. Thus $d(x, T^n x) < \varepsilon$ exactly when $d(y, T^n y) < \varepsilon$, for any $x, y \in X$. Since by Zorn's Lemma X contains minimal sets and hence almost periodic points, it follows that every point of X is almost periodic and $\{n : d(x, T^n x) < \varepsilon\}$ is a relatively dense subset of $\mathbb{Z}$ independent of x for each $\varepsilon > 0$.

Remark 1.4 *Dense orbits* Let us note also that T has dense orbits. If $\{U_1, U_2, \ldots\}$ is a countable base for the topology of X, let
$$X_i = \{x \in X : \text{there is } n \geq 0 \text{ with } T^n x \in U_i\}.$$
Since $T^{-1}X_i \subset X_i$ and $U_i \subset X_i$, so that $\mu(X_i) > 0$, we have $\mu(X_i) = 1$ for each i. Thus $\bigcap_i X_i$ has measure 1, and so in fact *almost every orbit is dense*. That T is an isometry, however, implies that *every orbit is dense*. For if $x_0 \in X$ has a dense orbit and $x, y \in X$ are given, then
$$d(T^n x, y) \leq d(T^n x, T^{n+m}x_0) + d(T^{n+m}x_0, y)$$
$$= d(x, T^m x_0) + d(T^{n+m}x_0, y),$$
which can be made arbitrarily small by appropriately choosing first m and then n. Thus our (X, T) is minimal.

Remark 1.5 *Uniform existence of mean values* We will show that

4. More about recurrence

for the ergodic isometry T,

$$\lim_{n\to\infty}\frac{1}{n}\sum_{k=0}^{n-1}f(T^k x)$$

exists uniformly for $x \in X$ for each $f \in \mathscr{C}(X)$. (This, together with the limit's being constant, is the condition that (X, T) be *uniquely ergodic* – there is exactly one T-invariant Borel probability measure on X.)

Since for each $f \in \mathscr{C}(X)$ the functions

$$f_n = \frac{1}{n}\sum_{k=0}^{n-1} f T^k$$

form an equicontinuous family which is uniformly bounded by $\|f\|_\infty$, by the Arzela–Ascoli Theorem some subsequence $\{f_{n_i}\}$ converges uniformly on X. Of course the limit must be $\int f \, d\mu$, since $f_n \to \int f \, d\mu$ pointwise.

This observation shows that (X, T) is uniquely ergodic: for if v is any T-invariant Borel probability measure on X, we have

$$\int f \, dv = \frac{1}{n_i}\sum_{k=0}^{n_i-1}\int f T^k \, dv = \int f_{n_i} \, dv \to \int \left[\int f \, d\mu\right] dv = \int f \, d\mu,$$

since $f_{n_i} \to \int f \, d\mu$ uniformly; therefore $v = \mu$.

Now if $\{f_n\}$ failed to converge to $\int f \, d\mu$ uniformly, we could find $\varepsilon > 0$, $m_i \to \infty$, and $x_i \in X$ with

$$\left|\frac{1}{m_i}\sum_{k=0}^{m_i-1} f(T^k x_i) - \int f \, d\mu\right| \geq \varepsilon.$$

Define continuous linear functionals μ_i on $\mathscr{C}(X)$ by

$$\mu_i(g) = \frac{1}{m_i}\sum_{k=0}^{m_i-1} g(T^k x_i) \text{ for } g \in \mathscr{C}(X),$$

and let v be a weak* limit point of $\{\mu_i\}$. Then

$$\left|\int f \, dv - \int f \, d\mu\right| \geq \varepsilon;$$

but also v is T-invariant, since for $g \in \mathscr{C}(X)$

$$(gT)_{m_i}(x_i) = \frac{1}{m_i}\sum_{k=0}^{m_i-1} g(T^{k+1} x_i) = g_{m_i}(x_i) + \frac{g(T^{m_i} x_i) - g(x_i)}{m_i},$$

so that

$$\mu_i(gT) = (gT)_{m_i}(x_i)$$

and

$$\mu_i(g) = g_{m_i}(x_i)$$

4.1. Construction of eigenfunctions

differ by a term which tends to 0 as $i \to \infty$. Since T is uniquely ergodic, this situation is impossible.

Remark 1.6 This proof shows that if $f \in \mathscr{C}(X)$, then
$$\frac{1}{n}\sum_{k=0}^{n-1} f(T^k x_n) \to \int f \, d\mu \quad \text{uniformly}$$
for any choice of $x_1, x_2, \ldots \in X$.

Remark 1.7 From 1.8 below, it will follow that each almost periodic sequence $\{a_n\}$ has a *mean value*
$$M\{a_n\} = \lim_{n \to \infty} \frac{1}{n} \sum_{k=0}^{n-1} a_{k+m},$$
the limit existing uniformly in m (even if m varies with n).

Remark 1.8 *A sequence $\{a_n : n \in \mathbb{Z}\}$ is almost periodic if and only if its orbit closure under translation, the set of all sequences*
$$\{a_{n+j} : n \in \mathbb{Z}\}, j \in \mathbb{Z},$$
is totally bounded in the supremum norm. This statement is nothing more than a paraphrase of the definition of almost periodicity: the ε-balls centered at K consecutive translates of a will cover the set of all translates of a, provided that every interval of length K contains an ε-period of f: given j, we find an ε-period p and a j_0 in the consecutive set such that $j = p + j_0$ and note that
$$|a_{n+j} - a_{n+j_0}| = |a_{n+p+j_0} - a_{n+j_0}| < \varepsilon \quad \text{for all } n.$$
(This characterization of almost periodicity (which Bochner (1927) showed to be equivalent to Bohr's for continuous functions on $\mathbb{R}$) was employed by Bochner and von Neumann (1935) to define almost periodic functions on a general group G: a bounded function $f : G \to \mathbb{R}$ is called *left almost periodic* if the set of all left translates of f is totally bounded in the supremum norm. One can prove that f is left almost periodic if and only if it is right almost periodic. Most of the properties of almost periodic sequences that we are considering here are shared by all almost periodic functions on groups.)

Now if $T : X \to X$ is an ergodic isometry, $x_0 \in X$, and $f \in \mathscr{C}(X)$, then $\{f(T^n x_0) : n \in \mathbb{Z}\}$ is an almost periodic sequence, since T is uniformly almost periodic. In fact, *all almost periodic sequences arise in this way*. For if $\{a_n\}$ is an almost periodic sequence, its orbit closure Ω under the shift σ is compact. Of course σ is an isometry with respect to the supremum norm, and (Ω, σ) is uniquely ergodic. (In fact, Ω is a compact monothetic group, since it contains a dense copy of $\mathbb{Z}$, and σ is multiplication by a fixed element of Ω.) The function f which evaluates 0th coordinates will recover the sequence a: $a(n) = f(\sigma^n a)$. (We could also form the function algebra generated by a given almost periodic sequence and its translates and consider the transformation on its maximal ideal space induced by translation.) We have established the following result.

Remark 1.9 *Almost periodic sequences are exactly those which come from ergodic isometries.*

Remark 1.10 *The almost periodic sequences form an algebra.* We will in fact prove a stronger result: *If $\{a_n\}$ and $\{b_n\}$ are almost periodic sequences, then $\{(a_n, b_n)\}$ forms an $\mathbb{R}^2$-valued almost periodic sequence.* For if $\{a_n\}$ comes from a pair (A, S) via a function f and a point a_0, and $\{b_n\}$ from (B, T) via g and b_0, then $S \times T$ is an isometry on $A \times B$. Let X be the orbit closure of (a_0, b_0) under $S \times T$. Then $(S \times T)^n(a_0, b_0) = (a_n, b_n)$, for $n \in \mathbb{Z}$, forms an $\mathbb{R}^2$-valued almost periodic sequence, by 1.9.

We also give the classical nondynamical proof of this observation. Given $\varepsilon > 0$, let P_a be the set of all $\frac{1}{2}\varepsilon$-periods of a. Choose K so that every interval of length K contains members of both P_a and P_b. In each interval $[(n-1)K, nK)$, choose $\alpha_n \in P_a$ and $\beta_n \in P_b$. We have

$$-K \leqslant \alpha_n - \beta_n \leqslant K \quad \text{for all } n.$$

Thus there are only finitely many possible values for the $\alpha_n - \beta_n$, and hence there is an n_0 such that for each $n \in \mathbb{Z}$ an n' can be found with

$$-n_0 \leqslant n' \leqslant n_0 \quad \text{and} \quad \alpha_n - \beta_n = \alpha_{n'} - \beta_{n'}.$$

Then

$$\alpha_n - \alpha_{n'} = \beta_n - \beta_{n'},$$

and each is an ε-period of its respective sequence. The spacing of these ε-periods is

$$|(\alpha_{n+1} - \alpha_{(n+1)'}) - (\alpha_n - \alpha_{n'})|$$
$$\leqslant |\alpha_{n+1} - \alpha_n| + |\alpha_{(n+1)'} - \alpha_{n'}|$$
$$\leqslant 2K + 2n_0.$$

4.1. Construction of eigenfunctions

Thus $\{\alpha_n - \alpha_{n'} : n = 1, 2, \ldots\}$ is a relatively dense set of ε-periods for each of a and b.

Now clearly if $\{(a_n, b_n)\}$ is an $\mathbb{R}^2$-valued almost periodic sequence and $f: \mathbb{R}^2 \to \mathbb{R}$ is continuous, then $\{f(a_n, b_n)\}$ is an almost periodic sequence. Since multiplication is continuous, 1.10 follows.

Remark 1.11 *The set AP of all almost periodic sequences is uniformly closed.*

C. Construction of the eigenfunction

Recall that T is an isometry on a compact metric space X consisting of more than one point and that T is ergodic with respect to an invariant Borel probability measure μ. We are trying to find a nonconstant $\phi \in L^2(X, \mu)$ for which there is $\xi \in \mathbb{C}$ with $\phi(Tx) = \xi \phi(x)$ a.e.

Choose any $f \in \mathscr{C}(X)$ which is not constant and for which $\int f \, d\mu = 0$, and fix $x_0 \in X$. Write

$$f(j) = f(T^j x_0) \quad (j \in \mathbb{Z}).$$

This is an almost periodic sequence. Since rotation of the circle through any fixed angle $2\pi\lambda$ is an isometry,

$$\{e^{2\pi i n \lambda} : n \in \mathbb{Z}\}$$

is an almost periodic sequence for each $\lambda \in \mathbb{R}$. By Remarks 1.10 and 1.7 there is a uniform mean value

$$\alpha_{x_0}(\lambda) = M\{f(j) e^{-2\pi i j \lambda}\} = \lim_{n \to \infty} \frac{1}{n} \sum_{j=0}^{n-1} f(j) e^{-2\pi i j \lambda}.$$

We define

$$h(x) = \alpha_x(\lambda) = \lim_{n \to \infty} \frac{1}{n} \sum_{j=0}^{n-1} f(T^j x) e^{-2\pi i j \lambda}.$$

Then h, being a uniform limit of continuous functions of x, is continuous, and

$$h(Tx) = \lim_{n \to \infty} \frac{1}{n} \sum_{j=0}^{n-1} f(T^{j+1} x) e^{-2\pi i j \lambda}$$

$$= e^{2\pi i \lambda} \lim_{n \to \infty} \frac{1}{n} \sum_{j=0}^{n-1} f(T^{j+1} x) e^{-2\pi i (j+1) \lambda} = e^{2\pi i \lambda} h(x).$$

We need then only *select* λ so that $h \not\equiv 0$. E.g., if we can find a λ such that

$$h(x_0) = \alpha_{x_0}(\lambda) \neq 0,$$

we will be through.

4. More about recurrence

We will be able to find such a λ by careful scrutiny of the almost periodic sequence $\{f(j)\}$. We will consider ε-periods (in fact $1/n$-periods) p_n of this sequence, and for each n pick a rational number k_n/p_n for which

$$f(j), \quad 0 \leqslant j \leqslant p_n - 1,$$

'best matches'

$$e^{2\pi i j k_n/p_n}, \quad 0 \leqslant j \leqslant p_n - 1,$$

in that

$$\left| \frac{1}{p_n} \sum_{j=0}^{p_n - 1} f(j) e^{-2\pi i j k_n/p_n} \right| \geqslant c > 0 \quad \text{for all } n.$$

The rational numbers k_n/p_n will converge to our eigenfrequency λ for which

$$\alpha_{x_0}(\lambda) = \lim_{n\to\infty} \frac{1}{p_n} \sum_{j=0}^{p_n - 1} f(j) e^{-2\pi i j \lambda} \neq 0.$$

Let $x_0 \in X$ and $f \in \mathscr{C}(X)$ be fixed. We will prove the following assertion:

Theorem 1.12 *If $\{f(j)\}$ is an almost periodic sequence for which $\alpha(\lambda) = M\{f(j) e^{-2\pi i j \lambda}\} = 0$ for all $\lambda \in \mathbb{R}$, then*

$$\lim_{n\to\infty} \frac{1}{n} \sum_{k=0}^{n-1} |f(k)|^2 = 0,$$

and hence $f(j) = 0$ for all j.

Since in our case $f(j) = f(T^j x_0)$ for a continuous function f on X, it will follow that if all $\alpha_{x_0}(\lambda)$ are 0, then $f(x) = 0$ for all $x \in X$, and so we will be finished. Theorem 1.12 is fundamental in the theory of almost periodic functions: it is the heart of Bohr's theorem characterizing AP as the uniform closure of the set of all trigonometric polynomials

$$a_n = \sum_{j=1}^{m} \alpha_j e^{2\pi i n \lambda_j}$$

(see Bohr 1925a, 1925b, 1926, 1932; Besicovitch 1954). Our proof reproduces the essential part of Bohr's original argument.

We will prove Theorem 1.12 by means of a sequence of lemmas. The idea is to show first that if $\alpha(\lambda) = 0$ for all λ, then, because of a certain continuity property of mean values (Lemma 1.13), in fact

$$\lim_{n\to\infty} \frac{1}{n} \sum_{j=0}^{n-1} f(j) e^{-2\pi i j \lambda} = 0 \quad \text{uniformly in } \lambda \in \mathbb{R}$$

(Remark 1.14).

4.1. Construction of eigenfunctions

Lemma 1.13 If $\{f(j)\}$ is an almost periodic sequence with $M\{f\} = 0$, then for each $\varepsilon > 0$ there are N and $\delta > 0$ such that

$$\left|\frac{1}{n}\sum_{j=0}^{n-1} f(j)e^{-2\pi i j \lambda}\right| < \varepsilon \quad \text{whenever } n \geq N \text{ and } |\lambda| < \delta.$$

Proof: Since $M\{f\} = 0$, given $\varepsilon > 0$ we can find n_0 such that

$$\left|\frac{1}{n}\sum_{j=0}^{n-1} f(k+j)\right| < \frac{\varepsilon}{2} \quad \text{for } n \geq n_0 \text{ and for all } k \in \mathbb{Z}.$$

For $n > n_0$, write $n = l n_0 + p$ with $p < n_0$. Choose δ so small that

$$|e^{-2\pi i j \lambda} - 1| < \frac{\varepsilon}{2\|f\|_\infty} \quad \text{for } |\lambda| < \delta \text{ and } j = 0, 1, \ldots, n_0 - 1.$$

Then

$$\left|\frac{1}{n}\sum_{j=0}^{n-1} f(j)e^{-2\pi i j \lambda}\right|$$

$$= \frac{1}{l n_0 + p}\left|\sum_{k=0}^{l-1}\sum_{j=kn_0}^{(k+1)n_0 - 1} f(j)e^{-2\pi i j \lambda} + \sum_{j=l n_0}^{n-1} f(j)e^{-2\pi i j \lambda}\right|$$

$$\leq \frac{1}{l}\left|\sum_{k=0}^{l-1} e^{-2\pi i k n_0 \lambda} \frac{1}{n_0}\sum_{j=0}^{n_0-1} f(j+kn_0)e^{-2\pi i j \lambda}\right|$$

$$+ \frac{1}{l}\left|\frac{1}{n_0}\sum_{j=l n_0}^{n-1} f(j)e^{-2\pi i j \lambda}\right|.$$

Now the second term is bounded by $l^{-1}\|f\|_\infty$, which tends to 0 as $l \to \infty$, while in the first term we have

$$\left|\frac{1}{n_0}\sum_{j=0}^{n_0-1} f(j+kn_0)\right| < \frac{\varepsilon}{2}$$

and

$$\left|\frac{1}{n_0}\sum_{j=0}^{n_0-1} f(j+kn_0) - \frac{1}{n_0}\sum_{j=0}^{n_0-1} f(j+kn_0)e^{-2\pi i j \lambda}\right|$$

$$\leq \frac{1}{n_0}\sum_{j=0}^{n_0-1} |f(j+kn_0)||1 - e^{-2\pi i j \lambda}| < \frac{\varepsilon}{2},$$

so that

$$\left|\frac{1}{n_0}\sum_{j=0}^{n_0-1} f(j+kn_0)e^{-2\pi i j \lambda}\right| < \varepsilon.$$

Thus the first term is less than ε, and hence the sum of the two is less than 2ε, if $n \geq l n_0$ and l is large enough.

144   4. More about recurrence

Remark 1.14 It follows readily that if
$$M\{f(j)e^{-2\pi ij\lambda}\} = 0 \quad \text{for all } \lambda \in \mathbb{R},$$
then
$$\lim_{n\to\infty} \frac{1}{n} \sum_{j=0}^{n-1} f(j)e^{-2\pi ij\lambda} = 0 \quad \text{uniformly for } \lambda \in \mathbb{R}.$$

Now we want to choose $k_n \in \mathbb{Z}$ and $1/n$-periods p_n such that
$$\left| \frac{1}{p_n} \sum_{k=0}^{p_n-1} f(j)e^{-2\pi ijk_n/p_n} \right| \geq c > 0 \quad \text{for all } n.$$

Then if $k_n/p_n \to \tau$, we will be able to argue that $\alpha_{x_0}(\tau) \neq 0$. If it is not possible to choose such k_n and p_n, then we will see that the $A_k^{(p)}$ of the following lemma tend to 0 uniformly in k as $p \to \infty$.

Lemma 1.15 Let
$$A_k^{(p)} = \frac{1}{p} \sum_{j=0}^{p-1} f(j)e^{-2\pi ijk/p} \text{ for } p = 1, 2, \ldots \text{ and } k = 0, 1, \ldots p-1.$$
If $\lim_{p\to\infty} A_k^{(p)} = 0$ uniformly in k, then
$$\lim_{n\to\infty} \frac{1}{n} \sum_{k=0}^{n-1} |f(k)|^2 = 0$$
(and hence $f \equiv 0$).

Assume henceforth that the hypotheses of Lemma 1.15 are satisfied. The following sequence of lemmas will constitute a proof of Lemma 1.15. Since AP is an algebra,
$$g(k) = \lim_{n\to\infty} \frac{1}{n} \sum_{j=0}^{n-1} f(k+j)\overline{f(j)}$$
exists uniformly in k. We need to prove that
$$g(0) = 0.$$
This will be accomplished by considering certain periodic functions related to f.

For a fixed integer p, let $F = F^{(p)}$ be the periodic sequence of period p for which
$$F(k) = f(k) \quad \text{for } k = 0, 1, \ldots p-1.$$
Fourier analysis on the finite abelian group $\mathbb{Z}_p$ shows that if
$$A_k = A_k^{(p)} = \frac{1}{p} \sum_{j=0}^{p-1} f(j)e^{-2\pi ijk/p},$$

4.1. Construction of eigenfunctions

then
$$F(k) = F^{(p)}(k) = \sum_{r=0}^{p-1} A_r e^{2\pi i r k/p}.$$

Further, either by direct calculation or knowledge of Parseval's formula, we see that

$$\frac{1}{p}\sum_{j=0}^{p-1} |f(j)|^2 = \frac{1}{p}\sum_{j=0}^{p-1} |F(j)|^2 = \sum_{j=0}^{p-1} |A_j|^2 \leq \|f\|_\infty^2.$$

Lemma 1.16 If $\sup_k |A_k^{(p)}| \to 0$ as $p \to \infty$, then
$$\sum_{k=0}^{p-1} |A_k^{(p)}|^4 \to 0 \quad \text{as } p \to \infty.$$

Proof:
$$\sum_{k=0}^{p-1} |A_k^{(p)}|^4 \leq \sup_k |A_k^{(p)}|^2 \sum_{k=0}^{p-1} |A_k^{(p)}|^2$$
$$\leq \|f\|_\infty^2 \sup_k |A_k^{(p)}|^2 \to 0 \quad \text{as } p \to \infty.$$

Lemma 1.17 If $G(k) = G^{(p)}(k) = 1/p \sum_{j=0}^{p-1} F(k+j)\overline{F(j)}$, then
$$\frac{1}{p}\sum_{j=0}^{p-1} |G^{(p)}(j)|^2 \to 0 \quad \text{as } p \to \infty.$$

Proof: Observe first that
$$\frac{1}{p}\sum_{j=0}^{p-1} G(j) e^{-2\pi i j k/p} = \frac{1}{p}\sum_{j=0}^{p-1} \frac{1}{p}\sum_{r=0}^{p-1} F(r+j)\overline{F(r)} e^{-2\pi i j k/p}$$
$$= \frac{1}{p}\sum_{r=0}^{p-1} \overline{F(r)} e^{2\pi i r k/p} \frac{1}{p}\sum_{j=0}^{p-1} F(r+j) e^{-2\pi i (r+j)k/p}$$
$$= \frac{1}{p}\sum_{r=0}^{p-1} \overline{F(r)} e^{2\pi i r k/p} \frac{1}{p}\sum_{j=0}^{p-1} F(j) e^{-2\pi i j k/p}$$
$$= \frac{1}{p}\sum_{r=0}^{p-1} \overline{F(r)} e^{2\pi i r k/p} A_k = A_k \frac{1}{p}\sum_{r=0}^{p-1} \overline{F(r)} e^{2\pi i r k/p}$$
$$= A_k \overline{A_k} = |A_k|^2,$$

so that, by Fourier analysis (or directly),
$$G(k) = \sum_{j=0}^{p-1} |A_j|^2 e^{2\pi i j k/p}.$$

Define
$$H(k) = H^{(p)}(k) = \frac{1}{p}\sum_{j=0}^{p-1} G(k+j)\overline{G(j)}.$$
A similar argument to the foregoing will show that
$$H(k) = \sum_{j=0}^{p-1} |A_j|^4 e^{2\pi i j k/p}.$$
By Lemma 1.16,
$$H^{(p)}(0) = \sum_{j=0}^{p-1} |A_j^{(p)}|^4 \to 0 \quad \text{as } p \to \infty.$$
In view of the definition of H, this says that
$$\frac{1}{p}\sum_{j=0}^{p-1} |G^{(p)}(j)|^2 \to 0 \quad \text{as } p \to \infty.$$

Recall that we are trying to obtain the conclusion of Lemma 1.15 for the almost periodic function g rather than the periodic function G, since this will immediately imply that $g \equiv 0$ and, in particular, $g(0) = 0$ as desired. However, if p is a large ε-period of f for a small $\varepsilon > 0$, there is not much difference between g and G.

Lemma 1.18 If $p = p_n$ is a (large) $1/n$-period of f, then
$$G^{(p)}(k) = g(k) + \theta_p,$$
where
$$|\theta_p| \le \varepsilon_p + \frac{\|f\|_\infty}{n}$$
with $\varepsilon_p \to 0$ as $p \to \infty$.

Proof: If $0 \le k \le p$, then
$$G^{(p)}(k) = \frac{1}{p}\left\{\sum_{j=0}^{p-k} F(k+j)\overline{F(j)} + \sum_{j=p-k+1}^{p-1} F(k+j-p)\,\overline{F(j)}\right\}$$
$$= \frac{1}{p}\left\{\sum_{j=0}^{p-k} f(k+j)\overline{f(j)} + \sum_{j=p-k+1}^{p-1} f(k+j-p)\overline{f(j)}\right\}$$
$$= \frac{1}{p}\sum_{j=0}^{p-1} f(k+j)\overline{f(j)} + \frac{1}{p}\sum_{j=p-k+1}^{p-1} \overline{f(j)}[f(k+j-p) - f(k+j)].$$
Now
$$|f(k+j-p) - f(k+j)| < \frac{1}{n} \quad \text{for all } k, j,$$

4.1. Construction of eigenfunctions

since p is a $1/n$-period of f, so that the second term is bounded by $\|f\|_\infty/n$. Since the first term converges to $g(k)$ uniformly in k as $p \to \infty$, the proof is complete.

Lemma 1.19 For each $n = 1, 2, \ldots$, let p_n be a $1/n$-period of f, and suppose that $p_n \to \infty$. Then

$$\left| \frac{1}{p_n} \sum_{k=0}^{p_n-1} |g(k)|^2 - \frac{1}{p_n} \sum_{k=0}^{p_n-1} |G^{(p_n)}(k)|^2 \right| \to 0 \quad \text{as } n \to \infty.$$

Proof: The discrepancy is

$$\frac{1}{p_n} \sum_{k=0}^{p_n-1} \{|g(k)| - |G^{(p_n)}(k)|\} \{|g(k)| + |G^{(p_n)}(k)|\}$$

$$\leq 2\|f\|_\infty^2 \frac{1}{p_n} \sum_{k=0}^{p_n-1} |g(k) - G^{(p_n)}(k)|,$$

which tends to 0 as $n \to \infty$, by Lemma 1.18.

Combining 1.17 and 1.19, we obtain

Lemma 1.20

$$\frac{1}{p_n} \sum_{k=0}^{p_n-1} |g(k)|^2 \to 0 \quad \text{as } n \to \infty.$$

and also

Lemma 1.21 $g \equiv 0$, and hence $g(0) = 0$.

Proof: $|g|^2$ is a nonnegative almost periodic function with mean 0.

Looking back a few pages, we realize that we have proved that $f \equiv 0$. Of course this was accomplished under the hypotheses of Lemma 1.15, namely assuming that

$$\lim_{p \to \infty} A_k^{(p)} = 0 \quad \text{uniformly in } k,$$

where

$$A_k^{(p)} = \frac{1}{p} \sum_{j=0}^{p-1} f(j) e^{-2\pi i j k/p}.$$

Since we selected $f \not\equiv 0$, this hypothesis cannot be satisfied. Then there is $c \geq 0$ and (by our proof) in fact $1/n$-periods $p_n \to \infty$ of f and integers k_n with

$$|A_{k_n}^{(p_n)}| \geq c \quad \text{for all } n.$$

We may assume that in fact $k_n/p_n \to \tau \in \mathbb{R}$.

Lemma 1.22 $\alpha_{x_0}(\tau) \neq 0$.

Proof: If
$$\alpha_{x_0}(\tau) = M\{f(j)e^{-2\pi i j \tau}\} = 0,$$
then by Lemma 1.13 there are $n_\tau \in \mathbb{Z}$ and $\delta_\tau > 0$ such that
$$\left|\frac{1}{m}\sum_{j=0}^{m-1} f(j)e^{-2\pi i j \lambda}\right| < \frac{c}{2} \quad \text{for } m \geq n_\tau \text{ and } |\lambda - \tau| < \delta_\tau.$$
However, we can choose $p_n \geq m$ and k_n with
$$|k_n/p_n - \tau| < \delta_\tau,$$
so this is impossible.

We have obtained, then, the nonzero eigenfunction $\alpha_x(\tau)$ with the nontrivial eigenfrequency τ. This completes our proof of the spectral characterization of weak mixing, avoiding invocation of the Spectral Theorem. Perhaps we should repeat that this was not just an idle exercise–besides providing a relatively concrete way to find eigenvalues, we have also learned something about almost periodic functions and in fact have established the main part of Bohr's fundamental theorem.

Corollary 1.23 The set of almost periodic sequences coincides with the uniform closure of the set of all trigonometric polynomials
$$P(k) = \sum_{j=1}^{n} a_j e^{2\pi i \lambda_j k} \quad (\lambda_j \in \mathbb{R})$$
on $\mathbb{Z}$.

Of course any uniform limit of such trigonometric polynomials is almost periodic, by 1.11. If $\{f(j)\}$ is an almost periodic sequence, then one can show that, in L^2,
$$f(k) = \sum_{n=1}^{\infty} a_n e^{2\pi i \lambda_n k}.$$
This is proved, using the above notation and machinery, by a sequence of steps familiar in Fourier analysis:

(*a*) There are at most countably many λ for which
$$\alpha(\lambda) = M\{f(j)e^{-2\pi i j \lambda}\} \neq 0.$$

(*b*) Bessel's Inequality: If $\alpha(\lambda_n) = a_n$ for each of the countably many λs, then
$$\Sigma |a_n|^2 \leq M\{|f(k)|^2\}.$$

4.1. Construction of eigenfunctions

(For if $\sigma_1, \ldots \sigma_N$ are arbitrary and $b_1, \ldots b_N \in \mathbb{C}$, then

$$M\left\{\left|f(k) - \sum_{j=1}^N b_j e^{2\pi i k \sigma_j}\right|^2\right\}$$

$$= M\{|f(k)|^2\} - \sum_{j=1}^N |\alpha(\sigma_j)|^2 + \sum_{j=1}^N |b_j - \alpha(\sigma_j)|^2,$$

which clearly is a minimum when $b_j = \alpha(\sigma_j)$, $j = 1, \ldots N$. This hint also helps to prove (a).)

(c) Parseval's Formula: For the $a_n = \alpha(\lambda_n)$,

$$M\{|f(k)|^2\} = \sum_{k=1}^\infty |a_k|^2.$$

Then the L^2 convergence of the above series to f follows readily.

That f is in fact a *uniform* limit of trigonometric polynomials is somewhat trickier to prove, although several arguments (especially a direct, elementary one due to Weyl 1926–27) are available. However, some *caution* is needed here, since it is not true that the above series converges uniformly to f: indeed f can be uniformly approximated to any desired degree of accuracy by a trigonometric polynomial, but these polynomials need *not* be partial sums of the above Bohr–Fourier series of f. In general, we can be sure only that eventually, as the approximations improve towards perfection, all the eigenfrequencies λ_n of f will appear in these approximating polynomials.

Many of the properties of almost periodic sequences that we have established here are shared by almost periodic functions on $\mathbb{R}$ and indeed by almost periodic functions on arbitrary groups. We have only introduced a large and beautiful subject which has important connections with and applications in many other parts of mathematics.

Exercises

1. Prove that an ergodic isometry on a compact metric space is isomorphic to a rotation (i.e. multiplication by a fixed element) on a compact abelian group. The group is *monothetic*, in that it contains a dense cyclic subgroup.
2. Prove 1.11.
3. Prove that if $\{a_n\}$ is an almost periodic sequence, then $M\{|a_n|^2\} = 0$ if and only if $a_n = 0$ for all n.
4. Fill in the details of the proof sketched above that an almost periodic sequence is, in L^2, the sum of its Bohr–Fourier series.

5. Can a sequence which is not almost periodic still have a uniform mean value?
6. Let $\{a_j\}$ be an almost periodic sequence and $A_n = \sum_{j=0}^{n-1} a_j$. Show that $\{A_n\}$ is almost periodic if and only if it is bounded.

4.2. Some topological dynamics

In the preceding section, in order completely to understand weakly mixing m.p.t.s it became necessary to consider topological aspects of dynamics such as minimality, the existence of dense orbits, and uniform almost periodicity. Here we will briefly indicate some aspects of the topological theory that parallels the metric theory of transformations which is our main concern. Some of these ideas will be used in the discussion of the Szemerédi–Furstenberg Theorem and related results in 4.3, and in 4.4 we will see that in fact every m.p.t. has a topological realization. The proofs of many of the facts we mention make reasonable exercises, and so they are numbered as such and assigned to the reader.

Topological dynamics is the study of groups acting on topological spaces by means of continuous maps. Rather than the most general situation, we concentrate on the case of a homeomorphism

$$T: X \to X$$

of a compact metric space X, in which case the acting group $\{T^n: n \in \mathbb{Z}\}$ may be identified with $\mathbb{Z}$; when we need to consider a general transformation group (X, G), we will rely on the ability of the reader to extend the concepts and definitions as required. The pair (X, T) is called a *cascade*. (We reserve the word *flow* for the case when the acting group is $\mathbb{R}$.) As usual the *orbit* of a point $x \in X$ is $\mathcal{O}(x) = \{T^n x : n \in \mathbb{Z}\}$.

A. Recurrence

Recall that a cascade (X, T) is called *minimal* in case X has no proper closed T-invariant subsets, or, equivalently, every point of X has a dense orbit. According to Gottschalk's Theorem (1.2), if (X, T) is minimal then every point of X has a very strong recurrence property (unfortunately called 'almost periodicity'): $T^k x$ returns to each neighborhood of x on a relatively dense set of k (i.e. with 'bounded gap'). Let us consider some weaker versions of topological recurrence, the first of which holds in any cascade.

Proposition 2.1 Every cascade contains a (nonempty) minimal set. (Exercise 1.)

4.2. Some topological dynamics

Theorem 2.2 Birkhoff Recurrence Theorem (1927) Every cascade (X, T) contains a point which is *recurrent under* T, i.e. a point x for which there are $n_k \to \infty$ with $T^{n_k} x \to x$.

Proof: Every point of a minimal set is recurrent under T.

This is a topological analogue of the Poincaré Recurrence Theorem, with compactness replacing the important condition of finite measure. In Section 4.3 we will consider an interesting extension of this result by Furstenberg and Weiss.

There is a weaker concept of recurrence than that of pointwise recurrence under T. We say that (X, T) is *regionally recurrent* if for any nonempty open $U \subset X$ there is $n \geq 1$ with $T^n U \cap U \neq \emptyset$. Within every (compact) cascade one can find a largest regionally recurrent cascade as follows. Call a point $x \in X$ *wandering* if there is a neighborhood U of x such that $T^n U \cap U = \emptyset$ for all $n \geq 1$. Then the *set of nonwandering points* Ω is closed and invariant, and (so long as X is compact) it is nonempty. It need not be true yet that (Ω, T) is regionally recurrent. Let Ω_1 be the set of nonwandering points of (Ω, T), and continue in this way. We obtain a descending sequence of closed invariant sets

$$X \supset \Omega \supset \Omega_1 \supset \Omega_2 \supset \ldots$$

such that if ξ is a successor ordinal, then Ω_ξ is the nonwandering set of $\Omega_{\xi-1}$; while if ξ is a limit ordinal, then

$$\Omega_\xi = \bigcap_{\eta < \xi} \Omega_\eta.$$

The intersection of all the Ω_ξ is denoted by Z and is called the *center* of (X, T).

Proposition 2.3 The center of (X, T) coincides with the closure of the set of points of X that are recurrent under T.

Proof: Exercise 2.

B. Topological ergodicity and mixing

(X, T) is *topologically ergodic* if every proper closed T-invariant subset of X is nowhere dense (i.e. its interior is empty).

(X, T) is *topologically weakly mixing* if $(X \times X, T \times T)$ is topologically ergodic.

(X, T) is *topologically strongly mixing* if given nonempty open $U, V \subset X$ there is $n_0 \in \mathbb{N}$ such that $T^n U \cap V \neq \emptyset$ whenever $n \geq n_0$.

There are several ways to prove that there exist T-invariant positive

Borel probability measures on X (Exercise 3); thus any cascade becomes a m.p.t. upon the choice of such a measure. A cascade (X, T) is called *uniquely ergodic* if there is only one such measure, and it is called *strictly ergodic* if it is minimal and uniquely ergodic.

Here are alternative characterizations of some of these properties.

Proposition 2.4 The following statements are equivalent:
  (i) (X, T) is topologically ergodic.
  (ii) Every point of X, with the possible exception of a set of first category, has an orbit which is dense in X.
  (iii) There is a point $x \in X$ which has a dense orbit.
  (iv) Given nonempty open $U, V \subset X$, there is $n \in \mathbb{Z}$ such that $T^n U \cap V \neq \emptyset$. (This property is sometimes called *regional transitivity*.)
Proof: Exercise 4.

Proposition 2.5 If (X, T) is a cascade and μ is a T-invariant Borel probability measure on X whose support is all of X, then each of metric ergodicity, weak mixing, and strong mixing implies its corresponding topological property.
Proof: Exercise 5.

Remark 2.6 The converses of these implications are not true (Exercise 6), even when (X, T) is uniquely ergodic. (Kolmogorov 1953, Petersen 1970a).

Remark 2.7 I apologize for the two distinct meanings of the word 'metric'.

Proposition 2.8 The following statements are equivalent:
  (i) (X, T) is uniquely ergodic.
  (ii) For each $f \in \mathscr{C}(X)$, $\{1/n \sum_{k=0}^{n-1} f T^k\}$ converges uniformly on X to a constant.
  (iii) For each $f \in \mathscr{C}(X)$, some subsequence of $\{1/n \sum_{k=0}^{n-1} f T^k\}$ converges pointwise on X to a constant.
Proof: Exercise 7.

Remark 2.9 Keynes and Robertson (1968) proved that (X, T) is topologically ergodic if and only if the only functions f on X which satisfy
  (i) the set of discontinuity points of f is residual,

4.2. Some topological dynamics

(ii) $fT = f$ except possibly on a set of first category

are those which are constant except possibly on a set of first category.

Similarly, a minimal cascade (X, T) is weakly mixing if and only if it has no continuous eigenfunctions (Keynes and Robertson 1968, 1969, Petersen 1970b).

Here are some typical examples of cascades to keep in mind.

1. *A translation on a compact group.* (e.g., an irrational rotation of the circle is topologically ergodic, uniquely ergodic, but not topologically weakly mixing – Exercise 8.)

2. *An automorphism of a compact group.* (e.g., an automorphism of an n-torus is not uniquely ergodic but is topologically strongly mixing whenever it is metrically strongly mixing with respect to Haar measure – Exercise 9.)

3. *Symbolic cascades.* The space $\Omega = \{0, 1, \ldots N - 1\}^{\mathbb{Z}}$ is metrizable, with metric

$$d(x, y) = \frac{1}{2^n}, \quad \text{where } n = \inf\{|k| : x_k \neq y_k\},$$

and the shift $\sigma : \Omega \to \Omega$ is a homeomorphism.

  (a) The orbit-closure of any point $\omega \in \Omega$ produces a topologically ergodic cascade $(\overline{\mathcal{O}(\omega)}, \sigma)$.
  (b) The full shift (Ω, σ) is topologically strongly mixing (Exercise 10).
  (c) A *subshift of finite type* (Σ_A, σ) is defined by an $N \times N$ matrix A with entries from $\{0, 1\}$ by

  $$x \in \Sigma_A \quad \text{if and only if} \quad A_{x_i x_{i+1}} = 1 \quad \text{for all } i \in \mathbb{Z}.$$

Thus Σ_A consists exactly of those sequences which contain only 'admissible transitions'.

4. *Diffeomorphisms of manifolds* (e.g., geodesic cascades and horocycle cascades).

C. Equicontinuous and distal cascades

Next we consider properties of cascades that are at the other extreme from mixing. With respect to a mixing transformation, subsets of X are extremely pliable; but transformations with either of these two properties will be akin to rigid motions. We will also mention the important fixed-point properties of groups of mappings of these kinds.

We say that (X, T) is *equicontinuous* in case $\{T^n : n \in \mathbb{Z}\}$ forms an equicontinuous family of maps $X \to X$. That is, given $\varepsilon > 0$ there is $\delta > 0$ such that if $d(x, y) < \delta$ then $d(T^n x, T^n y) < \varepsilon$ for all $n \in \mathbb{Z}$.

4. More about recurrence

We say that (X, T) is *distal* in case $x \neq y$ implies $\inf_n d(T^n x, T^n y) > 0$. Two points $x, y \in X$ are said to be *proximal* if there are $n_k \in \mathbb{Z}$ and $z \in X$ with $T^{n_k} x \to z$, $T^{n_k} y \to z$. Thus a distal cascade is one that has no pairs of distinct proximal points. Of course each equicontinuous cascade is distal.

Proposition 2.10 If (X, T) is equicontinuous, then $(\overline{\mathcal{O}(x)}, T)$ is minimal for each $x \in X$ (i.e. (X, T) is *pointwise almost periodic*). Therefore X is the union of its minimal subsets.
Proof: Exercise 11.

Theorem 2.11 The following statements about a minimal topological cascade (X, T) are equivalent:
  (i) (X, T) is equicontinuous.
  (ii) (X, T) is *uniformly almost periodic* in that given $\varepsilon > 0$ there is a relatively dense set $S \subset \mathbb{Z}$ such that $d(x, T^n x) < \varepsilon$ for all $n \in S$.
  (iii) X can be given a group structure which makes it a compact topological group, and there is an element $x_0 \in X$ such that $\{x_0^n : n \in \mathbb{Z}\}$ is dense in X (so that X is *monothetic*) and $Tx = x_0 x$ for all $x \in X$. (Thus with respect to Haar measure on X, T has discrete spectrum.)
Proof: Exercise 12.

The following fixed point theorem, which is frequently used in economics, easily implies, and is implied by, the existence of Haar measure on compact groups (Exercise 13).

Theorem 2.12 *Kakutani Fixed Point Theorem* (1938a) Let X be a nonempty compact convex subset of a locally convex topological vector space. Suppose that G is an equicontinuous group of *affine* maps $X \to X$: $g(\alpha x + (1 - \alpha) y) = \alpha g(x) + (1 - \alpha) g(y)$ for $0 \leq \alpha \leq 1$ and $x, y \in X$. Then G has a common fixed point in X.

Remark 2.13 F. Hahn (1967) showed that the Theorem is still true if the action of G is required only to be *distal* rather than equicontinuous: $g_i x \to z$ and $g_i y \to z$ implies $x = y$. We will give the clever proof of this stronger statement that is due to S. Glasner (1975, 1976).
Proof: By Zorn's Lemma, X contains a minimal nonempty compact convex G-invariant subset K. Let M denote the closure of the set of extreme points of K. ($M \neq \varnothing$ by the Krein–Milman Theorem.) I claim that M is the unique minimal subset of the system (K, G). (We are extending the terminology established for cascades to transformation groups in the obvious way.)

4.2. Some topological dynamics

For let $Y \subset K$ be a minimal subset of (K, G). Then $\overline{\text{co}}(Y)$ (the closed convex hull of Y) is closed, convex, and G-invariant and hence coincides with K. Therefore the extreme points of K are in Y, and it follows that $M \subset Y$ and $M = Y$.

I claim now that all pairs of points of M are proximal to one another. For if $x, y \in M$, then

$$\overline{G(\tfrac{1}{2}(x+y))} \supset M,$$

since M is the unique minimal set of (K, G). Thus there is an extreme point k of K and a net $\{g_\nu\}$ of elements of G such that

$$g_\nu(\tfrac{1}{2}(x+y)) \to k.$$

Using compactness, we may assume that $g_\nu x \to x_\infty$ and $g_\nu y \to y_\infty$. Then we find

$$\tfrac{1}{2}(x_\infty + y_\infty) = k,$$

and this is impossible, since k is an extreme point, unless $x_\infty = y_\infty$. Then x and y are proximal.

Since (M, G) is distal, this cannot happen unless M consists of just one point. The point is the one fixed by all elements of G.

This theorem was further generalized by Ryll–Nardzewski, who showed that it still held if X was assumed only to be *weakly* compact. We again give the remarkable simple proof found by Glasner (1975, 1976). There is no loss of generality in assuming that B is separable and a Banach space (see Ryll–Nardzewski 1967 for the reduction).

Theorem 2.14 *Ryll–Nardzewski Fixed Point Theorem* (1967) Let B be a separable Banach space, $X \subset B$ a weakly compact convex set, and G a group acting affinely and norm distally (and of course continuously with respect to the weak topology) on X. Then G has a common fixed point in X.

Proof: Repeat the proof of the Kakutani–Hahn Theorem for B with its weak topology. We find a minimal set (M, G), all pairs of points of which are proximal in the weak topology. Now we use norm distality to show that again M consists of just one point: given $x, y \in M$ and $\varepsilon > 0$, it is enough to find $g \in G$ such that $\|gx - gy\| < \varepsilon$, for then M will be simultaneously norm-distal and norm-proximal, hence a single point.

Let U be the closed ball of radius ε centered at the origin in B. Choose a sequence $b_1, b_2, \ldots$ of elements of B such that $\{b_i + U : i = 1, 2, \ldots\}$ covers B. Each $b_i + U$ is norm closed and convex, hence weakly closed. Then by the Baire Category Theorem, not every $(b_i + U) \cap M$ can be nowhere dense in the restriction of the weak topology to M. Thus there are an i and a weakly open subset W of B such that

$$\varnothing \neq W \cap M \subset (b_i + U) \cap M.$$

(Recall that the weak topology on a weakly compact subset of a separable Banach space is complete metric.)

Now x and y are proximal, so there are $g_v \in G$ and $z \in M$ such that $g_v x \to z$ and $g_v y \to z$, weakly. Choose $h \in G$ (using minimality of (M, G)) such that $hz \in W \cap M$. Then $W_0 = h^{-1}(W \cap M)$ is a weak neighborhood of z in M, so there is a v such that $g_v x \in W_0$ and $g_v y \in W_0$. Then $hg_v x$ and $hg_v y$ are both in $W \cap M \subset b_i + U$, and therefore $\|hg_v x - hg_v y\| < \varepsilon$.

D. Uniform distribution mod 1

Proposition 2.15 A minimal equicontinuous cascade is strictly ergodic.

Proof: Exercise 14.

Let X be the unit circle, here regarded as $[0, 1)$, and let $Tx = x + \alpha$ (mod 1), where α is irrational. There are many easy ways to see that T is a minimal isometry, and hence (X, T) is minimal and equicontinuous (Exercise 15). By the Proposition, (X, T) is strictly ergodic. Of course the unique invariant measure is Lebesgue measure m. Therefore

$$\frac{1}{n} \sum_{k=0}^{n-1} f(x + k\alpha) \to \int_0^1 f \, dm$$

uniformly on $[0, 1)$ for each continuous f on $[0, 1)$. In fact this statement holds for all *Riemann integrable* functions f on $[0, 1)$; this is one of the equidistribution theorems of H. Weyl (1916). (Exercise 16. Hint: Use monotone convergence. If necessary, see Remark 4.4(2).) In particular,

$$\frac{1}{n} \sum_{k=0}^{n-1} \chi_{[a,b)}(x + k\alpha) \to b - a$$

uniformly on $[0, 1)$ whenever $0 \leq a \leq b \leq 1$.

The fact that $\{k\alpha\}$ (mod 1) is dense in $[0, 1)$ when α is irrational has long been known. There are similar observations in the writings of Nicole Oresme (c. 1320–1382): In Proposition II, 4 of his *Tractatus de commensurabilitate vel incommensurabilitate motuum celi* (Grant 1971), for example, Oresme considers two bodies moving on a circle with uniform but incommensurable velocities and says, 'No sector of a circle is so small that two such mobiles could not conjuct in it at some future time, and could not

4.2. Some topological dynamics

have conjuncted in it sometime [in the past].' Oresme had an astonishing understanding of circular motion and rational and irrational numbers. In *De proportionibus proportionum*, Ch. III, Proposition V, he says, 'It is probable that two proposed unknown ratios are incommensurable because if many unknown ratios are proposed it is most probable that any [one] would be incommensurable to any [other].' (Oresme 1351). Could it be that a fourteenth-century scientist already knew that the set of rational numbers has measure zero (or at least measure less than that of the irrational numbers)? Apparently Oresme was also one of the first discoverers of the divergence of the harmonic series.

Oresme's *Ad pauca respicientes* was written to demonstrate that astrology is futile because the future is essentially unpredictable. His ideas are surprisingly close to P. Bohl's (1909) treatment of the 'problem of mean motion,' a question that had already been discussed by Lagrange. For a given Keplerian element $\rho(t)$, such as the longitude of the perihelion of a planet orbiting the sun, are there a constant c and a bounded function h such that $\rho(t) = ct + h(t)$? (We suppose that at each time t the planet is on a Kepler ellipse, but that the ellipse is changing with time because of the perturbing influences of the other bodies in the solar system.) If there were, the perihelion would essentially revolve regularly about the sun, up to bounded anomalies. Lagrange (1870) gave an affirmative answer in case the equations of motion happened to take certain special forms, and it was generally believed (e.g. Cavällin and Gyldén 1895) that there always did exist a mean motion. However, Bohl showed that in one situation the mean motion existed if and only if the expression $\sum_{k=0}^{n-1} \chi_{[a,b)}(x + k\alpha) - n(b - a)$ is bounded in n. This happens if and only if $b - a$ is an integral multiple of α mod 1. (Cf. Petersen 1973a, Furstenberg, Keynes and Shapiro 1973.) Since practically speaking such a determination could never be made by actual experimental measurement, the problem of existence of mean motion was one that by its very nature could never be settled. Similarly Oresme (*Tractatus de Commensurabilitate*, Part III), concluded that although the celestial motions are probably incommensurable, the final answer is necessarily unknowable, and indeed it is good that this is so.

An elementary argument using the 'pigeonhole principle' due to Dirichlet, shows that if α is irrational then there are even infinitely many k with

$$k\alpha \pmod 1 < \frac{1}{k}.$$

(Exercise 17.)

4. More about recurrence

There is a similar theory of uniform distribution on the n-torus $\mathbb{K}^N = [0, 1)^n$, a compact abelian group under coordinatewise addition mod 1. Let $\alpha = (\alpha_1, \alpha_2, \ldots, \alpha_n)$ be rationally independent. Then it is a theorem of Kronecker (1884) (see also Tchebychef 1866) that $\{k\alpha\}$ is dense in $\mathbb{K}^n$. (Exercise 18.) That $\{k\alpha\}$ is equidistributed in $\mathbb{K}$ is another theorem of Weyl (1916).
(Exercise 19.)

Further, if P is any polynomial with at least one irrational coefficient, then $\{P(k)\}$ is equidistributed mod 1 (Weyl 1916; for dynamical proofs, see Furstenberg 1960, Hahn 1965, and Postnikov 1966).

E. Structure of distal cascades

The two most basic facts about distal cascades are that they can be characterized in terms of pointwise almost periodicity and that they can be resolved into towers composed of equicontinuous pieces.

The sequences and nets that arise in discussions of proximality are most easily handled by means of the *Ellis semigroup* $E(X, T)$ (also called the *enveloping semigroup*), which is the closure in the product topology (i.e. the topology of pointwise convergence) of $\{T^n : n \in \mathbb{Z}\}$ in the compact space X^X. It is not hard to show that for any cascade (X, T),

(i) E is a semigroup;
(ii) right multiplication is a continuous operation in E;
(iii) left multiplication by a continuous element of E is a continuous operation in E;
(iv) $x_1, x_2 \in X$ are proximal if and only if there is $p \in E$ with $px_1 = px_2$ (note that we write the action of the functions in E on the *left*);
(v) $\overline{\mathcal{O}(x)} = E(X, T)x$ for each $x \in X$.

Proposition 2.16 (Ellis 1958) Every compact semigroup in which multiplication on at least one side is continuous contains an idempotent.
Proof: Let us suppose that right multiplication is continuous in the compact semigroup E. By Zorn's Lemma, there is a minimal compact subset $K \subset E$ with $K^2 \subset K$. If $u \in K$ then $Ku \subset K$ is compact and $(Ku)^2 = KuKu \subset K^2u \subset Ku$, so by minimality $Ku = K$. Let $L = \{v \in K : vu = u\}$. Since $Ku = K$, $L \neq \emptyset$. However, L is closed and $L^2 = L$, so we must have $L = K$. Thus $Ku = u$, and in particular $u^2 = u$.

Proposition 2.17 (Ellis 1958) (X, T) is distal if and only if $E(X, T)$ is a group.

4.2. Some topological dynamics

Proof: Suppose that $E(X, T)$ is a group. If x_1 and x_2 are proximal, then $px_1 = px_2$ for some $p \in E(X, T)$. Multiplying by $p^{-1} \in E(X, T)$ gives $x_1 = x_2$. Hence (X, T) is distal.

Conversely, suppose that (X, T) is distal. I claim that then $E(X, T)$ has left cancellation. If $p, p_1, p_2 \in E(X, T)$ and $pp_1 = pp_2$, then for any fixed $x \in X$ we have $p(p_1 x) = p(p_2 x)$, which by distality implies $p_1 x = p_2 x$; therefore $p_1 = p_2$.

Of course $E(X, T)$ contains an identity element $e = T^0$. By left cancellation, the identity element is unique.

Inverses can be found as follows. Given $p \in E(X, T)$, $K = E(X, T)p$ is compact and satisfies

$$K^2 = (E(X, T)pE(X, T))p \subset E(X, T)p = K.$$

Thus K is a compact semigroup and, by the preceding Proposition, K contains an idempotent u. Using left cancellation, $u \cdot u = u = u \cdot e$ implies that $u = e$. Thus $e = qp$ for some $q \in E(X, T)$.

Repeating this argument for q, we find $r \in E(X, T)$ with $e = rq$. Then $p = ep = rqp = re = r$, so we have $qp = pq = e$.

Of course inverses will be unique by left cancellation.

Proposition 2.18 (*Ellis* 1958) A distal cascade is pointwise almost periodic.

Proof: Let $x \in X$. We need to show that $\overline{\mathcal{O}(x)} = E(X, T)x$ is a minimal subset of X. For this purpose it is enough to show that if $y \in \overline{\mathcal{O}(x)}$, then $x \in \overline{\mathcal{O}(y)}$. But if $y \in \overline{\mathcal{O}(x)}$, then $y = px$ for some $p \in E(X, T)$, and hence $x = p^{-1}y \in E(X, T)y = \overline{\mathcal{O}(y)}$.

Theorem 2.19 (*Ellis* 1958) (X, T) is distal if and only if $(X \times X, T \times T)$ is pointwise almost periodic.

Proof: If (X, T) is distal, then so is $(X \times X, T \times T)$, and hence it is pointwise almost periodic by the preceding Proposition.

For the converse, suppose that x and y are proximal. Let $\Delta \subset X \times X$ denote the diagonal. Then $\overline{\mathcal{O}(x, y)} \cap \Delta \neq \emptyset$, and hence, since Δ is closed and invariant and $\overline{\mathcal{O}(x, y)}$ is minimal, we must have $(x, y) \in \Delta$.

The fundamental theorem on the structure of distal flows was proved by Furstenberg. Similar ideas, except for m.p.t.s, played a role in his proof of the Szemerédi Theorem, which we will discuss in Section 4.3.

Let (X, T) and (Y, S) be cascades, and let $\phi: X \to Y$ be a continuous onto map such that $\phi T = S\phi$. We say that ϕ is a *homomorphism*, that (Y, S) is a *factor* of (X, T), and (X, T) is an *extension* of (Y, S). Also, the entire diagram $\phi:(X, T) \to (Y, S)$ is called a factor, homomorphism, or extension.

An extension $\phi:(X, T) \to (Y, S)$ is called *isometric* in case there is a continuous metric $\rho(x_1, x_2)$ defined on each fiber $\phi^{-1}\{y\}$ such that $\rho(Tx_1, Tx_2) = \rho(x_1, x_2)$ whenever $\phi(x_1) = \phi(x_2)$. According to Furstenberg's Theorem, every distal cascade is composed of a (possibly transfinite) string of isometric extensions. In order to state this theorem, one more concept, that of *inverse limit*, is necessary.

Let $\{(X_\alpha, T_\alpha): \alpha \in A\}$ be a family of cascades indexed by a partially ordered set A such that whenever $\alpha, \beta \in A$ and $\alpha \leq \beta$, there is a homomorphism $\phi_{\beta\alpha}:(X_\beta, T_\beta) \to (X_\alpha, T_\alpha)$. The *inverse limit* (X, T) of the system $\{(X_\alpha, T_\alpha)\}$ is defined by

$$X = \{x = (x_\alpha) \in \prod_{\alpha \in A} X_\alpha : \phi_{\beta\alpha} x_\beta = x_\alpha \text{ for all } \alpha, \beta \in A \text{ with } \alpha \leq \beta\}$$

and

$$Tx = T(x_\alpha) = (T_\alpha x_\alpha).$$

There are natural homomorphisms

$$\phi_\alpha:(X, T) \to (X_\alpha, T_\alpha)$$

for each $\alpha \in A$.

Theorem 2.20 Furstenberg Structure Theorem (1963) Let (X, T) be a minimal distal cascade. Then there is an ordinal number η and a family of factors $\{(X_\xi, T_\xi): \xi \leq \eta\}$ of X such that:

(i) X_0 consists of a single point,
(ii) $(X_\eta, T_\eta) = (X, T)$;
(iii) if $\xi < \tau \leq \eta$, then there is a homomorphism $\phi_{\tau\xi}:(X_\tau, T_\tau) \to (X_\xi, T_\xi)$;
(iv) if $\xi \leq \eta$ is a successor ordinal, then (X_ξ, T_ξ) is an isometric extension of $(X_{\xi-1}, T_{\xi-1})$;
(v) if $\xi \leq \eta$ is a limit ordinal, then (X_ξ, T_ξ) is the inverse limit of $\{(X_\tau, T_\tau): \tau < \xi\}$.

Corollary 2.21 A nontrivial minimal distal cascade has nontrivial equicontinuous factors, and hence nontrivial continuous eigenfunctions. *Proof*: Exercise 20.

This theorem has been extended by Veech (1970) and Ellis (1973) so that no countability assumptions on X or the acting group G are needed

4.2. Some topological dynamics

and one must only assume that X is *point-distal*: at least one point of X is not proximal to any other points. For further reading about this and the many other topics in topological dynamics that we have not been able even to mention we recommend the following references.

The origins of the subject are in

Henri Poincaré *Les méthodes nouvelles de la mécanique céleste*, I (1892), II (1893), and III (1899), Gauthiers–Villars, Paris. Also Dover, New York, 1957; and NASA TTF 450–452, Washington, D.C., 1967.

G. D. Birkhoff (1927) and (1966). *Dynamical Systems*, A.M.S. Colloquium Publications **9**, Providence, R.I.

Systematic and progressively more recent accounts are

V. V. Nemytskii and V. V. Stepanov (1960). *Qualitative Theory of Differential Equations*, Princeton University Press.

W. H. Gottschalk and G. A. Hedlund (1955). *Topological Dynamics*, A.M.S. Colloquium Publications **36**, Providence, R.I.

Robert Ellis (1969). *Lectures on Topological Dynamics*, W. A. Benjamin, Inc., New York.

I. U. Bronstein (1975). *Extensions of Minimal Transformation Groups*, Sijthoff & Noordhoff, Alphen an den Rijn, The Netherlands.

For the connections with ergodic theory,

Manfred Denker, Christian Grillenberger, and Karl Sigmund, (1976). *Ergodic Theory on Compact Spaces*, Springer Lectures Number 527.

For the complete story of uniform distribution,

L. Kuipers and H. Niederreiter (1974). *Uniform Distribution of Sequences*, John Wiley & Sons, New York.

For the most recent (as well as an unbelievably thorough and novel) treatment of recurrence in both the topological and metric contexts,

H. Furstenberg (1981). *Recurrence in Ergodic Theory and Combinatorial Number Theory*, Princeton University Press.

Finally, a survey of much of the newer work in the field, including many open problems:

William A. Veech (1977). Topological dynamics, *Bull. Amer. Math. Soc.* **83**, 775–830.

Exercises

21. Show that the space $\mathcal{M}$ of positive Borel probability measures on a cascade (X, T) is convex and compact in the weak* topology and the set $\mathcal{E}$ of ergodic measures in $\mathcal{M}$ is exactly the set of extreme points of $\mathcal{M}$. (In fact $\mathcal{M}$ is metrizable: Prohorov 1956; Varadarajan 1958, 1962; cf. Parthasarathy 1967.)

4. More about recurrence

22. Give necessary and sufficient combinatorial conditions on a sequence $x \in \{0, 1\}^{\mathbb{Z}}$ in order that its orbit closure under the shift be (a) minimal (b) uniquely ergodic (c) strongly mixing.
23. Construct explicitly a point which has a dense orbit in $\{0, 1\}^{\mathbb{Z}}$. Do the same for a given irreducible subshift of finite type.
24. Give an example of a distal cascade that is not equicontinuous. Now give a minimal example. (Hint: Try a skew product.)
25. If $\beta = j\alpha$ mod 1 for some integer j, then $\sum_{k=0}^{n-1} \chi_{[0,\beta)}(k\alpha) - n\beta$ is bounded in n. (Hecke 1922).
26. Show that a cascade which has the structure specified in the conclusion of the Furstenberg Structure Theorem is distal.
27. Show that the classes of distal cascades and minimal cascades are each closed under the formation of factors and inverse limits.

4.3. The Szemerédi Theorem

A. Furstenberg's approach to the Szemerédi and van der Waerden Theorems

Baudet (a Dutch mathematician working in Göttingen) conjectured that if the positive integers are divided into two disjoint classes, then one of them at least must contain arbitrarily long arithmetic progressions. In 1927 van der Waerden published the proof of a stronger statement:

Theorem 3.1 *Van der Waerden's Theorem* (1927). If $\mathbb{N} = C_1 \cup C_2 \cup \ldots \cup C_r$ (disjoint), then there is a j, $j = 1, \ldots r$, such that C_j contains arbitrarily long arithmetic progressions: given k, an a and n can be found such that

$$a, a+n, \ldots a+(k-1)n \in C_j.$$

Van der Waerden (1971) describes at great length how he, Artin, and Schreier found the proof; Khintchine (1948) includes this result as one of his 'three pearls of number theory', although he has forgotten Baudet's name; and Rado's doctoral dissertation consisted of an extension of this result. It is possible that the conjecture is actually due to Schur, and that Baudet only propagated it – See Schur (1973), I, p. xiii.

Later, Erdös and Turán (1936) conjectured that in fact any subset of $\mathbb{N}$ which has positive upper density must contain arbitrarily long arithmetic progressions. This is of course a stronger statement than the van der Waerden Theorem. Roth (1952) obtained the result for progressions of length 3 and Szemerédi established first (1969) the case $k = 4$ and then (1975) the general statement.

4.3. The Szemerédi Theorem

Theorem 3.2 Szemerédi's Theorem (1975) Any set of positive integers with positive upper density contains arbitrarily long arithmetic progressions.

By a set $\mathscr{C}$ with *positive upper density* we mean one for which N_i, M_i with $N_i - M_i \to \infty$ can be found such that

$$\lim_{i \to \infty} \frac{\text{card}\,[\mathscr{C} \cap (M_i, N_i)]}{N_i - M_i} = \delta > 0.$$

The Theorem asserts that for such a set $\mathscr{C}$, given any k it is possible to find a and n such that

$$a, a + n, \ldots a + (k-1)n \in \mathscr{C}.$$

In 1977, Furstenberg proved an ergodic-theoretic version of Szemerédi's Theorem. Although this Theorem can be proved on the basis of Szemerédi's Theorem, Furstenberg's proof uses ergodic theory and does not depend on the van der Waerden, Roth, or Szemerédi results; this approach provides a new and most interesting proof of the Szemerédi Theorem.

Theorem 3.3 Ergodic Szemerédi Theorem (Furstenberg 1977) If $T: X \to X$ is a measure-preserving transformation on a finite measure space $(X, \mathscr{B}, \mu)$, $A \in \mathscr{B}$ with $\mu(A) > 0$, and $k > 0$, then there is an $n > 0$ such that

$$\mu(A \cap T^n A \cap T^{2n} A \cap \ldots \cap T^{(k-1)n} A) > 0.$$

Thus every m.p.t. on a finite measure space is *multiply recurrent*. Of course for $k = 2$ the statement reduces to the Poincaré Recurrence Theorem.

When we take up the proof of this theorem, we will first consider the case when T is weakly mixing. There the result will follow from an even stronger one: *every weakly mixing transformation is 'weakly mixing of all orders along multiples'*.

Theorem 3.4 If $T: X \to X$ is a weakly mixing m.p.t. on a finite measure space $(X, \mathscr{B}, \mu)$, then for any $A_1, A_2, \ldots, A_k \in \mathscr{B}$,

$$\lim_{N-M \to \infty} \frac{1}{N-M} \sum_{n=M+1}^{N} |\mu(A_1 \cap T^n A_2 \cap \ldots \cap T^{(k-1)n} A_k)$$
$$- \mu(A_1)\mu(A_2)\ldots\mu(A_k)| = 0.$$

Of course taking $A_1 = A_2 = \ldots = A_k = A$ will yield the preceding Theorem as an easy corollary, in case T is weakly mixing.

Combining this Theorem with the observation of Kakutani and Jones

164 4. More about recurrence

that we discussed in Section 2.6, we can prove that *each weakly mixing transformation is in fact 'strongly mixing of all orders along multiples' provided that we exclude all those powers of T which lie in a single set of density* 0.

Theorem 3.5 If $T: X \to X$ is weakly mixing, then there is a set $J \subset \mathbb{N}$ with density 0 such that given $A_1, \ldots, A_k \in \mathcal{B}$,

$$\lim_{\substack{n \to \infty \\ n \notin J}} \mu(A_1 \cap T^n A_2 \cap \ldots \cap T^{(k-1)n} A_k) = \mu(A_1)\mu(A_2)\ldots\mu(A_k).$$

For the proof, for each k and each choice of $A_1, \ldots, A_k$ from a countable generating set for $\mathcal{B}$ we choose a set $J \subset \mathbb{Z}^+$ of density 0 (whose existence is guaranteed by Theorem 3.4 and Lemma 2.6.2) such that the statement holds for $A_1, \ldots, A_k$ and this J. This way we obtain countably many such sets J. The rest of the proof proceeds as before.

Furstenberg and Weiss have observed that the topological analogue of the Ergodic Szemerédi Theorem is much easier to prove and also yields van der Waerden's Theorem as an easy corollary.

Theorem 3.6 (*Furstenberg and Weiss* 1978) Let X be a compact Hausdorff space and $T: X \to X$ a minimal homeomorphism. If $U \subset X$ is open and nonempty and $k > 1$, then there is an n such that

$$U \cap T^n U \cap \ldots \cap T^{(k-1)n} U \neq \varnothing.$$

In fact, Furstenberg and Weiss found it not much harder to prove a similar theorem for any k commuting homeomorphisms, not just $I, T, T^2, \ldots T^{k-1}$.

Theorem 3.7 (*Furstenberg and Weiss* 1978) Let X be a compact metric space and $T_1, \ldots T_k$ commuting homeomorphisms of X. Then there is a point $x \in X$ which is *multiply recurrent*: there are $n_l \to \infty$ with $T_i^{n_l} x \to x$ for each $i = 1, \ldots k$.

Of course this Theorem implies the one preceding it. By minimality, $T^m x \in U$ for some m. Letting $T_i = T^{-i}$ for $i = 0, 1, \ldots k-1$, we have

$$T_i^{n_k} x \to x \quad \text{as } k \to \infty \quad \text{for all } i,$$

that is,

$$T^{-n_k i} x \to x$$

and hence

$$T^{-n_k i} T^m x \to T^m x \quad \text{for all } i.$$

4.3. The Szemerédi Theorem

Then clearly a single n can be found such that
$$T^{-ni}x \in U \quad \text{for all } i = 0, 1, \ldots k-1,$$
and this implies that
$$x \in U \cap T^n U \cap \ldots \cap T^{(k-1)n} U.$$

The existence of this 'commuting' version of the topological Szemerédi theorem suggested to Furstenberg and Katznelson that a similar extension might be possible in the ergodic setting as well.

Theorem 3.8 (*Furstenberg and Katznelson* 1978) Let $T_1, T_2, \ldots, T_k$ be commuting measure-preserving transformations on a finite measure space $(X, \mathcal{B}, \mu)$. If $A \in \mathcal{B}$ with $\mu(A) > 0$, then
$$\liminf_{N \to \infty} \frac{1}{N} \sum_{n=1}^{N} \mu(T_1^n A \cap T_2^n A \cap \ldots \cap T_k^n A) > 0.$$
In particular, there is an $n > 0$ with $\mu(T_1^n A \cap T_2^n A \cap \ldots \cap T_k^n A) > 0$.

This strengthening of the ergodic version leads to a strengthening of the original Szemerédi Theorem. In order to indicate the connection between the combinatorial and ergodic settings, we will prove this Theorem immediately, assuming the Furstenberg–Katznelson Theorem.

Theorem 3.9 *Multidimensional Szemerédi Theorem* (*Furstenberg and Katznelson* 1978) Let $S \subset \mathbb{Z}^r$ be a subset with positive upper density (calculated with respect to any sequence of cubes $[a_n^1, b_n^1] \times \ldots \times [a_n^r, b_n^r]$ with $b_n^j - a_n^j \to \infty$ as $n \to \infty$) and let $F \subset \mathbb{Z}^r$ be any finite configuration. Then there are an integer d and a vector $u \in \mathbb{Z}^r$ such that $u + dF \subset S$ (i.e., S contains a figure *similar* to F).

Proof: Let $S \subset \mathbb{Z}^r$ have positive upper density. Let
$$\Omega_r = \{0, 1\}^{\mathbb{Z}^r}$$
with the product topology; an element ω of Ω_r can be thought of as a 0, 1-valued function $\omega(n_1, n_2, \ldots, n_r)$ on $\mathbb{Z}^r$. On Ω_r we have r commuting transformations $\sigma_1, \sigma_2, \ldots, \sigma_r, \sigma_i$ being the shift in the ith coordinate: for example
$$(\sigma_1 \omega)(n_1, n_2, \ldots, n_r) = \omega(n_1 + 1, n_2, \ldots, n_r).$$
Let X be the orbit closure of $\chi_S \in \Omega_r$ under the transformations $\sigma_1, \ldots, \sigma_r$:
$$X = \mathrm{cl}\,\{\sigma_i^n \chi_S : n \in \mathbb{Z}, i = 1, 2, \ldots, r\}.$$
Let
$$A = \{\omega \in X : \omega(0, 0, \ldots 0) = 1\},$$

so that $\sigma_1^{m_1}\sigma_2^{m_2}\ldots\sigma_r^{m_r}\chi_S \in A$ if and only if $(m_1, m_2, \ldots, m_r) \in S$. We want first to find a Borel probability measure μ on X invariant under all the σ_i such that $\mu(A) > 0$.

Since $S \subset \mathbb{Z}^r$ has positive upper density, there are cubes $[a_n^1, b_n^1] \times \ldots \times [a_n^r, b_n^r]$ with each $b_n^j - a_n^j \to \infty$ as $n \to \infty$ such that the density of S calculated with respect to these cubes is positive. This is to say that if for each n we define a measure μ_n on Ω_r by

$$\mu_n(f) = \frac{1}{(b_n^1 - a_n^1)(b_n^2 - a_n^2)\ldots(b_n^r - a_n^r)}$$
$$\times \sum_{j=1}^{r} \sum_{m_j = a_n^j}^{b_n^j - 1} f(\sigma_1^{m_1}\sigma_2^{m_2}\ldots\sigma_r^{m_r}\chi_S) \quad (f \in \mathscr{C}(\Omega_r)),$$

then

$$\lim_{n \to \infty} \mu_n(A) > 0.$$

Let μ be any weak* limit point of $\{\mu_n : n = 1, 2, \ldots\}$. Then, μ is invariant under each σ_i, and also $\mu(A) > 0$.

Let $F \subset \mathbb{Z}^r$ be any finite configuration; we need to prove that S contains a translate of a dilate of F. For this purpose it is enough to show that given any $K \in \mathbb{Z}^+$, there are $u \in \mathbb{Z}^r$ and $d \in \mathbb{Z}^+$ such that

$$u + d(k_1, k_2, \ldots k_r) \in S \quad \text{for all } 0 \leq k_i \leq K.$$

Since the transformations

$$\{\sigma_1^{-k_1}\sigma_2^{-k_2}\ldots\sigma_r^{-k_r} : 0 \leq k_i \leq K, i = 1, \ldots, r\}$$

form a commuting family of transformations of (X, μ) and $\mu(A) > 0$, by the Furstenberg–Katznelson Theorem we can find $d > 0$ with

$$\mu\left(\bigcap_{\substack{\{k_i\} \\ 0 \leq k_i \leq K}} (\sigma_1^{-k_1}\sigma_2^{-k_2}\ldots\sigma_r^{-k_r})^d A\right) > 0.$$

The definition of $\mu = \lim \mu_n$ shows that if $\mu(E) > 0$ then there is $(u_1, u_2, \ldots, u_r) \in \mathbb{Z}^r$ such that

$$\sigma_1^{u_1}\sigma_2^{u_2}\ldots\sigma_r^{u_r}\chi_S \in E.$$

(In fact there are enough such $\{u_i\}$ that their limiting 'frequency' is $\mu(E) > 0$.) Letting E be the above set of positive measure, we see that there is $u \in \mathbb{Z}^r$ such that for each choice of $k_1, \ldots, k_r$ with $0 \leq k_i \leq K$ for all i,

$$\sigma_1^{u_1}\sigma_2^{u_2}\ldots\sigma_r^{u_r}\chi_S \in \sigma_1^{-dk_1}\sigma_2^{-dk_2}\ldots\sigma_r^{-dk_r}A.$$

This says that

$$\sigma_1^{u_1 + dk_1}\sigma_2^{u_2 + dk_2}\ldots\sigma_r^{u_r + dk_r}\chi_S \in A,$$

or

$$(u_1 + dk_1, u_2 + dk_2, \ldots, u_r + dk_r) \in S.$$

There are further conjectures along these lines (due to Erdös (1964) who has offered monetary rewards for their solution), some of which may also be amenable to ergodic-theoretic analysis. Perhaps the most important one is the following, a positive solution of which would imply that the sequence of primes contains arbitrarily long arithmetic progressions.
Problem 1: If $\{n_k\}$ is a sequence of positive integers for which

$$\sum_k \frac{1}{n_k} = \infty,$$

then $\{n_k\}$ contains arbitrarily long arithmetic progressions.
Problem 2: If ω is any infinite sequence on the symbols 1 and -1, then for every k an n and N can be found for which

$$|\omega(n) + \omega(2n) + \ldots + \omega(Nn)| > k.$$

While in Problem 2 less is required than in van der Waerden's Theorem – in that we don't ask that either $\{\omega = 1\}$ or $\{\omega = -1\}$ contain arbitrarily long arithmetic progressions but seek only progressions along which either the 1s or -1s are predominant – at the same time the progressions considered are restricted, in that translation is not allowed.

In the following section we will prove the topological Szemerédi Theorem and its consequence, van der Waerden's Theorem. The same approach yields a dynamical proof of Hindman's Theorem. Next, we will prove Furstenberg's ergodic Szemerédi theorem for weakly mixing transformations (i.e., weak mixing implies weak mixing of all orders). Only then, once the basics and the direction are clear, will we sketch the proof of the Furstenberg–Katznelson Theorem. (For the details of the proof, and much more information about recurrence, see Furstenberg 1981).

We should remark also that equally important with establishing a proof of the Szemerédi Theorem is the development of the new ergodic-theoretic ideas which the process entails. Furstenberg has contributed new insights which can be applied to many other problems in ergodic theory besides this one.

B. *Topological multiple recurrence, van der Waerden's Theorem, and Hindman's Theorem*

The fundamental recurrence result that will be generalized and applied in this section is the following one.

Theorem 3.10 (*Birkhoff* 1927) Let $T: X \to X$ be a homeomorphism on a compact metric space X. Then there is a point $x \in X$ which is recurrent under T: that is, there are $n_k \to \infty$ with $T^{n_k}x \to x$.

Proof: By Zorn's Lemma, there are minimal sets in X (nonempty closed invariant sets M, each point of which has an orbit dense in M). Every point of a minimal set is recurrent, by Gottschalk's Theorem (1.2).

We will say that a pair (X, T), where X is compact metric and $T: X \to X$ is a homeomorphism, is *homogeneous* if there is a group G of homeomorphisms of X commuting with T such that (X, G) is a minimal transformation group. A closed subset $A \subset X$ is called *homogeneous in* (X, T) if there is a group G of homeomorphisms of X commuting with T such that $GA = A$ and (A, G) is a minimal transformation group.

Proposition 3.11 Let $T: X \to X$ be a homeomorphism of a compact metric space X and $A \subset X$ a closed homogeneous subset. Suppose that for each $\varepsilon > 0$ there are $x, y \in A$ and $n \geq 1$ with $d(T^n x, y) < \varepsilon$. Then for each $\varepsilon > 0$ there are $z \in A$ and $n \geq 1$ with $d(T^n z, z) < \varepsilon$. (Notice that A need not be T-invariant).

Proof (Bowen): We will prove first that the point $y \in A$ mentioned in the hypotheses is arbitrary.

Suppose then that (A, G) is minimal and $\varepsilon > 0$. We can find $g_1, \ldots, g_n \in G$ with

(1) $\min_i d(g_i x, y) < \dfrac{\varepsilon}{2}$ for all $x, y \in A$:

for cover A by finitely many sets V_j of diameter less than $\varepsilon/2$. Then for each j,

$$\{g^{-1} V_j : g \in G\}$$

is an open cover of A, and hence has a finite subcover

$$\{g_{1,j}^{-1} V_j, g_{2,j}^{-1} V_j, \ldots, g_{n,j}^{-1} V_j\}.$$

Then given any $x, y \in A$, $y \in V_j$ for some j and $x \in g_{i,j}^{-1} V_j$ for some i and the same j. Consequently

$$d(g_{i,j} x, y) < \frac{\varepsilon}{2},$$

proving (1).

Now choose $\delta > 0$ small enough that $d(x, x') < \delta$ implies $d(g_i x, g_i x') < \varepsilon/2$ for all i (note that the g_i are fixed to satisfy (1)). By hypothesis, there are $x_0, y_0 \in A$ and $n_0 \geq 1$ such that

$$d(T^{n_0} x_0, y_0) < \delta.$$

4.3. The Szemerédi Theorem

Then for each i

$$d(g_i T^{n_0} x_0, g_i y_0) = d(T^{n_0} g_i x_0, g_i y_0) < \frac{\varepsilon}{2},$$

and hence, since (1) allows us to choose i so that $d(g_i y_0, y) < \varepsilon/2$,

$$\min_i d(T^{n_0} g_i x_0, y) < \varepsilon \quad \text{for each } y \in A.$$

This proves that for each $y \in A$ there are $x \in A$ and $n \geq 1$ such that

$$d(T^n x, y) < \varepsilon.$$

Now fix an arbitrary $z_0 \in A$ and choose $z_1 \in A$ and $n_1 \geq 1$ with

(2) $\quad d(T^{n_1} z_1, z_0) < \dfrac{\varepsilon}{2}.$

Choose $z_2 \in A$ and $n_2 \geq 1$ with $T^{n_2} z_2$ very close to z_1 and $T^{n_1 + n_2} z_2$ very close to z_0: that is, choose z_2 and n_2 so that

$$d(T^{n_2} z_2, z_1) < \varepsilon_2,$$

where $\varepsilon_2 < \varepsilon/2$ is chosen small enough to guarantee that (2) still holds when z_1 is replaced by $T^{n_2} z_2$.

Continue in this manner. If $z_0, z_1, \ldots, z_r \in A, n_1, n_2, \ldots, n_r \in \mathbb{N}$, and $\varepsilon_2, \ldots, \varepsilon_r \in (0, \varepsilon/2)$ have already been chosen so that

(3) $\quad d(T^{n_j} z_j, z_{j-1}) < \varepsilon_j, \quad j = 1, \ldots, r,$

find $\varepsilon_{r+1} < \varepsilon/2$ so that (3) still holds when z_r is replaced by any point at distance less than ε_{r+1} from it. Then choose $z_{r+1} \in A$ and $n_{r+1} \in N$ with

$$d(T^{n_{r+1}} z_{r+1}, z_r) < \varepsilon_{r+1}.$$

We have that $i < j$ implies

$$d(T^{n_j + n_{j-1} + \ldots + n_{i+1}} z_j, z_i) < \frac{\varepsilon}{2}.$$

Simply by compactness of A, there are i, j with $i < j$ and

$$d(z_i, z_j) < \frac{\varepsilon}{2}.$$

Choosing $n = n_j + n_{j-1} + \ldots + n_i$, we have

$$d(T^n z_j, z_j) < \varepsilon.$$

Proposition 3.12 Under the hypotheses of the preceding Proposition, there is $x \in A$ which is recurrent under T.

Proof: We will use a category argument. For each $n = 1, 2, \ldots$, let

$$E_n = \left\{ x \in A : \inf_k d(T^k x, x) \geq \frac{1}{n} \right\}.$$

170   4. *More about recurrence*

If A contains no points recurrent under T, then

$$A \subset \bigcup_{n=1}^{\infty} E_n.$$

We will show that the interior E_n^0 of each of the (closed) sets E_n is empty, and this will contradict the Baire Category Theorem.

If $E_n^0 \neq \emptyset$ for some n, then since (A, G) is minimal we have $A \subset GE_n^0$ and, by compactness,

$$A \subset g_1^{-1} E_n^0 \cup \ldots \cup g_r^{-1} E_n^0$$

for some $g_1, \ldots, g_r \in G$.

Choose $\delta > 0$ such that $d(x, x') < \delta$ implies $d(g_i x, g_i x') < 1/n$ for all i. We claim that if $x \in g_j^{-1} E_n^0$, then $\inf_k d(T^k x, x) \geqslant \delta$. This is so because if there is a k such that $d(T^k x, x) < \delta$, then

$$d(T^k g_j x, g_j x) < \frac{1}{n} \quad \text{for all } j,$$

or

$$d(T^k y, y) < \frac{1}{n} \quad \text{for some } y \in E_n^0.$$

Since each $x \in A$ is in $g_j^{-1} E_n^0$ for some j, we have proved that

$$\inf_k d(T^k x, x) \geqslant \delta \quad \text{for all } x \in A.$$

This conclusion, however, contradicts the preceding Proposition.

Proof 3.13 *Proof of the Furstenberg–Weiss Theorem* Recall that $T_1, \ldots, T_k$ are commuting homeomorphisms of the compact metric space X, and we are to find $x \in X$ such that for each $\varepsilon > 0$ there is $n \geqslant 1$ with $d(T_i^n x, x) < \varepsilon$ for all $i = 1, \ldots, k$. (Notice that in Proposition 3.11, the z could change with ε.)

The case $k = 1$ is just Birkhoff's Theorem. We proceed by induction. Suppose then that the Theorem holds for any $k - 1$ commuting transformations.

Let G be the group generated by $T_1, \ldots, T_k$. By restricting to a minimal set if necessary, we may assume without loss of generality that (X, G) is minimal.

Let $\Delta \subset X^k$ be the diagonal, and let $T = T_1 \times \ldots \times T_k$. The elements g of G act on X^k by $g(x_1, \ldots, x_k) = (gx_1, \ldots, gx_k)$; i.e., g corresponds to $g \times g \times \ldots \times g$. Clearly these maps commute with T, and (Δ, G) (being isomorphic to (X, G)) is minimal; therefore Δ is homogeneous in (X^k, T).

4.3. The Szemerédi Theorem

We verify the common hypothesis of the two preceding Propositions. We need to show that for each $\varepsilon > 0$ there are $x^*, y^* \in \Delta$ and $n \in \mathbb{N}$ such that $d(T^n x^*, y^*) < \varepsilon$. Here we use the induction hypothesis. Let
$$R_i = T_i T_k^{-1} \quad \text{for } i = 1, \ldots, k-1,$$
and find $x \in X$ and $n_m \to \infty$ such that
$$R_i^{n_m} x \to x \quad \text{for } i = 1, \ldots, k-1.$$

Suppose $\varepsilon > 0$, and let
$$y^* = (x, x, \ldots, x) \text{ and } x^* = (T_k^{-n_m} x, T_k^{-n_m} x, \ldots, T_k^{-n_m} x).$$
Then
$$\begin{aligned} d(T^{n_m} x^*, y^*) &= d(T_1^{n_m} \times T_2^{n_m} \times \ldots \times T_k^{n_m} x^*, y^*) \\ &= d((T_1^{n_m} T_k^{-n_m} x, \ldots, T_{k-1}^{n_m} T_k^{-n_m} x, x), (x, x, \ldots, x)) \\ &= d((R_1^{n_m} x, \ldots, R_{k-1}^{n_m} x, x), (x, x, \ldots, x)), \end{aligned}$$
which can be made less than ε by an appropriate choice of m.

By Proposition 3.12, then, there is a point $(x, x, \ldots, x) \in \Delta$ which is recurrent under $T = T_1 \times T_2 \times \ldots \times T_k$. This is exactly the conclusion of the Theorem.

Proof 3.14 *Proof of van der Waerden's Theorem* Suppose that we are given a partition
$$\mathbb{N} = C_1 \cup C_2 \cup \ldots \cup C_r.$$
We form the sequence space
$$\Omega = \Lambda^{\mathbb{Z}}$$
on the alphabet $\Lambda = \{1, 2, \ldots, r\}$. Λ has the discrete topology, and Ω has the product topology. Ω may be given the metric $d(\omega_1, \omega_2) = 1/2^k$, where $k = \inf\{|n| : \omega_1(n) \neq \omega_2(n)\}$. $\sigma : \Omega \to \Omega$ is the shift, defined by $\sigma\omega(n) = \omega(n+1)$.

Define $\omega_0 \in \Omega$ by
$$\omega_0(n) = \begin{cases} j & \text{if } n \geq 1 \text{ and } n \in C_j \\ 1 & \text{if } n \leq 0. \end{cases}$$
Let X be the set of limit points of $\{\sigma^n \omega_0 : n \geq 1\}$. Then X is closed and invariant under σ.

Given $k \geq 1$, let $T_i = \sigma^i$ for $i = 1, 2, \ldots, k$. Applying the Furstenberg–Weiss Theorem with $\varepsilon = 1$, we find $x \in X$ and $n \geq 1$ with
$$d(T_i^n x, x) < 1 \quad \text{for all } i = 1, 2, \ldots, k.$$

This implies that $x, \sigma^n x, \sigma^{2n} x, \ldots, \sigma^{kn} x$ all agree in the 0th coordinate, so that
$$x(0) = x(n) = x(2n) = \ldots = x(kn).$$
Now the central $(2kn + 1)$-block of x must appear somewhere in the sequence ω_0, say starting at the mth place. Then
$$\omega_0(m) = \omega_0(m + n) = \omega_0(m + 2n) = \ldots = \omega_0(m + kn).$$
Therefore $C_{\omega_0(m)}$ contains an arithmetic progression of length $k + 1$.

We have shown that for every k there is a j such that C_j contains an arithmetic progression of length k. Of course then there must be a single j such that C_j contains arbitrarily long arithmetic progressions.

Remark 3.15 For this application, the 'ε first' version of the Furstenberg–Weiss Theorem would have sufficed: For each $\varepsilon > 0$, there is $x \in X$ with $d(T_i^n x, x) < \varepsilon$ for all $i = 1, 2, \ldots, k$. The complete stronger version, based on Proposition 3.12, was not needed.

The topological multiple recurrence theorem leads to a higher-dimensional version of van der Waerden's Theorem. We leave the proof as an exercise.

Theorem 3.16 *Grünwald's Theorem* (see Rado 1943) For any finite partition of $\mathbb{N}^r$,
$$\mathbb{N}^r = C_1 \cup C_2 \ldots \cup C_m,$$
and any $K = 1, 2, \ldots$, there are C_j, $d \in \mathbb{N}$, and $b \in \mathbb{N}^r$ such that
$$b + d(k_1, k_2, \ldots, k_m) \in C_j \quad \text{for } 1 \leq k_i \leq K, 1 \leq i \leq m.$$
Thus for any finite configuration $F \subset \mathbb{N}^r$ some C_j contains a figure *similar* to F (i.e., a translate of a dilate of F).

Graham and Rothschild (1971) conjectured that van der Waerden's Theorem could be extended to apply also to certain infinite configurations, and this was proved by N. Hindman.

Theorem 3.17 *Hindman's Theorem* (1974) If
$$\mathbb{N} = C_1 \cup C_2 \cup \ldots \cup C_r \quad \text{(disjoint)},$$
then there is a $j = 1, \ldots, r$ such that C_j contains a sequence $p_1, p_2, \ldots$ for which all finite sums $p_{i_1} + p_{i_2} + \ldots + p_{i_n}$ ($i_1 < i_2 < \ldots < i_n, n = 1, 2, \ldots$) also belong to C_j.

A subset $C \subset \mathbb{N}$ having this property i.e., containing a sequence p_1,

4.3. The Szemerédi Theorem

$p_2, \ldots$ for which

$$C = \{p_{i_1} + \ldots + p_{i_n} : i_1 < i_2 < \ldots < i_n, n = 1, 2, \ldots\},$$

is called an *IP-sequence*. The terminology arises from the fact that the set of all such sums forms an 'infinite-dimensional parallelepiped'

$$\{0, p_1\} \cup \{p_2, p_1 + p_2\} \cup \{p_3, p_1 + p_3, p_2 + p_3, p_1 + p_2 + p_3\} \cup \ldots$$

(each set is a translate of the union of the preceding ones). Thus Hindman's Theorem says that for any partition of the positive integers, at least one of the partitioning sets contains an infinite-dimensional parallelepiped. Earlier, Hilbert (1892) had proved that for any partition of the positive integers and any $n \geq 1$, at least one of the partitioning sets has to contain infinitely many translates of a parallelepiped of dimension n (i.e. one for which the above union terminates after n sets).

Furstenberg and Weiss showed how to base the proof of Hindman's Theorem on the notion of *proximality* in topological dynamics. Recall that if X is a compact metric space and $T: X \to X$ is a continuous map, then two points $x_1, x_2 \in X$ are said to be *proximal* in the system (X, T) if there is a sequence $\{n_k\}$ of positive integers with $d(T^{n_k} x_1, T^{n_k} x_2) \to 0$. (If T were a homeomorphism, this would be the definition of *positive* proximality). Recall also that the *Ellis semigroup* $E(X, T)$ is the closure in the product topology of $\{T^n : n > 0\}$ in X^X.

Proposition 3.18 Let X be a compact metric space, $T: X \to X$ a continuous map, $x_0 \in X$, $L(x_0)$ the set of all limit points of $\{T^n x_0 : n \geq 0\}$, and $M \subset L(x_0)$ a minimal set. Then there is a point $m_0 \in M$ such that x_0 and m_0 are proximal.

Proof: We have $E(X, T) x_0 = L(x_0)$. Thus if

$$F = \{p \in E : p x_0 \in M\},$$

then $F x_0 = M$, F is closed, and $F^2 \subset EF \subset F$. By Proposition 2.16 F contains an idempotent u. Then $u x_0 \in M$ and is proximal to x_0.

Proof 3.19 **Proof of Hindman's Theorem** Given a partition $\mathbb{N} = C_1 \cup C_2 \cup \ldots \cup C_r$, again form the (compact metric) sequence space $\Omega = \Lambda^{\mathbb{Z}}$ on the alphabet $\Lambda = \{1, 2, \ldots, r\}$, and define

$$\omega_0(n) = \begin{cases} j & \text{if } n \geq 1 \text{ and } n \in C_j \\ 1 & \text{if } n \leq 0. \end{cases}$$

Let L be the set of limit points of $\{\sigma^n\omega_0 : n \geq 0\}$. Choose a minimal set $M \subset L$ and a point $m_0 \in M$ which is proximal to ω_0.

Recall that two points ω_1 and ω_2 of Ω are close together if they agree on a long central block: $\omega_1(j) = \omega_2(j)$ for $|j| \leq k$ (some large k). Also, by Gottschalk's theorem (Theorem 1.2) m_0 is almost periodic; in this context, this means that each block in m_0 reappears with bounded gap as one moves out along the sequence m_0. Let $m_0(0) = j_0$; we will show that

$$C_{j_0} = \{n \geq 1 : \omega_0(n) = j_0\}$$

contains an IP-sequence $\{p_1, p_2, \ldots\}$. This is possible because the symbol j_0 recurs in the sequence m_0 with bounded gap, while the sequence ω_0 agrees with m_0 on arbitrarily long blocks; thus the occurrences of j_0 in ω_0 can be controlled.

More precisely, we may define the sequence $\{p_1, p_2, \ldots\}$ inductively as follows. Let B_0 be the block j_0, and find $p_1 \geq 1$ such that B_0 appears at the p_1th place in both ω_0 and m_0:

$$\omega_0(p_1) = m_0(p_1) = B_0 = j_0.$$

Then define $B_1 = m_0(0)m_0(1)\ldots m_0(p_1)$. Since any sufficiently long block in m_0 contains B_1, there is $p_2 \geq 1$ such that B_1 appears at the p_2th place in both ω_0 and m_0:

$$\omega_0(p_2)\omega_0(p_2+1)\ldots\omega_0(p_2+p_1) = m_0(p_2)m_0(p_2+1)\ldots m_0(p_2+p_1)$$
$$= B_1 = m_0(0)m_0(1)\ldots m_0(p_1).$$

In particular,

$$\omega_0(p_2+p_1) = m_0(p_2+p_1) = m_0(p_1) = j_0,$$
$$\omega_0(p_2) = m_0(p_2) = m_0(0) = j_0,$$
$$\omega_0(p_1) = m_0(p_1) = j_0,$$

so that $p_1, p_2, p_1 + p_2 \in C_{j_0}$.

If $p_1, p_2, \ldots, p_n$ have been selected with

$$\omega_0(p_{i_1} + p_{i_2} + \ldots + p_{i_r}) = m_0(p_{i_1} + p_{i_2} + \ldots + p_{i_r}) = j_0$$

for any choice of $1 \leq i_1 < i_2 < \ldots < i_r \leq n$,

let

$$B_n = m_0(0)m_0(1)\ldots m_0(p_1 + p_2 + \ldots + p_n)$$

and find a $p_{n+1} > 0$ such that B_n appears at the p_{n+1}th place in both ω_0 and m_0. Then

$$\omega_0(p_{n+1})\omega_0(p_{n+1}+1)\ldots\omega_0(p_{n+1}+(p_1+\ldots+p_n))$$
$$= m_0(p_{n+1})m_0(p_{n+1}+1)\ldots m_0(p_{n+1}+(p_1+\ldots+p_n))$$
$$= m_0(0)m_0(1)\ldots m_0(p_1+\ldots+p_n).$$

4.3. The Szemerédi Theorem

This implies that

$$p_{i_1} + p_{i_2} + \ldots + p_{i_r} \in C_{j_0} \text{ for any choice of}$$
$$1 \leq i_1 < i_2 < \ldots < i_r \leq n+1.$$

In Section 4.4 this result will be used in the proof of the Jewett–Krieger Theorem on the representation of arbitrary ergodic transformations by uniquely ergodic subsystems of (Ω, σ).

C. Weak mixing implies weak mixing of all orders along multiples

We will prove in this section that if $T: X \to X$ is a weakly mixing m.p.t. on a probability space $(X, \mathcal{B}, \mu)$, and $A_1, A_2, \ldots, A_k \in \mathcal{B}$, then

$$\lim_{N-M \to \infty} \frac{1}{N-M} \sum_{n=M+1}^{N} |\mu(A_1 \cap T^n A_2 \cap \ldots \cap T^{(k-1)n} A_k) - \mu(A_1)\mu(A_2) \ldots \mu(A_k)| = 0.$$

Recall that an ergodic m.p.t. $T: X \to X$ is weakly mixing if and only if $T \times S$ is ergodic on $X \times Y$ for each ergodic $(Y, \mathcal{C}, \nu, S)$. This has the following immediate consequence.

Proposition 3.20 If $T: X \to X$ is weakly mixing, then $T_k = T \times T^2 \times \ldots \times T^k$ is ergodic on X^k for each $k \geq 1$.

Proof: Each (X, T^j) is weakly mixing. (Consider the characterization of weak mixing involving sets $J \subset \mathbb{Z}^+$ of density 0.) Since (X, T) is ergodic and (X, T^2) is weakly mixing, $(X \times X, T \times T^2)$ is ergodic. Since (X, T^3) is weakly mixing $(X \times X \times X, T \times T^2 \times T^3)$ is ergodic. We may continue in this manner for as many steps as needed, thereby establishing the result for all k.

At the heart of the ergodic Szemerédi Theorem for weakly mixing transformations is a relationship between measures that generalizes that between point measures and the given invariant ergodic measure μ: averages of translates of one approximate the other.

Definition 3.21 Let $T: X \to X$ be an ergodic m.p.t. on a finite measure space $(X, \mathcal{B}, \mu)$ and let ν be another measure on $(X, \mathcal{B})$. Let $\mathcal{A} \subset L^\infty(X, \mathcal{B}, \mu)$ be a T-invariant (i.e. $T\mathcal{A} = \mathcal{A}$) self-conjugate algebra of bounded measurable functions on X, and let $\{M_k, N_k\}$ be a sequence of pairs of integers with $N_k - M_k \to \infty$. We say that ν *is generic for* μ *with respect to* $\mathcal{A}$ *and* $\{M_k, N_k\}$ if

$$\frac{1}{N_k - M_k} \sum_{n=M_k+1}^{N_k} \int f(T^n x) \, d\nu(x) \to \int f \, d\mu \quad \text{for all } f \in \mathcal{A}.$$

4. More about recurrence

Remarks 3.22 (1) μ is always generic for μ.

(2) If $\mathscr{A}$ is countably generated (with respect to μ), then for almost all x, δ_x (the point mass at x) is generic for μ with respect to $\mathscr{A}$ and $\{0,k\}$. This follows from the Ergodic Theorem.

(3) Let T be ergodic, $X_2 = X \times X$, $\mathscr{B}_2 =$ the product σ-algebra, and $T_2 = T \times T^2$. Then the *diagonal measure* v_2 on $(X, \mathscr{B}_2)$ defined by

$$\int f(x_1, x_2) dv_2(x_1, x_2) = \int f(x, x) d\mu(x) \quad (f \in L^\infty(X_2, \mathscr{B}_2))$$

is generic for $\mu_2 = \mu \times \mu$ with respect to any sequence $\{M_k, N_k\}$ for which $N_k - M_k \to \infty$ and the algebra

$$\mathscr{A}_2 = \left\{ \sum_{i=1}^n f_i(x_1) g_i(x_2) : f_i, g_i \in L^\infty(X, \mathscr{B}, \mu) \right\}.$$

(Functions of the above form will be abbreviated $\sum f_i \otimes g_i$.)

Proof: By the Mean Ergodic Theorem, if $f \in L^2(X, \mathscr{B}, \mu)$, then

$$\frac{1}{N_k - M_k}(Tf + T^2 f + \ldots + T^{N_k - M_k} f) \to \int f \, d\mu \quad \text{in } L^2.$$

Using the invariance of μ, we see that

$$\frac{1}{N_k - M_k} \sum_{j = M_k + 1}^{N_k} T^j f \to \int f \, d\mu \quad \text{in } L^2.$$

(For we may fix a k with the approximation in the first statement better than ε, and then change variables, replacing x by $T^{M_k} x$.)

This implies that if $f, g \in L^\infty$, then

$$\left(g, \frac{1}{N_k - M_k} \sum_{j = M_k + 1}^{N_k} T^j f \right) \to \left(g, \int f \, d\mu \right),$$

and hence

$$\frac{1}{N_k - M_k} \sum_{j = M_k + 1}^{N_k} \int g \cdot T^j f \, d\mu \to \int f \, d\mu \cdot \int g \, d\mu.$$

Changing variables in each integral gives

$$\frac{1}{N_k - M_k} \sum_{n = M_k + 1}^{N_k} \int g(T^n x) f(T^{2n} x) \, d\mu \to \int f \, d\mu \cdot \int g \, d\mu,$$

or

$$\frac{1}{N_k - M_k} \sum_{n = M_k + 1}^{N_k} \int T_2^n(g \otimes f) \, dv_2 \to \int g \otimes f \, d\mu_2.$$

4.3. The Szemerédi Theorem

We want to extend Remark 3.22 (3) from the case of $(X_2, \mathscr{B}_2, T_2)$ to $(X_k, \mathscr{B}_k, T_k)$ ($T_k = T \times T^2 \times \ldots \times T^k$) for any $k \geq 2$, in case T is weakly mixing; weak mixing of all orders along multiples will then follow easily. The argument is similar to the one just given for the case $k = 2$, and it depends on an extension of the Mean Ergodic Theorem.

Theorem 3.23 If $(X, \mathscr{B}, \mu, T)$ is ergodic and v is generic for μ with respect to $\{M_k, N_k\}$ and an algebra $\mathscr{A}$, then for each $f \in \mathscr{A}$,

$$\frac{1}{N_k - M_k} \sum_{n=M_k+1}^{N_k} T^n f \to \int f \, d\mu \quad \text{in } L^2(X, \mathscr{B}, v).$$

(In case $v = \mu$, this is the ordinary Mean Ergodic Theorem).

Proof: Let $f \in \mathscr{A}$; we assume without loss of generality that $\int f \, d\mu = 0$. Given $\varepsilon > 0$, choose Q so large that

$$\int \left| \frac{Tf + T^2 f + \ldots + T^Q f}{Q} \right|^2 d\mu < \varepsilon.$$

Since

$$g = \left| \frac{Tf + T^2 f + \ldots + T^Q f}{Q} \right|^2 \in \mathscr{A},$$

by applying the definition of 'generic', we see that for large k

$$\frac{1}{N_k - M_k} \sum_{n=M_k+1}^{N_k} \int \left| \frac{T^{n+1} f + \ldots + T^{n+Q} f}{Q} \right|^2 dv < \varepsilon.$$

By the Schwarz Inequality ($|\sum_{k=1}^{n}(a_k/n)| \leq \sqrt{\sum_{k=1}^{n} a_k^2} \sqrt{1/n}$), it follows that also

$$I_{Q,k} = \int \left| \frac{1}{N_k - M_k} \sum_{n=M_k+1}^{N_k} \frac{T^{n+1} f + \ldots + T^{n+Q} f}{Q} \right|^2 dv$$

$$= \int |A_{Q,k}|^2 dv < \varepsilon.$$

But

$$A_{Q,k} = \frac{1}{N_k - M_k} \sum_{n=M_k+1}^{N_k} \frac{T^{n+1} f + \ldots + T^{n+Q} f}{Q}$$

is not far from

$$\frac{1}{N_k - M_k} \sum_{n=M_k+1}^{N_k} T^n f;$$

in fact, the difference is (sluffing the k subscripts for the moment)

$$\frac{1}{N-M} \sum_{n=M+1}^{N} \sum_{j=1}^{Q} \frac{T^{n+j}f - T^n f}{Q}$$

$$= \frac{1}{Q} \sum_{j=1}^{Q} \frac{1}{N-M} \sum_{n=M+1}^{N} (T^{n+j}f - T^n f)$$

$$= \frac{1}{Q} \sum_{j=1}^{Q} \frac{1}{N-M} \left(\sum_{n=N-j+1}^{N} T^{n+j}f - \sum_{n=M+1}^{M+j} T^n f \right),$$

which has $L^2(v)$ norm less than or equal to

$$\frac{2Q\|f\|_\infty}{N_k - M_k},$$

and this tends to 0 as $k \to \infty$. We are able to conclude, then, that

$$\limsup_{k\to\infty} \int \left| \frac{1}{N_k - M_k} \sum_{n=M_k+1}^{N_k} T^n f \right|^2 dv \leq \varepsilon,$$

as required.

Theorem 3.24 If $(X, \mathcal{B}, \mu, T)$ is weakly mixing, then for each $k \geq 2$ the *diagonal measure* v_k on $(X_k, \mathcal{B}_k)$ ($X_k = X^k$, $\mathcal{B}_k = k$-fold product σ-algebra of $\mathcal{B}$) defined by

$$\int f(x_1, x_2, \ldots, x_k) dv_k = \int f(x, x, \ldots, x) d\mu(x) \quad (f \in L^\infty(X_k, \mathcal{B}_k))$$

is generic for $\mu_k = \mu \times \mu \times \ldots \times \mu$ with respect to $T_k = T \times T^2 \times \ldots \times T^k$, any sequence $\{M_k, N_k\}$ with $N_k - M_k \to \infty$, and the algebra $\mathscr{A}_k$ of all functions of the form

$$\sum_{j=1}^{J} f_1^j \otimes \ldots \otimes f_k^j(x_1, \ldots x_k) = \sum_{j=1}^{J} f_1^j(x_1) f_2^j(x_2) \ldots f_k^j(x_k)$$

(each $f_i^j \in L^\infty(X)$).

That is,

$$\frac{1}{N} \sum_{n=1}^{N} \int f_1(x) f_2(T^n x) \ldots f_k(T^{(k-1)n} x) d\mu(x) \to \int f_1 d\mu \ldots \int f_k d\mu$$

for $f_1, \ldots, f_k \in L^\infty(X)$.

Proof: The proof is by induction on k. By Remark 3.22 (3), the statement holds for $k = 2$. Assume now that the Theorem is true for k, and we will prove it for $k+1$. Since $(X_k, \mathcal{B}_k, \mu_k, T_k)$ is ergodic (Proposition 3.20), Theorem 3.23 implies that, given $f_i \in L^\infty(X)$, $i = 1, \ldots, k$,

$$\frac{1}{N_k - M_k} \sum_{n=M_k+1}^{N_k} T_k^n (f_1 \otimes f_2 \ldots \otimes f_k) \to \int f_1 \otimes \ldots \otimes f_k d\mu_k$$

in $L^2(v_k)$.

4.3. The Szemerédi Theorem

This says that

$$\frac{1}{N_k - M_k} \sum_{n=M_k+1}^{N_k} T^n f_1 \otimes T^{2n} f_2 \otimes \ldots \otimes T^{kn} f_k \to \prod_{i=1}^{k} \int f_i \, d\mu$$

in $L^2(\nu_k)$,

or (in light of the definition of ν_k),

$$\frac{1}{N_k - M_k} \sum_{n=M_k+1}^{N_k} f_1(T^n x) f_2(T^{2n} x) \ldots f_k(T^{kn} x) \to \prod_{i=1}^{k} \int f_i \, d\mu$$

in $L^2(\mu)$.

Multiply by $f_0(x)$ and integrate to find that

$$\frac{1}{N_k - M_k} \sum_{n=M_k+1}^{N_k} \int f_0(x) f_1(T^n x) \ldots f_k(T^{kn} x) \, d\mu \to \prod_{i=0}^{k} \int f_i \, d\mu.$$

Changing variables in each integral ($x \to T^n x$) gives

$$\frac{1}{N_k - M_k} \sum_{n=M_k+1}^{N_k} \int f_0(T^n x) f_1(T^{2n} x) \ldots f_k(T^{(k+1)n} x) \, d\mu \to \prod_{i=0}^{k} \int f_i \, d\mu,$$

i.e.

$$\frac{1}{N_k - M_k} \sum_{n=M_k+1}^{N_k} \int T_{k+1}^n (f_0 \otimes f_1 \otimes \ldots \otimes f_k) \, d\nu_{k+1} \to$$

$$\int f_0 \otimes f_1 \otimes \ldots \otimes f_k \, d\mu_{k+1}.$$

Corollary 3.25 If $(X, \mathscr{B}, \mu, T)$ is weakly mixing and $A_1, A_2, \ldots, A_k \in \mathscr{B}$, then

$$\lim_{N-M \to \infty} \frac{1}{N-M} \sum_{n=M+1}^{N} \mu(A_1 \cap T^n A_2 \cap \ldots \cap T^{(k-1)n} A_k)$$
$$= \mu(A_1) \mu(A_2) \ldots \mu(A_k).$$

Proof: Apply the Theorem to $f_i = \chi_{A_i}, i = 1, \ldots, k$.

Corollary 3.26 If $(X, \mathscr{B}, \mu, T)$ is weakly mixing and $A_1, A_2, \ldots, A_k \in \mathscr{B}$, then

$$\lim_{N-M \to \infty} \frac{1}{N-M} \sum_{n=M+1}^{N} |\mu(A_1 \cap T^n A_2 \cap \ldots \cap T^{(k-1)n} A_k)$$
$$- \mu(A_1) \mu(A_2) \ldots \mu(A_k)| = 0.$$

Proof: Apply Corollary 3.25 to the weakly mixing transformation $T \times T$ on $X \times X$ and the sets $A_1 \times A_1, \ldots, A_k \times A_k$; we obtain

$$\frac{1}{N-M} \sum_{n=M+1}^{N} \mu \times \mu [(A_1 \times A_1) \cap (T \times T)^n (A_2 \times A_2) \cap \ldots$$
$$\cap (T \times T)^{(k-1)n}(A_k \times A_k)]$$
$$\to (\mu \times \mu)(A_1 \times A_1) \ldots (\mu \times \mu)(A_k \times A_k),$$

i.e.
$$\frac{1}{N-M}\sum_{n=M+1}^{N}[\mu(A_1\cap T^nA_2\cap\ldots\cap T^{(k-1)n}A_k)]^2$$
$$\to [\mu(A_1)\mu(A_2)\ldots\mu(A_k)]^2.$$

If we let
$$a_n = \mu(A_1\cap T^nA_2\cap\ldots\cap T^{(k-1)n}A_k)$$
and
$$\alpha = \mu(A_1)\mu(A_2)\ldots\mu(A_k),$$
then
$$\frac{1}{N-M}\sum_{n=M+1}^{N}(a_n-\alpha)^2$$
$$=\frac{1}{N-M}\sum_{n=M+1}^{N}a_n^2 - 2\alpha\frac{1}{N-M}\sum_{n=M+1}^{N}a_n + \alpha^2 \to 0,$$

since the first term tends to α^2 and the second to $-2\alpha^2$. By the Schwarz Inequality, then also
$$\frac{1}{N-M}\sum_{n=M+1}^{N}|a_n-\alpha| \leq \sqrt{\sum_{n=M+1}^{N}|a_n-\alpha|^2}\sqrt{\frac{1}{N-M}} \to 0.$$

Remark 3.27 In connection with Corollary 3.25, notice that $(X,\mathcal{B},\mu,T)$ is weakly mixing if and only if
$$\lim_{N\to\infty}\frac{1}{N}\sum_{n=1}^{N}\mu(A\cap T^nB\cap T^{2n}C) = \mu(A)\mu(B)\mu(C) \text{ for all } A,B,C\in\mathcal{B}.$$
For this statement implies, as usual, that
$$\lim_{N\to\infty}\frac{1}{N}\sum_{n=1}^{N}\int f(x)g(T^nx)h(T^{2n}x)\,d\mu$$
$$=\int f\,d\mu\cdot\int g\,d\mu\cdot\int h\,d\mu \quad \text{for all } f,g,h\in L^2(X).$$

But if T is not weakly mixing, then it has a nontrivial eigenfunction ϕ with eigenvalue $\lambda \neq 1$, and taking $f=\phi, g=\phi^{-2}, h=\phi$ would give
$$f(x)g(T^nx)h(T^{2n}) = \phi(x)\lambda^{-2n}\overline{\phi(x)}^2\lambda^{2n}\phi(x) \equiv 1;$$
on the other hand, $\int f\,d\mu = 0$, so that the condition cannot hold. Of course the statement of Corollary 3.25 for $k=2$ is equivalent to ergodicity.

4.3. The Szemerédi Theorem

D. Outline of the proof of the Furstenberg–Katznelson Theorem

At the other extreme, with respect to dynamical behavior, from the weakly mixing transformations are the group rotations. In this section we will show first that the multiple recurrence result holds for such transformations also. The proof of the full multiple recurrence theorem, which we will then outline, depends on strengthening this result and the one of the previous section so that they will apply to the case of extensions.

Suppose then that X is a compact metric abelian group, $\mathscr{B}$ is the Borel σ-algebra of X, μ is Haar measure, and $Tx = x_0 x$, where $x_0 \in X$ is a point such that $\{x_0^n : n \in \mathbb{Z}\}$ is dense in X. Recall that every ergodic isometry has a model of this kind.

Lemma 3.28 If $f_1, f_2, \ldots, f_k \in L^\infty(X)$, then

$$\psi(x) = \int_X f_1(y) f_2(xy) \ldots f_k(x^{k-1} y) \, d\mu(y)$$

is a continuous function of x.

Proof: Each of the functions $x \to f_i(x^{i-1} y)$ is a continuous function from X into $L^1(X)$. (For example, if $f \in L^1$, given $\varepsilon > 0$, we may choose $\phi \in \mathscr{C}(X)$ with $\|\phi - f\|_1 < \varepsilon$ and, since multiplication is continuous, a $\delta > 0$ such that $d(x, x') < \delta$ implies $|\phi(xy) - \phi(x' y)| < \varepsilon$ for all $y \in X$. Writing $f_x(y) = f(xy)$, we then have

$$\|f_x - f_{x'}\|_1 \leq \|f_x - \phi_x\|_1 + \|\phi_x - \phi_{x'}\|_1 + \|\phi_{x'} - f_{x'}\|_1 < 3\varepsilon$$

if $d(x, x') < \delta$.) Then if $x_j \to x$ we have

$$\int f_1(y) f_2(x_j y) \ldots f_k(x_j^{k-1} y) \, d\mu(y) - \int f_1(y) f_2(xy) \ldots f_k(x^{k-1} y) \, d\mu(y)$$

$$= \int f_1(y) f_2(x_j y) \ldots f_{k-1}(x_j^{k-2} y) [f_k(x_j^{k-1} y) - f_k(x^{k-1} y)] \, d\mu(y)$$

$$+ \int f_1(y) \ldots f_{k-2}(x_j^{k-3} y) [f_{k-1}(x_j^{k-2} y) - f_{k-1}(x^{k-2} y)]$$

$$\times f_k(x^{k-1} y) \, d\mu(y) + \ldots$$

$$+ \int f_1(y) [f_2(x_j y) - f_2(xy)] f_3(x^2 y) \ldots f_k(x^{k-1} y) \, d\mu(y),$$

and, using the boundedness of the f_i, each of these integrals tends to 0 as $j \to \infty$.

4. More about recurrence

Theorem 3.29 If $(X, \mathcal{B}, \mu, T)$ is an ergodic group rotation and $f \in L^\infty(X, \mathcal{B}, \mu)$ is nonnegative but not identically 0, then

$$\lim_{N-M\to\infty} \frac{1}{N-M} \sum_{n=M+1}^{N} \int f(y) f(T^n y) \ldots f(T^{(k-1)n} y) \, d\mu(y)$$

exists and is positive.

Proof: Given such an $f \in L^\infty$, let

$$\psi(x) = \int_X f(y) f(xy) \ldots f(x^{k-1} y) \, d\mu(y)$$

for $x \in X$. Since ψ is continuous, by the Lemma, and T is minimal and uniquely ergodic,

$$\lim_{N-M\to\infty} \frac{1}{N-M} \sum_{n=M+1}^{N} \psi(x_0^n) = \int_X \psi(y) \, d\mu(y).$$

This is, however, just the limit mentioned in the Theorem. Since $\psi \geq 0$, $\psi(e) > 0$, and ψ is continuous, $\int_X \psi \, d\mu > 0$.

Suppose that $(X, \mathcal{B}, \mu)$ is a probability space and Γ is an abelian group which acts on X by m.p.t.s. A function $f \in L^2(X)$ is called an *eigenfunction* in case there is a character ξ of Γ (i.e. a homomorphism $\xi: \Gamma \to \mathbb{C}$) such that

$$f(\gamma x) = \xi(\gamma) f(x) \quad \text{for all } \gamma \text{ and a.a. } x.$$

The system $(X, \mathcal{B}, \mu, \Gamma)$ is called *almost periodic* (or a *Kronecker system*) in case the eigenfunctions span L^2. Theorem 3.29 shows, then, that the ergodic Szemerédi Theorem holds for Kronecker systems $(X, \mathcal{B}, \mu, \mathbb{Z})$.

Remarks 3.30 (1) Every measure-preserving system $(X, \mathcal{B}, \mu, \Gamma)$ has a maximal almost periodic factor.

(2) If Γ is the group generated by k commuting m.p.t.s $T_1, T_2, \ldots, T_k$ on X and $(X, \mathcal{B}, \mu, \Gamma)$ is almost periodic, then for each $0 \leq f \in L^\infty, f \not\equiv 0$,

$$\lim_{N-M\to\infty} \frac{1}{N-M} \sum_{n=M+1}^{N} \int f(T_1^n y) f(T_2^n y) \ldots f(T_k^n y) \, d\mu(y) > 0.$$

We will now consider the relativized versions of ergodicity, weak mixing and almost periodicity. Let Γ be a fixed countable abelian group which acts by m.p.t.s on the measure spaces mentioned in the following discussion. If $(X, \mathcal{B}, \mu)$ and $(Y, \mathcal{C}, \nu)$ are probability spaces and $\pi: X \to Y$ is a measurable onto map which commutes with the action of each $T \in \Gamma$ and for which $\mu \pi^{-1} = \nu$, then we write $X \xrightarrow{\pi} Y$ and say that Y is a *factor* of X, or X is an *extension* of Y.

4.3. The Szemerédi Theorem

The measure μ on X has a *disintegration* in terms of *fiber measures* indexed by the points of Y. We always deal with Lebesgue spaces, so let us assume that $X = Y = [0, 1)$, and $\pi:(X, \mathscr{B}, \mu) \to (Y, \mathscr{C}, \nu)$ is an extension. Let $\mathscr{B}_Y = \pi^{-1}\mathscr{C}$. Let $\{\phi_n\}$ be a countable dense set in $\mathscr{C}(X)$. For each n choose a version of the conditional expectation $E(\phi_n | \mathscr{B}_Y)$. For almost all $x \in X$, the map

$$\lambda_x(\phi_n) = E(\phi_n | \mathscr{B}_Y)(x)$$

is uniformly continuous on $\{\phi_n\}$, and hence λ_x extends to all of $\mathscr{C}(X)$. The extension is linear and positive with $\lambda_x(1) = 1$, so there is a positive Borel measure $\mu_{\pi x}$ (depending only on $y = \pi x$) such that $\lambda_x(\phi) = \int \phi \, d\mu_{\pi x}$ for all $\phi \in \mathscr{C}(X)$. Then

$$\int f \, d\mu = \int_Y \int_X f(x) \, d\mu_y(x) \, d\nu(y)$$

for all bounded Borel functions f on X, so we write

$$\mu = \int \mu_y \, d\nu(y).$$

An extension $X \xrightarrow{\pi} Y$ is called *relatively ergodic with respect to* $T \in \Gamma$ if every T-invariant L^2 function on X is a.e. a function on Y, in that there is $f_0 \in L^2(Y)$ with $f = f_0 \pi$ a.e.

The relativized version of the cross product is the *fiber product* $X \times_Y X'$, which is defined as follows. If $X \xrightarrow{\pi} Y$ and $X' \xrightarrow{\pi'} Y$ are extensions of Y, let $X \times_Y X' = \{(x, x'): \pi x = \pi' x'\}$. The σ-algebra, $\tilde{\mathscr{B}}$, of $X \times_Y X'$ is the one it inherits by restriction from $X \times X'$, and its measure, $\tilde{\mu}$, is the one determined by the fiber measure $\tilde{\mu}_y = \mu_y \times \mu'_y$:

$$\int f(x, x') \, d\tilde{\mu}(x, x') = \int \int \int f(x, x') \, d\mu_y(x) \, d\mu'_y(x') \, d\nu(y).$$

Now we can say that an extension $X \xrightarrow{\pi} Y$ is *relatively weakly mixing with respect to* $T \in \Gamma$ if $X \times_Y X \xrightarrow{\pi} Y$ is a relatively ergodic extension for T.

The extension of the definition of almost periodicity is a little more complicated.

It can be proved that the following two properties of an extension $X \xrightarrow{\pi} Y$, for a subgroup $\Lambda \subset \Gamma$ generated by $T_1, \ldots, T_k$, are equivalent:

(1) There is a dense set $\mathscr{D} \subset L^2(X, \mathscr{B}, \mu)$ consisting of functions whose Λ-orbits have finite rank over Y, in that for each $f \in \mathscr{D}$ and $\varepsilon > 0$, there is a finite set $g_1, \ldots, g_n \in L^2(X, \mathscr{B}, \mu)$ such that for each $T \in \Lambda$ there is a j with $\|Tf - g_j\|_{L^2(X, \mathscr{B}, \mu_y)} < \varepsilon$ for a.e. $y \in Y$.

(2) The autocorrelation

$$\lim_{n\to\infty} \frac{1}{(2n+1)^k} \sum_{|i_1,i_2,\ldots,i_k|\leq n} f(T_1^{i_1}T_2^{i_2}\cdots T_k^{i_k}x_1)\overline{f(T_1^{i_1}T_2^{i_2}\cdots T_k^{i_k}x_2)}$$

of a function $f \in L^2(X, \mathcal{B}, \mu)$ is 0 a.e. in $L^2(X \times_Y X, \tilde{\mathcal{B}}, \tilde{\mu})$ if and only if $f = 0$ a.e. in X.

An extension $X \xrightarrow{\pi} Y$ having one of these properties will be called *relatively compact*. (1) and (2) are relativized versions of almost periodicity: (1) corresponds to the fact that the eigenfunctions of an almost periodic system span L^2 (see also the remarks at the end of Section 4.1.C), and (2) corresponds to a consequence of the recurrence property of almost periodic functions. Other characterizations of relative compactness analogous to properties of almost periodic systems, for example in terms of vector-valued eigenfunctions with unitary-matrix eigenvalues, are also possible.

Remark 3.31 If $X \xrightarrow{\pi} Y$ is a relatively compact extension for each of $\Lambda_1, \Lambda_2 \subset \Gamma$, then it is relatively compact for $\Lambda_1 \Lambda_2$.

Now we extend the characterization of weak mixing $T \times \mathcal{E} \subset \mathcal{E}$ and Theorem 3.24 to the case of relatively weakly mixing extensions. Note that in 3.24 and 3.33, the presence or absence of the exponent 2 is immaterial – by increasing k, as in the proof of Corollary 3.26, it can be added when desired.

Proposition 3.32 If $X \xrightarrow{\pi} Y$ is relatively weakly mixing and $X' \xrightarrow{\pi'} Y$ is relatively ergodic, then $X \times_Y X' \to Y$ is relatively ergodic.

Theorem 3.33 Let $(X, \mathcal{B}, \mu) \xrightarrow{\pi} (Y, \mathcal{C}, \nu)$ be a relatively weakly mixing extension, for each $T \in \Gamma, T \neq 1$. If $f_1, f_2, \ldots, f_k \in L^\infty(X, \mathcal{B}, \mu)$ and $T_1, \ldots, T_k$ are distinct elements of Γ, then

$$\lim_{N\to\infty} \frac{1}{N} \sum_{n=1}^{N} \int \left[E\left(\prod_{j=1}^{k} T_j^n f_j \Big| \mathcal{B}_Y\right) - \prod_{j=1}^{k} T_j^n E(f_j|\mathcal{B}_Y) \right]^2 d\mu = 0.$$

The next theorem will imply that any system $(X, \mathcal{B}, \mu, T)$ can be resolved into relatively weakly mixing and relatively compact extensions, beginning with a one-point space.

Theorem 3.34 If a proper extension $(X, \mathcal{B}, \mu) \xrightarrow{\pi} (Y, \mathcal{C}, \nu)$ is not relatively weakly mixing for $T \in \Gamma$, then there is a factor $(Z, \mathcal{D}, \tau)$ of $(X, \mathcal{B}, \mu)$

4.3. The Szemerédi Theorem

which is a nontrivial relatively compact extension of $(Y, \mathscr{C}, v)$ for the group generated by T.

Using Remark 3.31, the following resolution can now be established.

Theorem 3.35 Let Γ be finitely generated and $(Y, \mathscr{C}, v)$ a proper factor of $(X, \mathscr{B}, \mu)$. Then there are (1) a factor $(Z, \mathscr{D}, \tau)$ of $(X, \mathscr{B}, \mu)$ which is a proper extension of $(Y, \mathscr{C}, v)$ and (2) a decomposition $\Gamma = \Gamma_W \times \Gamma_C$ of Γ into the direct product of two subgroups such that

(i) $(Z, \mathscr{D}, \tau) \to (Y, \mathscr{C}, v)$ is relatively weakly mixing for each $T \neq 1$ in Γ_W, and

(ii) $(Z, \mathscr{D}, \tau) \to (Y, \mathscr{C}, v)$ is relatively compact for Γ_C.

We say that the action of Γ on $(X, \mathscr{B}, \mu)$ is SZ if given $T_1, \ldots, T_k \in \Gamma$, the conclusion of the Furstenberg-Katznelson Theorem holds for $T_1, \ldots, T_k$: given $A \in \mathscr{B}$ with $\mu(A) > 0$, there is $\delta > 0$ such that

$$\mu(T_1^n A \cap T_2^n A \cap \ldots \cap T_k^n A) > \delta$$

on a set of lower density (calculated with respect to the intervals $[1, N]$ as $N \to \infty$) greater than δ. We want to show that the action of $\mathbb{Z}^k$ on a finite measure space is always SZ.

In order to prove the Furstenberg-Katznelson Theorem, one assumes that the commuting m.p.t.s $T_1, \ldots T_k$ on $(X, \mathscr{B}, \mu)$ are given, and that each is an element of $\Gamma \approx \mathbb{Z}^k$. The action of Γ on the maximal almost periodic factor $(X, \mathscr{D}, v)$ of $(X, \mathscr{B}, \mu)$ is SZ, by Remark 3.30(2), above. The aim is to show that the maximal SZ factor of $(X, \mathscr{B}, \mu)$ is $(X, \mathscr{B}, \mu)$ itself by using the resolution of this extension into relatively weakly mixing and relatively compact extensions.

If $\{(Y_\alpha, \mathscr{C}_\alpha, v_\alpha) : \alpha \in A\}$ is a totally-ordered family of factors of $(X, \mathscr{B}, \mu)$, we define their *supremum* to be the factor determined by the closure of the union of the corresponding sub-σ-algebras of $\mathscr{B}$. It can be proved that if the action of Γ on each $(Y_\alpha, \mathscr{C}_\alpha, v_\alpha)$ is SZ, then the action of Γ on $\sup_\alpha (Y_\alpha, \mathscr{C}_\alpha, v_\alpha)$ is SZ. Therefore, by Zorn's Lemma, $(X, \mathscr{B}, \mu)$ has maximal factors on which the action of Γ is SZ. Let us suppose that $(Y, \mathscr{C}, v)$ is such a maximal factor. If the extension $(X, \mathscr{B}, \mu) \to (Y, \mathscr{C}, v)$ is proper, it can be resolved as in Theorem 3.35 into relatively weakly mixing and relatively compact parts. The core of the argument consists, then, in showing that the SZ property lifts through every extension of this particular form. Significantly, the key combinatorial fact required to accomplish this turns out to be Grünwald's Theorem, the multidimensional van der Waerden Theorem (3.16). In this way the delicate multidimensional Szemerédi Theorem is reduced to a much simpler special case, and the quantitative Furstenberg–Katznelson Theorem is reduced to a recurrence result in topological dynamics.

Exercises
1. If $T: X \to X$ is a m.p.t. on $(X, \mathcal{B}, \mu)$ and $\mu(A) > 0$, then for each $r = 1, 2, \ldots$ a.e. point of A is *r-recurrent with respect to* A: if $\rho_r(x, A) = \inf\{n \geq 1 : x, T^n x, T^{2n} x, \ldots, T^{rn} x \in A\}$, then $\rho_r(x, A) < \infty$ for a.e. $x \in A$.

4.4. The topological representation of ergodic transformations

Because much of the motivation for the study of ergodic measure-preserving transformations comes from Liouville's Theorem and Boltzmann's Ergodic Hypothesis, many of the important systems for the study of which ergodic theory was created are physical. They are governed by differential equations, and their evolution is described by diffeomorphisms of manifolds which preserve measures with smooth densities. Poincaré and Birkhoff initiated the study of the purely topological and metric (i.e. measure-theoretic) properties of such maps in order to analyze their most basic properties, and this led to the abstract study of the dynamics of homeomorphisms of compact topological spaces and measure-preserving transformations on measure spaces.

How far are these now vastly-developed subjects from their origins? More precisely, one may ask, for example:

(a) Can every ergodic m.p.t. on a Lebesgue space be realized as (i.e., is it metrically isomorphic to) a minimal, uniquely ergodic homeomorphism of a compact metric space?

(b) Can every ergodic m.p.t. on a Lebesgue space be realized as a diffeomorphism which preserves a smooth measure on a manifold?

The answer to (a) is Yes, as we shall now see. Work on (b) is in progress, and the answer may also be Yes, subject to certain known restrictions (e.g. finite entropy).

Robert I. Jewett (1970) gave the then astonishing positive answer to (a) for weakly mixing transformations. Before then substantial effort had gone into the construction of uniquely ergodic systems which had positive entropy (Furstenberg 1967, Hahn and Katznelson 1967), were weakly or

4.4. Topological representation

strongly mixing (Jacobs 1970b, Kakutani 1973, Petersen 1970a etc.). The existence of such systems is now, of course, a direct corollary of Jewett's Theorem. Subsequently, it remained for some time an interesting problem to give explicit constructions of uniquely ergodic K-systems and Bernoulli systems. This was finally accomplished by Grillenberger (1973a, b) and Grillenberger and Shields (1975).

Krieger (1972) established the positive answer to (a) for ergodic transformations, using his theorem on the existence of generators and the theory of entropy. Other proofs, improvements, and extensions were due to Hansel (1974), Hansel and Raoult (1973), Jacobs (1970a), Denker (1973), and Denker and Eberlein (1974).

In realizing an ergodic m.p.t. T topologically, the topological space may always be taken to be the Cantor set. If the entropy of the transformation is finite, the space may be taken to be a closed shift-invariant uniquely ergodic (even strictly ergodic – that is, uniquely ergodic and minimal) subset of the space of all bilateral sequences on N symbols, where N is the smallest integer larger than $2^{h(T)}$ ($h(T)$ is the entropy of T), and the transformation may be taken to be the shift. (We must have the entropy of T strictly less than the entropy of the full shift on N symbols.)

When one is working with an ergodic m.p.t. T, it may be convenient to assume, as this Theorem allows us to do, that T is a strictly ergodic subshift. It is hard to see, though, that such a supposition, comforting though it may be, can have any real usefulness. The Jewett–Krieger Theorem is much like the Isomorphism Theorem for Lebesgue spaces: its theoretical importance is tremendous, but its practical effects are essentially nil. Indeed, the theorem itself says that this must be so, since it guarantees that *the property of being realizable as a uniquely ergodic homeomorphism tells us nothing at all about a given m.p.t.* Still, just as some measure-theoretic constructions are more easily described on the unit interval, so some dynamics may more conveniently be done in a topological setting, even in a subshift.

A. Bellow and H. Furstenberg (1979) extended Jewett's argument from the weakly mixing case to the ergodic case. They noticed that weak mixing was used at exactly one point, and that the proof could be carried past this point in the ergodic case by using Hindman's Theorem (which, symmetrically Furstenberg and Weiss had proved by dynamical methods in 1978 – see Section 4.3.B). There is no need to bring in the machineries of entropy and generators; the argument, close and clever as it is, uses only combinatorics, basic ergodic theory, and soft analysis. It is this argument that we will present here.

188   4. *More about recurrence*

Theorem (*Jewett–Krieger*) Let $T: X \to X$ be an ergodic m.p.t. on a Lebesgue space $(X, \mathscr{B}, \mu)$. Then T is metrically isomorphic to a strictly ergodic homeomorphism S of the Cantor set C.

A. Preliminaries

First we reduce the problem as far as possible by applying soft analysis and also record a few observations which will be useful later.

Remark 4.1 *A functional–analytic condition for isomorphism.* We may assume that $(X, \mathscr{B}, \mu)$ is the ordinary measure space of the unit interval $[0, 1)$. According to the theorems of von Neumann (1.4.6 and 1.4.7), in the case of m.p.t.s on complete, separable metric spaces, every (set) isomorphism between the maps of the measure algebras arises from a (point) isomorphism between the m.p.t.s. Such a set isomorphism between T and S can be obtained from a Banach algebra isomorphism

$$\phi: L^\infty(X, \mathscr{B}, \mu) \xrightarrow{\text{onto}} L^\infty(C, \mathscr{F}, \nu),$$

such that

$$\phi(fT) = \phi(f)S \quad \text{and} \quad \int \phi(f) \, d\nu = \int f \, d\mu \quad \text{for all } f \in L^\infty(X, \mathscr{B}, \mu).$$

(Here $\mathscr{F}$ is the Borel field of C and ν is the unique S-invariant Borel probability measure on C.) Of course we put $\tilde{\phi}(E) = \phi(\chi_E)$ for $E \in \mathscr{B}$ to define the set isomorphism. Then clearly

$$\tilde{\phi}(TE) = S\tilde{\phi}(E), \quad \nu(\tilde{\phi}E) = \mu(E), \quad \tilde{\phi}(E^c) = \tilde{\phi}(E)^c,$$

and

$$\tilde{\phi}(A \cup B) = \phi(\chi_{A \cup B}) = \phi(\chi_A + \chi_B - \chi_A \chi_B) = \tilde{\phi}(A) \cup \tilde{\phi}(B).$$

For countable disjoint unions, we have

$$\tilde{\phi} \bigcup_{n=1}^\infty E_n = \phi \sum_{n=1}^\infty \chi_{E_n} = \sum_{n=1}^N \phi(\chi_{E_n}) + \phi\left(\sum_{n=N+1}^\infty \chi_{E_n} \right),$$

which converges in $L^1(C, \mathscr{F}, \nu)$ to $\sum_{n=1}^\infty \phi(\chi_{E_n})$:

$$\int_C \left| \phi\left(\sum_{n=N+1}^\infty \chi_{E_n} \right) - \sum_{n=N+1}^\infty \phi(\chi_{E_n}) \right| d\nu$$

$$\leq \int_C \phi\left(\sum_{n=N+1}^\infty \chi_{E_n} \right) d\nu + \sum_{n=N+1}^\infty \int_C \phi(\chi_{E_n}) \, d\nu \to 0.$$

Therefore

$$\tilde{\phi} \bigcup_{n=1}^\infty E_n = \bigcup_{n=1}^\infty \tilde{\phi} E_n.$$

4.4. Topological representation

Remark 4.2 Construction of the Cantor set The Cantor set C is obtained by specifying its algebra of continuous functions $\mathscr{C}(C)$. This will be the closed subalgebra $\mathrm{Alg}_T(h)$ of $L^\infty(X, \mathscr{B}, \mu)$ generated by $\{hT^n : n \in \mathbb{Z}\}$ and the constants, where $h \in L^\infty(X, \mathscr{B}, \mu)$ is carefully chosen.

We must have h *total* in that $\{h^{-1}A : A \subset \mathbb{R} \text{ is Borel}\}$ is dense in the measure algebra $\mathscr{B}$, in order to guarantee that $L^\infty(C, \mathscr{F}, v)$ is isomorphic to $L^\infty(X, \mathscr{B}, \mu)$. And we must be sure that each function $g \in \mathrm{Alg}_T(h)$ is *uniform*, in that

$$\frac{1}{n}\sum_{k=0}^{n-1} gT^k \to \int g \, d\mu \text{ in } L^\infty(X, \mathscr{B}, \mu),$$

in order to guarantee that S is uniquely ergodic. (Recall that $S : C \to C$ will be uniquely ergodic if and only if

$$\frac{1}{n}\sum_{k=0}^{n-1} gS^k$$

converges uniformly to a constant for each $g \in \mathscr{C}(C)$.) There will be much more discussion of these two properties in what follows.

It will also be important that $h(X) \subset D$ for some compact totally disconnected set $D \subset \mathbb{R}$, for this will imply that the simple continuous functions are dense in $\mathscr{C}(C)$ and hence C is totally disconnected. These assertions follow from the three facts about totally disconnected sets to be mentioned forthwith.

Remarks 4.3 Facts about totally disconnected sets

(1) Suppose every function in a set $S \subset L^\infty(X, \mathscr{B}, \mu)$ has range contained in a fixed compact totally disconnected set $D \subset \mathbb{R}$. Then $\mathrm{cl}_{L^\infty} \mathrm{Alg}(S)$ contains a dense set of simple functions. (We write $\mathrm{Alg}(S)$ for the algebra generated by S and the constants.)

Proof: By the Gelfand theory,

$$\mathrm{cl}_{L^\infty} \mathrm{Alg}(S) = \mathscr{C}(M)$$

for some compact Hausdorff space M. Thus if $g \in \mathrm{cl}_{L^\infty} \mathrm{Alg}(S)$ and $f : \mathbb{R} \to \mathbb{R}$ is continuous, then $f \circ g \in \mathrm{cl}_{L^\infty} \mathrm{Alg}(S)$.

Now if we are given $g_0 \in \mathrm{cl}_{L^\infty} \mathrm{Alg}(S)$ and $\varepsilon > 0$, we may choose $g \in \mathrm{Alg}(S)$ with $\|g - g_0\|_\infty < \varepsilon/2$. Let $\delta > 0$, and select a continuous simple function $\phi : D \to D$ such that if $\phi(D) = \{d_1, d_2, \ldots, d_n\}$, then $\phi(d_i) = d_i$ and

$$\mathrm{diam}\, \phi^{-1}\{d_i\} < \delta \text{ for } i = 1, 2, \ldots, n.$$

The function g is a polynomial in members of S. Form a new function g' in $\mathrm{Alg}(\phi \circ S)$ by replacing each monomial

$$S_{i_1} S_{i_2} \ldots S_{i_k} \text{ by } \phi \circ S_{i_1} \cdot \phi \circ S_{i_2} \ldots \phi \circ S_{i_k}.$$

Then g' is in $\mathrm{cl}_{L^\infty}\mathrm{Alg}(S)$ (since each $\phi \circ S_{i_j}$ is), g' is simple, and $\|g'-g\|_\infty < \varepsilon/2$ if δ is small enough.

(2) *Suppose that M is a compact metric space such that the simple continuous functions are dense in $\mathscr{C}(M)$. Then M is totally disconnected.*
Proof: Let F be a component of M. Then F is closed. Suppose $x, y \in F$, $x \neq y$. There is a continuous function f with $f(x) \neq f(y)$ and hence a continuous simple function f_0 with $f_0(x) \neq f_0(y)$. We may assume that f_0 is two-valued. Then $f_0^{-1}\{f_0(x)\} \cup f_0^{-1}\{f_0(y)\}$ splits F into two disjoint closed sets.

(3) *If M is compact, metric, totally disconnected, and perfect (i.e. without isolated points), then M is homeomorphic to the ordinary Cantor set C.* (Hocking and Young 1961, p. 100).

Remarks 4.4 Facts about uniform functions
(1) *The set of uniform functions is a closed subspace of $L^\infty(X, \mathscr{B}, \mu)$.*
Proof: For each function f on X and integer n, write

$$A_n f = \frac{1}{n}\sum_{k=0}^{n-1} fT^k.$$

Then if $f_k \to f$ in L^∞ and the f_k are all uniform,

$$\left\|\int f\,d\mu - A_n f\right\|_\infty \leq \left|\int f\,d\mu - \int f_k\,d\mu\right| + \left\|\int f_k\,d\mu - A_n f_k\right\|_\infty$$
$$+ \left\|A_n f_k - A_n f\right\|_\infty.$$

First choose k, then n, to make this small.

(2) If $g_1 \leq g_2 \leq \ldots$ and $h_1 \geq h_2 \geq \ldots$ are uniform functions in $L^\infty(X, \mathscr{B}, \mu)$ with $g_n \to f$, $h_n \to f$ a.e., then f is uniform.
Proof: For each $n, k \geq 1$,

$$A_n g_k - \int f\,d\mu \leq A_n f - \int f\,d\mu \leq A_n h_k - \int f\,d\mu.$$

Given $\varepsilon > 0$, choose k with

$$\int (f - g_k)\,d\mu < \varepsilon/2, \quad \int (h_k - f)\,d\mu < \varepsilon/2,$$

and n_0 with

$$A_n g_k > \int g_k\,d\mu - \varepsilon/2 \quad \text{a.e.}$$

$$A_n h_k < \int h_k\,d\mu + \varepsilon/2 \quad \text{a.e., for } n \geq n_0.$$

Then
$$-\varepsilon \leq A_n f - \int f \, d\mu \leq \varepsilon \quad \text{a.e. for } n \geq n_0.$$

Remarks 4.5 *Facts about total functions*

(1) If $h \in L^\infty(X, \mathcal{B}, \mu)$ is total, then $\text{Alg}(h)$ is dense in $L^1(X, \mathcal{B}, \mu)$.

Proof: For any polynomial P and Borel $A \subset \mathbb{R}$,
$$\int_X |P \circ h - \chi_{h^{-1}A}| \, d\mu = \int_\mathbb{R} |P - \chi_A| \, d(\mu \circ h^{-1}).$$

Given A, this can be made arbitrarily small by an appropriate choice of P, since the measure $\mu \circ h^{-1}$ has compact support. Because the sets $h^{-1}A$ are dense in $\mathcal{B}$, the result follows.

(2) There are total functions on X, e.g.
$$\sum_{n=1}^\infty \frac{\chi_{B_n}}{3^n},$$
where $\{B_n\}$ is a countable dense set in $\mathcal{B}$.

(3) If $f_1, f_2, \ldots \in L^\infty(X, \mathcal{B}, \mu)$ are total and
$$\sum_{n=1}^\infty \mu\{f_n \neq f_{n+1}\} < \infty,$$
then $\{f_n\}$ converges a.e. and the limit function f is total.

Proof: $\{f_n\}$ converges on the complement of
$$\bigcap_{n=1}^\infty \bigcup_{k \geq n} \{f_k \neq f_{k+1}\},$$
which has measure 0 by the easy half of the Borel–Cantelli Lemma.

Given $B \in \mathcal{B}$ and $\varepsilon > 0$, we want to find a Borel set $A \subset \mathbb{R}$ with
$$d(B, f^{-1}A) = \mu(B \triangle f^{-1}A) < \varepsilon.$$

Find an n with
$$\mu \bigcup_{k \geq n} \{f_k \neq f_{k+1}\} < \varepsilon/2$$
and a Borel set $A \subset \mathbb{R}$ with
$$d(B, f_n^{-1}A) < \varepsilon/2.$$

Then
$$d(B, f^{-1}A) \leq d(B, f_n^{-1}A) + d(f_n^{-1}A, f^{-1}A).$$

The first term is less than $\varepsilon/2$, while the second is
$$\mu(f_n^{-1}A \triangle f^{-1}A) \leq \mu\{f_n \neq f\} \leq \mu \bigcup_{k \geq n} \{f_k \neq f_{k+1}\} < \varepsilon/2.$$

(4) If $f, g \colon X \to (0, 1]$ are in $L^1(X, \mathscr{B}, \mu)$ with $\|f - g\|_1 < \varepsilon$, and f is total, then there is a total $h \colon X \to (0, 1]$ such that
$$\|h - g\|_\infty < \sqrt{\varepsilon}$$
and
$$\mu\{h \neq f\} < \sqrt{\varepsilon}.$$

Proof: Choose $\varepsilon_1 > 0$ and $n \in \mathbb{N}$ such that
$$\|f - g\|_1 < \varepsilon_1 < \varepsilon \quad \text{and} \quad n \geq 1/\sqrt{\varepsilon_1}.$$
Let
$$E = \{|f - g| \geq \sqrt{\varepsilon_1}\}.$$
Fix a δ with $0 < \delta < 1/n$, to be specified later. Using the usual Lebesgue Ladder technique, for each $k = 0, 1, \ldots, n - 1$, let
$$S_k = \left\{ x \notin E \colon \frac{k}{n} + \delta < f(x) \leq \frac{k+1}{n} \right\},$$
$$S = \bigcup_{k=0}^{n-1} S_k,$$
$$Q_k = \left\{ x \notin S \colon \frac{k}{n} < g(x) \leq \frac{k+1}{n} \right\}.$$
Then $\mathscr{S} = \{S_0, S_1, \ldots, S_{n-1}, Q_0, Q_1, \ldots, Q_{n-1}\}$ forms a partition of X. We define h on each cell of the partition by
$$h = f \text{ on } S$$
$$h = \frac{k}{n} + \delta f \text{ on } Q_k, \quad k = 0, \ldots, n - 1.$$

The claim is that if δ is small enough, then this h satisfies the statement.

First, why is h total? We have
$$h(S_k) \subset \left(\frac{k}{n} + \delta, \frac{k+1}{n} \right]$$
and
$$h(Q_k) \subset \left(\frac{k}{n}, \frac{k}{n} + \delta \right]$$
(since $f \colon X \to (0, 1]$) for each k. These $2n$ intervals form a partition of $(0, 1]$. Now if $A \subset (0, 1]$ is a Borel set, then
$$S_k \cap f^{-1} A = h^{-1}\left(A \cap \left(\frac{k}{n} + \delta, \frac{k+1}{n} \right] \right)$$

4.4. Topological representation

and
$$Q_k \cap f^{-1}A = h^{-1}\left(\frac{k}{n} + \delta A\right).$$

Thus given an arbitrary $B \in \mathcal{B}$, we can choose a Borel set A_0 with $d(B, f^{-1}A_0)$ small, and then construct

$$A = \bigcup_{k=0}^{n-1}\left(A_0 \cap \left[\frac{k}{n} + \delta, \frac{k+1}{n}\right]\right) \cup \left(\frac{k}{n} + \delta A_0\right).$$

Then $h^{-1}(A)$ will be very close to B, since they will nearly agree on each cell of $\mathcal{S}$.

Now
$$|h - g| \leqslant \sqrt{\varepsilon_1} < \sqrt{\varepsilon} \quad \text{on each } S_k \subset E^c,$$
and
$$|h - g| \leqslant \frac{1}{n} \leqslant \sqrt{\varepsilon_1} < \sqrt{\varepsilon} \quad \text{on each } Q_k,$$
so we have
$$\|h - g\|_\infty < \sqrt{\varepsilon}.$$

Finally,
$$\mu\{h \neq f\} \leqslant \mu(S^c).$$

However,
$$\mu(S^c) \to \mu(E) \quad \text{as } \delta \to 0,$$
and
$$\sqrt{\varepsilon_1}\mu(E) \leqslant \|f - g\|_1 \leqslant \varepsilon_1.$$

Therefore
$$\mu(E) \leqslant \sqrt{\varepsilon_1}$$
and
$$\mu\{h \neq f\} \leqslant \mu(S^c) < \sqrt{\varepsilon}$$

for sufficiently small δ.

(5) If $f: X \to [0, 1]$ is total and $\phi: [0, 1] \to [0, 1]$ is strictly increasing, then $\phi \circ f$ is total.

Proof: $(\phi \circ f)^{-1}(A) = f^{-1}(\phi^{-1}A)$, and $\phi^{-1}A$ runs through all Borel subsets of $[0, 1]$ as A does.

Remark 4.6 Reduction of the problem In order to prove the Jewett–Krieger Theorem, it is enough, given $(X, \mathcal{B}, \mu, T)$, to find a compact totally disconnected $D \subset [0, 1]$ and a total function $h: X \to D$ such that

the algebra generated by $\{hT^n : n \in \mathbb{Z}\}$ consists entirely of uniform functions.
Proof: Recall that $\text{Alg}_T(h)$ denotes the L^∞ closure of the algebra generated by the $hT^n, n \in \mathbb{Z}$, and the constant functions. By 4.4(1), $\text{Alg}_T(h)$ consists entirely of uniform functions. As promised in 4.2, we let M be the maximal ideal space of $\text{Alg}_T(h)$, so that

$$\text{Alg}_T(h) = \mathscr{C}(M).$$

By the Gelfand theory, the action of T on this algebra arises from a homeomorphism $S: M \to M$. S is uniquely ergodic because all functions in $\mathscr{C}(M)$ are uniform: $\{A_n g\}$ converges uniformly on M to a constant for each $g \in \mathscr{C}(M)$. Let v be the unique invariant Borel probability measure on M.

We have a Banach algebra isomorphism

$$\phi : \text{Alg}_T(h) \to \mathscr{C}(M)$$

such that

$$\int_X f \, d\mu = \int_M \phi(f) \, dv \quad \text{for all } f \in \text{Alg}_T(h).$$

Now $\text{Alg}_T(h) \subset L^\infty(X)$ and $\mathscr{C}(M) \subset L^\infty(M)$ are each dense in the respective L^1 norms. Thus ϕ extends to a Banach algebra isomorphism

$$\phi : L^\infty(X, \mathscr{B}, \mu) \to L^\infty(M, \mathscr{F}, v)$$

such that

$$\phi(fT) = \phi(f)S \quad \text{and} \quad \int_X f \, d\mu = \int_M \phi(f) \, dv \quad \text{for all} f \in L^\infty(X, \mathscr{B}, \mu).$$

By 4.1, $(X, \mathscr{B}, \mu, T)$ is metrically isomorphic to $(M, \mathscr{F}, v, S)$.

By discarding an open set of measure 0 from M if necessary, we may assume that v has full support. Then (M, S) must be minimal, and so it is in fact *strictly ergodic*. Notice that this step corresponds to discarding a set of μ-measure 0 from X, so that $\text{Alg}_T(h)$ remains unchanged.

By 4.3(1), $\mathscr{C}(M)$ contains a dense set of simple functions. M is metrizable because $\mathscr{C}(M)$ is separable. Thus 4.3(2) applies, and M is totally disconnected. M must be perfect, since the set of isolated points is invariant and open, hence is either empty or all of M; but if it were all of M, M would have to consist of finitely many points, which is impossible because $(M, \mathscr{F}, v) \approx (X, \mathscr{B}, \mu)$, a Lebesgue space. By 4.3(3), M is homeomorphic to the Cantor set C.

B. Recurrence along IP-sets

This section presents the extension, due to Bellow and Fursten-

4.4. Topological representation

berg, to ergodic m.p.t.s of a property that is obvious for weakly mixing m.p.t.s. This property is essential in the proof of Jewett's Uniform Perturbation Lemma, which is used to produce the function h.

Recall *Hindman's Theorem*: If $\mathbb{N} = C_1 \cup C_2 \cup \ldots \cup C_r$ (disjoint), then at least one C_j contains an IP-set, i.e. a set $\{p_1, p_2, \ldots\}$ which is closed under the formation of finite sums $p_{i_1} + p_{i_2} + \ldots + p_{i_n} (i_1 < i_2 < \ldots < i_n)$. In 4.3.B we gave a proof of this result by means of dynamics. Before applying it, some simple observations about ergodic theory and IP-sets are needed.

Lemma 4.7 Let $T: X \to X$ be a not necessarily ergodic m.p.t. on a probability space $(X, \mathcal{B}, \mu)$. Let $S \subset \mathbb{N}$ be an IP-set and $B \in \mathcal{B}$ with $\mu(B) > 0$. Then for each $\varepsilon > 0$ there are infinitely many $s \in S$ for which

$$\mu(B \cap T^{-s}B) > \mu(B)^2 - \varepsilon.$$

Proof: Suppose that S is generated by $\{p_1, p_2, \ldots\}$, i.e. S consists of all finite sums $p_{i_1} + p_{i_2} + \ldots + p_{i_n}$, where $i_1 < i_2 < \ldots < i_n$. Replacing S by a subset if necessary, we may assume that

$$p_{k+1} > p_1 + p_2 + \ldots + p_k = n_k \quad \text{for all } k.$$

Note that

$$i < j \text{ implies } n_j - n_i \in S$$

and

$$i' < j', i < j < j' \text{ implies } n_{j'} - n_{i'} > n_j - n_i.$$

From Khintchine's Theorem (2.3.3) we already know that

$$\{k \in \mathbb{N} : \mu(B \cap T^{-k}B) > \mu(B)^2 - \varepsilon\}$$

is relatively dense, hence infinite. However, we claim that there are infinitely many k of the form $n_j - n_i$ in this set.

For if not, we would have

$$\mu(T^{-n_i}B \cap T^{-n_j}B) \leq \mu(B)^2 - \varepsilon \quad \text{for all } j > i \geq \text{some } I.$$

If $f = \chi_B - \mu(B)$, this would imply that for all $p \in \mathbb{N}$,

$$0 \leq \int (fT^{n_I} + fT^{n_{I+1}} + \ldots + fT^{n_{I+p-1}})^2 d\mu$$

$$= \int (\chi_{T^{-n_I}B} + \ldots + \chi_{T^{-n_{I+p-1}}B} - p\mu(B))^2 d\mu$$

$$\leq p(p-1)[\mu(B)^2 - \varepsilon] + p\mu(B) - 2p^2\mu(B)^2 + p^2\mu(B)^2$$

$$= p\mu(B)(1 - \mu(B)) - \varepsilon p(p-1).$$

However, this tends to $-\infty$ as $p \to \infty$, so this situation is impossible.

4. More about recurrence

Proposition 4.8 Let $T: X \to X$ be a not necessarily ergodic m.p.t. on a probability space $(X, \mathscr{B}, \mu)$. Let $A, B \in \mathscr{B}$, let $S \subset \mathbb{N}$ be an IP-set, and suppose that

$$\mu(A \cap T^{-s} B) = \varepsilon(s) > 0 \quad \text{for all } s \in S.$$

Then we cannot have $\varepsilon(s) \to 0$ as $s \to \infty$.

Proof: Suppose that S is generated by $p_1, p_2, \ldots$. Apply the Lemma to

$$A' = A \cap T^{-p_1} B, \quad \varepsilon = \tfrac{1}{2}\mu(A')^2,$$

and the IP-set S' generated by $p_2, p_3, \ldots$. We find that there are infinitely many $s' \in S'$ with

$$\mu(A' \cap T^{-s'} A') > \varepsilon,$$

i.e.

$$\mu(A \cap T^{-p_1} B \cap T^{-s'} A \cap T^{-s'-p_1} B) \geq \varepsilon.$$

Hence

$$\mu(A \cap T^{-(s'+p_1)} B) \geq \varepsilon$$

for infinitely many $s' \in S'$. □

This Proposition together with Hindman's Theorem will provide the desired steppingstone. For a finite partition $\mathscr{P}$ of X, define the *gauge* $\delta(\mathscr{P})$ of $\mathscr{P}$ by

$$\delta(\mathscr{P}) = \inf\{\mu(P) : P \in \mathscr{P}, \mu(P) > 0\}.$$

Theorem 4.9 (Bellow–Furstenberg 1979) Let $T: X \to X$ be a m.p.t. on a probability space $(X, \mathscr{B}, \mu)$, and let $\mathscr{F}$ be a finite measurable partition of X. Then

$$\limsup_{n \to \infty} \delta(\mathscr{F} \vee T^{-n} \mathscr{F}) > 0.$$

Proof: If $\mathscr{F} = \{F_1, F_2, \ldots, F_r\}$, let

$$\mathscr{A} = \{(i,j) : 1 \leq i, j \leq r \text{ and there is } n \in \mathbb{N} \text{ with}$$

$$\mu(F_i \cap T^{-n} F_j) = \delta(\mathscr{F} \vee T^{-n} \mathscr{F})\}$$

be the set of gauge-minimizing indices. For each $(i,j) \in \mathscr{A}$, let

$$H(i,j) = \{n \in \mathbb{N} : \delta(\mathscr{F} \vee T^{-n}\mathscr{F}) = \mu(F_i \cap T^{-n} F_j)\}$$

be the set of times at which this pair of indices scores a minimum measure. We then have a finite covering

$$\mathbb{N} = \bigcup_{(i,j) \in \mathscr{A}} H(i,j).$$

By Hindman's Theorem, some $H(i,j)$ contains an IP-set. By Proposi-

4.4. Topological representation

tion 4.8 applied to F_i, F_j, and this set,

$$0 < \delta(\mathscr{F} \vee T^{-n}\mathscr{F}) = \mu(F_i \cap T^{-n}F_j) = \varepsilon(n) \not\to 0 \quad \text{as } n \to \infty.$$

C. Perturbation to uniformity

The key step in Jewett's argument consists in perturbing a given simple function to a nearly uniform simple function. This is accomplished within a suitable tower in X. Henceforth $T: X \to X$ is ergodic.

Recall that for $E \subset X$,

$$n_E(x) = \inf\{n \geq 1 : T^n x \in E\}.$$

We use the notation

$$A_n f(x) = \frac{1}{n} \sum_{k=0}^{n-1} f(T^k x),$$

$$A_{-n} f(x) = \frac{1}{n} \sum_{k=0}^{n-1} f(T^{-k} x) \quad \text{for } n \geq 1.$$

Lemma 4.10 *Tower Lemma* Let $f \in L^1(X, \mathscr{B}, \mu)$, $\varepsilon > 0$, and $p \in \mathbb{N}$. Then there are a measurable set $E \subset X$ and $q > p$ such that

(i) $p \leq n_E(x) \leq q$ for all $x \in E$;
(ii) if $x \in E$ and $n_E(x) > p$, then $\left| \int_X f d\mu - A_{n_E(x)} f(x) \right| < \varepsilon$;
(iii) $\mu\{x \in E : n_E(x) = p\} < \varepsilon/p$

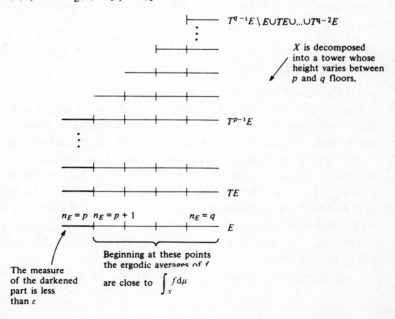

X is decomposed into a tower whose height varies between p and q floors.

Beginning at these points the ergodic averages of f are close to $\int_X f d\mu$

The measure of the darkened part is less than ε

4. More about recurrence

Proof: Assume that $\varepsilon < 1$. For each $n = 1, 2, \ldots$ let

$$G_n = \left\{ x \in X : \left| \int_X f \, d\mu - A_k f(x) \right| < \varepsilon \text{ for } |k| \geq n \right\}.$$

Then $G_1 \subset G_2 \subset \ldots$ and, by the Ergodic Theorem and because on a set of finite measure a.e. convergence implies convergence in measure, $\mu(G_n) \nearrow 1$.

Choose $N > p$ with $\mu(G_N) > 1 - \varepsilon/p$, and choose $D \subset TG_N$ with $0 < \mu(D) < 1/N$. Let

$$E_0 = D \setminus \bigcup_{k=1}^{N} T^k D.$$

Then $\mu(E_0) > 0$, since otherwise, up to sets of measure 0,

$$D \subset \bigcup_{k=1}^{N} T^k D,$$

so that

$$T^{-1} D \subset D \cup TD \cup \ldots \cup T^{N-1} D,$$

and hence

$$T^{-1}(D \cup TD \cup \ldots \cup T^{N-1} D) \subset D \cup TD \cup \ldots \cup T^{N-1} D;$$

this is impossible because T is ergodic and $D \cup TD \cup \ldots \cup T^{N-1} D$ has measure strictly between 0 and 1.

Now form the tower decomposition of X with respect to E_0; a certain choice procedure will be used to construct the set E. Let

$$E_0^i = \{ x \in E_0 : n_{E_0}(x) = i \}, \quad \text{for } i = 1, 2, \ldots.$$

Then

$$E_0^i \cup T E_0^i \cup \ldots \cup T^{i-1} E_0^i$$

forms the *i*th column of the tower.

Start at the roof of the skyscraper, in a column of height at least $2N$, and count down $2N$ floors. When our choice procedure lands us above this boundary, nothing above the landing cell will be included in E.

Start with a point $x \in E_0$, and look at its successive images under T, moving up the tower (but only up to the first point beyond the boundary level $n_{E_0}(x) - 2N$). Whenever we encounter a point of G_N^c, we include it in E and skip p steps; whenever we encounter a point of G_N, we include it in E and skip N steps:

Thus $E_0 \subset E$, and if we are looking at $T^k x$ for some $x \in E_0$ and $k \leq n_{E_0}(x) - 1$, then

(a) if $T^k x \in G_N^c$, we put $T^k x \in E$, $T^{k+1} x, \ldots, T^{k+p-1} x \notin E$,

4.4. Topological representation

and turn our attention to $T^{k+p}x$ (so long as $k+p \leq n_{E_0}(x)-1$);
  (b) if $T^k x \in G_N$, we put $T^k x \in E$, $T^{k+1}x, \ldots, T^{k+N-1}x \notin E$, and turn our attention to $T^{k+N}x$ (so long as $k+N \leq n_{E_0}(x)-1$);
  (c) if $k \geq n_{E_0}(x) - 2N$, we put $T^k x \in E$ but $T^{k+1}x, \ldots, T^{n_{E_0}(x)-1}x \notin E$.

Now if $x \in E$, either

(1) $x \in G_N$ and $n_E(x) = N$,
(2) $x \notin G_N$ and $n_E(x) = p$, or
(3) $T^{m-1}x \in G_N$ for some m with $N < m \leq 2N$.

(The entire top floor of the tower must be a subset of G_N, since $D \subset TG_N$). It is clear then that

$$p \leq N_E(x) \leq 2N \quad \text{for all } x \in E,$$

so (i) holds if we take $q = 2N$.

To verify (ii), notice that if $x \in E$ and $n_E(x) > p$, then $n_E(x) \geq N$. If $n_E(x) = N$, then $x \in G_N$ and (ii) holds by definition of G_N. If $n_E(x) > N$, then

$$y = T^{n_E(x)-1}x \in G_N$$

and

$$f(x) + f(Tx) + \ldots + f(T^{n_E(x)-1}x) = f(T^{-n_E(x)+1}y) + \ldots + f(y),$$

so that

$$A_{n_E(x)}f(x) = A_{-n_E(x)}f(y),$$

and again (ii) holds by the definition of G_N.

Finally, if $x \in E$ and $n_E(x) = p$, then $x \in G_N^c$. This implies that

$$p\mu\{x \in E : n_E(x) = p\} \leq p\mu(G_N^c) < p \cdot \frac{\varepsilon}{p} < \varepsilon,$$

so that (iii) is also true.

Lemma 4.11 Uniform Perturbation Lemma Let $f: X \to \mathbb{R}$ be a simple (measurable) function (that is, f assumes only finitely many values), $\varepsilon > 0$, and $m \in \mathbb{N}$. Then there is a simple function $g: X \to \mathbb{R}$ such that

(i) $\mu\{x : f(x) \neq g(x)\} < \varepsilon$,
(ii) $\| \int_X f \, d\mu - A_n g \|_\infty < \varepsilon$ for all large enough n,
(iii) $\{(g(x), g(Tx), \ldots, g(T^{m-1}x)) : x \in X\} \subset \{(f(x), f(Tx), \ldots, f(T^{m-1}x)) : x \in X\}$.

Proof: The function $F : X \to \mathbb{R}^m$ defined by

$$F(x) = (f(x), f(Tx), \ldots, f(T^{m-1}x))$$

assumes only finitely many values; call them $a_1, a_2, \ldots, a_r$. Then the sets

$F^{-1}\{a_1\},\ldots,F^{-1}\{a_r\}$ form a finite partition $\mathscr{F}$ of X, so by the Bellow–Furstenberg Theorem,

$$\limsup_{n\to\infty} \delta(\mathscr{F} \vee T^{-n}\mathscr{F}) > 0.$$

This means that there is an α with $0 < \alpha < 1$ for which there are infinitely many $p \in \mathbb{N}$ such that

(1) $\mu(F^{-1}\{a_i\} \cap T^{-p}F^{-1}\{a_j\}) > 0$ implies $\mu(F^{-1}\{a_i\} \cap T^{-p}F^{-1}\{a_j\}) \geq \alpha$.

Again, if

$$G_n = \left\{ x \in X : \left| \int_X f\,d\mu - A_k f(x) \right| < \varepsilon/2 \text{ for } k \geq n \right\},$$

then $\mu(G_n) \nearrow 1$, so we can find $N > m$ with $\mu(G_N) > 1 - \alpha$. Choose and fix $p > N$ for which (1) holds.

Let

$$\mathscr{K} = \{(i,j) : 1 \leq i, j \leq r \text{ and } \mu(F^{-1}\{a_i\} \cap T^{-p}F^{-1}\{a_j\}) > 0\},$$

and

$$E_{ij} = F^{-1}\{a_i\} \cap T^{-p}F^{-1}\{a_j\} \quad \text{for } (i,j) \in \mathscr{K}.$$

Then $\{E_{ij} : (i,j) \in \mathscr{K}\}$ forms a finite partition of X into sets of measure at least α. We must have, therefore,

$$\mu(G_N \cap E_{ij}) > 0 \quad \text{for all } (i,j) \in \mathscr{K}.$$

Thus for each $(i,j) \in \mathscr{K}$, it is possible to choose

$$y_{ij} \in G_N \cap E_{ij}.$$

This completes the hard part, for which the Bellow–Furstenberg Theorem is required. Now the function g is easily constructed by means of skyscraper architecture.

Apply the Tower Lemma to f, $\varepsilon/2$, and p; we obtain an E and q satisfying (i), (ii) and (iii) of that Lemma. Refer to the Figure (p. 197). We will redefine g only on the leftmost column of the tower, leaving $g = f$ elsewhere: if $E_p = \{x \in E : n_E(x) = p\}$, then g will be equal to f everywhere except maybe at some points of $E_p \cup TE_p \cup \ldots \cup T^{p-1}E_p$. Thus (iii) of the Tower Lemma gives (i) of the present lemma.

Now if $x \in E_p$, then $x \in E_{ij}$ for a unique $(i,j) \in \mathscr{K}$. We will redefine g at x and its images $Tx, T^2x, \ldots, T^{p-1}x$ up the column by

$$(g(x), g(Tx), \ldots, g(T^{p-1}x)) = (f(y_{ij}), f(Ty_{ij}), \ldots, f(T^{p-1}y_{ij})).$$

Let $G(x) = (g(x), g(Tx), \ldots, g(T^{m-1}x))$ for $x \in X$. In order to verify (iii), we have to show that range$(G) \subset$ range(F). It is clear, since $m < n < p \leq n_E(x)$, that if $x \in E$ then $G(x) \in$ range(F). (Note $F(y_{ij}) = a_i$.)

4.4. Topological representation

The difficulty comes from the possibility that $x, Tx, \ldots, T^{m-1}x$ slop over the top of the tower, i.e. that $m > n_E(x)$. We overcome this by showing that for each $x \in E$ there is $y \in X$ such that

$$(g(x), g(Tx), \ldots, g(T^{m+n_E(x)-1}x)) = (f(y), f(Ty), \ldots, f(T^{m+n_E(x)-1}y)).$$

Then m-step agreement, starting at any point in the tower, will follow from this agreement on each column plus m steps, starting from a point in the base.

Suppose then that $x \in E$. There are several cases to consider, depending on where x starts from in E and where it first returns to E.

Consider first the case when $x \in E_p$ (i.e., $n_E(x) = p$). Assume that $x \in E_{ij} = F^{-1}\{a_i\} \cap T^{-p}F^{-1}\{a_j\}$. Then $T^p x, T^p y_{ij} \in F^{-1}\{a_j\}$.

If $x \in E_p$ and $T^p x \in E_p$, then

$$(g(x), g(Tx), \ldots, g(T^{p-1}x)) = (f(y_{ij}), f(Ty_{ij}), \ldots, f(T^{p-1}y_{ij})).$$

Now $T^p x \in F^{-1}\{a_j\}$; suppose $T^p x \in E_{jk} = F^{-1}\{a_j\} \cap T^{-p}F^{-1}\{a_k\}$. Then

$$(g(T^p x), \ldots, g(T^{p+m-1}x)) = (f(y_{jk}), \ldots, f(T^{m-1}y_{jk}))$$
$$= a_j = (f(T^p y_{ij}), \ldots, f(T^{p+m-1}y_{ij})),$$

since $T^p y_{ij} \in F^{-1}\{a_j\}$. Recall that $m < p$, so this case is settled.

If $x \in E_p$ and $T^p x \notin E_p$, then

$$(g(T^p x), \ldots, g(T^{p+m-1}x)) = (f(T^p x), \ldots, f(T^{p+m-1}x))$$
$$= a_j = (f(T^p y_{ij}), \ldots, f(T^{p+m-1}y_{ij})),$$

so that again we can take $y = y_{ij}$.

If $x \notin E_p$ and $T^{n_E(x)}x \notin E_p$, then clearly

$$(g(x), \ldots, g(T^{m+n_E(x)-1}x)) = (f(x), \ldots, f(T^{m+n_E(x)-1}x)).$$

The only case left to consider, then, is when $x \notin E_p$ but $T^{n_E(x)}x \in E_p$. We have

$$(g(x), g(Tx), \ldots, g(T^{n_E(x)-1}x)) = (f(x), f(Tx), \ldots, f(T^{n_E(x)-1}x)).$$

Suppose that $T^{n_E(x)}x \in E_{ij}$. Then

$$(g(T^{n_E(x)}x), \ldots, g(T^{n_E(x)+p-1}x)) = (f(y_{ij}), f(Ty_{ij}), \ldots, f(T^{p-1}y_{ij})),$$

so that

$$(g(T^{n_E(x)}x), \ldots, g(T^{n_E(x)+m-1}x)) = F(y_{ij}) = a_i = F(T^{n_E(x)}x)$$
$$= (f(T^{n_E(x)}x), \ldots, f(T^{n_E(x)+m-1}x)).$$

To verify (ii), note that from the Tower Lemma if $x \in E \setminus E_p$ then

$$\left| \int_X f \, d\mu - A_{n_E(x)} f(x) \right| < \varepsilon/2.$$

The same conclusion holds also for $x \in E_p$, because each $y_{ij} \in G_N$ and $p > N$.

Because g and the return time to E are bounded, this observation is sufficient to yield (ii). For any average

$$A_n g(x) = \frac{1}{n} \sum_{k=0}^{n-1} g(T^k x)$$

can be broken up into blocks according to the entrances of $T^k x$ into E: choose $0 \leq k_1 < k_2 < \ldots < k_r \leq n-1$ with $T^{k_i} x \in E$ for all i, $T^k x \notin E$ for other values of k between 0 and $n-1$. Then

$$A_n g(x) = \frac{1}{n} \sum_{k=0}^{k_1-1} g(T^k x) + \frac{1}{n} \sum_{j=1}^{r-1} \sum_{k=k_j}^{k_{j+1}-1} g(T^k x) + \frac{1}{n} \sum_{k=k_r}^{n-1} g(T^k x).$$

The first and last sums each contain at most q terms and therefore tend uniformly to 0 as $n \to \infty$. The central term may be written as

$$\frac{1}{n}\left\{ n_E(T^{k_1}x) A_{n_E(T^{k_1}x)} f(x) + \ldots + n_E(T^{k_{r-1}}x) A_{n_E(T^{k_{r-1}}x)} f(x) \right\}.$$

For large n, this expression approximates a convex combination of $r-1$ terms, each of which is within $\varepsilon/2$ of $\int_X f \, d\mu$; the result must be within ε of $\int_X f \, d\mu$.

D. Uniform polynomials

According to the preceding Lemma any simple function f can be changed on a set of arbitrarily small measure to obtain a nearly uniform simple function g which runs through the same m-tuples of values as f. Next, given a polynomial P we construct the simple g so that $P \circ g$ is nearly uniform, and then we take a limit of such functions g.

Lemma 4.12 Let $f: X \to \mathbb{R}$ be simple, $\varepsilon > 0$, $m \geq 1$, and P a polynomial in m variables. Then there is a simple function $g: X \to \mathbb{R}$ such that

(i) $\mu\{x: f(x) \neq g(x)\} < \varepsilon$,
(ii) $\left\| \int_X P(g, gT, \ldots, gT^{m-1}) d\mu - A_n P(g, gT, \ldots, gT^{m-1}) \right\|_\infty < \varepsilon$ for $n \geq$ some n_0,
(iii) $\{(g(x), g(Tx), \ldots, g(T^{m-1}x)): x \in X\}$
$\subset \{(f(x), f(Tx), \ldots, f(T^{m-1}x)): x \in X\}$.

Proof: We will apply the Uniform Perturbation Lemma to a simple function $\bar{f}$ obtained from f as follows. Again let

$$F(x) = (f(x), f(Tx), \ldots, f(T^{m-1}x)).$$

and suppose that the values assumed by F are $a_1, a_2, \ldots, a_r$.

4.4. Topological representation

Choose distinct $\bar{a}_1, \bar{a}_2, \ldots, \bar{a}_r$ with

$$|P(a_i) - \bar{a}_i| < \varepsilon/4 \quad \text{for all } i,$$

and define

$$\bar{f} \equiv \bar{a}_i \text{ on } F^{-1}\{a_i\}.$$

Use the Uniform Perturbation Lemma to find a simple $\bar{g}$ with

(a) $\mu\{x : \bar{f}(x) \neq \bar{g}(x)\} < \varepsilon/4$,
(b) $\|\int_X \bar{f} d\mu - A_n \bar{g}\|_\infty < \varepsilon/4$ for $n \geq n_0$,
(c) $\{(\bar{g}(x), \bar{g}(Tx), \ldots, \bar{g}(T^{m-1}x)) : x \in X\}$
$\subset \{(\bar{f}(x), \bar{f}(Tx), \ldots, \bar{f}(T^{m-1}x)) : x \in X\}.$

We define

$$g \equiv \text{first entry of } a_i \text{ on } \bar{g}^{-1}\{\bar{a}_i\}.$$

For (i), notice that if $\bar{f}(x) = \bar{g}(x) = \bar{a}_i$, then $x \in F^{-1}\{a_i\}$ and $g(x) = f(x)$. Thus

$$\{x : f(x) \neq g(x)\} \subset \{x : \bar{f}(x) \neq \bar{g}(x)\}.$$

For (iii), given $x \in X$ there is $y_x \in X$ such that

$$(\bar{g}(x), \ldots, \bar{g}(T^{m-1}x)) = (\bar{f}(y_x), \ldots, \bar{f}(T^{m-1}y_x)).$$

Then for each $k = 0, 1, \ldots, m-1$, $\bar{g}(T^k x) = \bar{f}(T^k y_x)$; thus if both are $\bar{a}_i$, then $g(T^k x)$ is the first coordinate of a_i while $T^k y_x \in F^{-1}\{a_i\}$ so that $a_i = (f(T^k y_x), \ldots)$.

For (ii), given $x \in X$ find y_x as above such that $\bar{g}(x) = \bar{f}(y_x)$ and

$$(g(x), g(Tx), \ldots, g(T^{m-1}x)) = (f(y_x), f(Ty_x), \ldots, f(T^{m-1}y_x)).$$

Suppose that $y_x \in F^{-1}\{a_i\}$. Then

$$|\bar{g}(x) - P(g(x), g(Tx), \ldots, g(T^{m-1}x))|$$
$$= |\bar{f}(y_x) - P(f(y_x), f(Ty_x), \ldots, f(T^{m-1}y_x))|$$
$$= |\bar{a}_i - P(a_i)| < \varepsilon/4.$$

Therefore

$$-\varepsilon/4 < A_n P(g, gT, \ldots, gT^{m-1}) - A_n \bar{g} < \varepsilon/4$$

everywhere, and hence, using (b),

$$-\varepsilon/2 < A_n P(g, gT, \ldots, gT^{m-1}) - \int_X \bar{f} d\mu < \varepsilon/2$$

for all large enough n. Now integrate the inequalities

$$\int_X \bar{f} d\mu - \varepsilon/2 < A_n P(g, gT, \ldots, gT^{m-1}) < \int_X \bar{f} d\mu + \varepsilon/2$$

to find that
$$\left| \int_X P(g, gT, \ldots, gT^{m-1}) d\mu - \int_X \bar{f} d\mu \right| < \varepsilon/2.$$
The triangle inequality gives
$$\left\| \int_X P(g, gT, \ldots, gT^{m-1}) d\mu - A_n P(g, gT, \ldots, gT^{m-1}) \right\|_\infty < \varepsilon$$
for all large n.

Lemma 4.13 There is a total function $f \in L^\infty(X)$ such that the (not necessarily closed) algebra generated by the functions fT^n, $n \in \mathbb{Z}$, consists entirely of uniform functions.

Proof: It is enough to find a total f such that each monomial in $f, fT, \ldots, fT^{m-1}$ is uniform. (Compose with a negative power of T to obtain all monomials in the fT^k, $k \in \mathbb{Z}$.) Begin by writing a sequence of monomials $M_n : \mathbb{R}^n \to \mathbb{R}$,
$$M_n(t_1, \ldots, t_n) = t_1^{e_1(n)} t_2^{e_2(n)} \cdots t_n^{e_n(n)},$$
in such a way that every possible monomial appears infinitely many times in this list: for any choice of nonnegative integers $e_1, \ldots, e_k$, there are infinitely many n with
$$M_n(t_1, \ldots, t_n) = t_1^{e_1} \cdots t_k^{e_k}.$$
(e.g., the list $1; 1, t_1, t_2; 1, t_1, t_2, t_1 t_2, t_1^2 t_2, t_1 t_2^2, t_1^2 t_2^2; \ldots$ will do.)

Use uniform continuity of each M_m on $[0,1]^m$ to choose $\varepsilon_{mn} < 1/2^n$ such that
$$|s_k - t_k| < \varepsilon_{mn} \quad \text{for } k = 1, 2, \ldots, m \text{ implies}$$
$$|M_m(s_1, \ldots, s_m) - M_m(t_1, \ldots, t_m)| < \frac{1}{2^{n+1}}.$$
Let us simplify the notation $M_m(\phi, \phi T, \ldots, \phi T^{m-1})$ to $M_m(\phi)$.

The aim is to set up Remark 4.5(3) about total functions by defining inductively total $f_n : X \to (0, 1)]$ and $r_n \geqslant n$ such that
$$\mu\{x : f_{n+1}(x) \neq f_n(x)\} < \frac{1}{2^n}$$
and
$$(1) \quad \left\| \int_X M_m(f_n) d\mu - A_{r_m} M_m(f_n) \right\|_\infty$$
$$\leqslant 2 \left(\frac{1}{2^m} + \frac{1}{2^{m+1}} + \ldots + \frac{1}{2^n} \right) \quad \text{for } 1 \leqslant m \leqslant n \text{ and all } n.$$

4.4. Topological representation

Then 4.5(3) will give that $f_n \to f$ a.e., f is total, and

(2) $\left\| \int_X M_m(f) d\mu - A_{r_m} M_m(f) \right\|_\infty \leq 4 \cdot \frac{1}{2^m}$ for all m.

To begin, let $f_1 : X \to (0, 1]$ be any total function and let $r_1 = 1$. If $n > 1$ and $f_1, \ldots, f_{n-1}$ and $r_1, \ldots, r_{n-1}$ have been chosen, let

$$p = n + \max_{1 \leq m \leq n-1} r_m$$

and

$$\varepsilon = \frac{1}{2} \min_{1 \leq m \leq n} \varepsilon_{mn}^2.$$

Since the simple functions are dense in $L^\infty(X)$, there is a simple measurable $g : X \to (0, 1]$ such that $\|g - f_{n-1}\|_\infty < \varepsilon$. By restricting attention to an invariant subset of full measure if necessary, we may assume that for each $(t_1, \ldots, t_n) \in \mathbb{R}^n$,

$$\{x : (g(x), \ldots, g(T^{n-1}x)) = (t_1, \ldots, t_n)\}$$

is either empty or has positive measure.

Now apply Lemma 4.12 to g, ε, p, and M_n (which may be considered a polynomial in p variables). We obtain a simple function $h : X \to (0, 1]$ and an $r_n \geq n$ such that

(i) $\mu\{x : h(x) \neq g(x)\} < \varepsilon$,
(ii) $\|\int_X M_n(h) d\mu - A_r M_n(h)\|_\infty < \varepsilon$ for $r \geq r_n$,
(iii) $\{(h(x), h(Tx), \ldots, h(T^{p-1}x)) : x \in X\}$
$\subset \{(g(x), g(Tx), \ldots, g(T^{p-1}x)) : x \in X\}$.

Next get ready to use Remark 4.5(4) about total functions: f_{n-1} is total and

$$\|h - f_{n-1}\|_1 \leq \|h - g\|_1 + \|g - f_{n-1}\|_1 \leq \mu\{x : h(x) \neq g(x)\}$$
$$+ \|g - f_{n-1}\|_\infty < \varepsilon + \varepsilon = 2\varepsilon,$$

so there is a total $f_n : X \to (0, 1]$ such that

$$\|f_n - h\|_\infty < \sqrt{2\varepsilon}$$

and

$$\mu\{x : f_n(x) \neq f_{n-1}(x)\} < \sqrt{2\varepsilon} < \frac{1}{2^n}.$$

Note next that since $\sqrt{2\varepsilon} \leq \varepsilon_{nn}$, we have

$$\|M_n(f_n) - M_n(h)\|_\infty < \frac{1}{2^{n+1}},$$

4. More about recurrence

so that also $|\int_X M_n(f_n)\mathrm{d}\mu - \int_X M_n(h)\mathrm{d}\mu| < 1/2^{n+1}$; since $\varepsilon < 1/2^{n+1}$, (ii) gives

$$\left\|\int_X M_n(f_n)\mathrm{d}\mu - A_r M_n(f_n)\right\|_\infty < 2\cdot\frac{1}{2^n} \quad \text{for } r \geqslant r_n,$$

proving (1) for the case $m = n$.

Suppose now that $1 \leqslant m \leqslant n-1$. Since $p \geqslant n + r_m \geqslant m$, we have, by (iii), that

$$\left\|\int_X M_m(f_n)\mathrm{d}\mu - A_{r_m}M_m(h)\right\|_\infty \leqslant \left\|\int_X M_m(f_n)\mathrm{d}\mu - A_{r_m}M_m(g)\right\|_\infty.$$

Remembering that $\varepsilon < \varepsilon_{mn}$ and $\sqrt{2\varepsilon} < \varepsilon_{mn}$, we see that

$$\|M_m(g) - M_m(f_{n-1})\|_\infty < \frac{1}{2^{n+1}}$$

and

$$\|M_m(f_n) - M_m(h)\|_\infty < \frac{1}{2^{n+1}}.$$

Thus

$$\left\|\int_X M_m(f_n)\mathrm{d}\mu - A_{r_m}M_m(f_n)\right\|_\infty \leqslant \|A_{r_m}M_m(f_n) - A_{r_m}M_m(h)\|_\infty$$

$$+ \left\|A_{r_m}M_m(h) - \int_X M_m(f_n)\mathrm{d}\mu\right\|_\infty$$

$$< \frac{1}{2^{n+1}} + \left\|A_{r_m}M_m(g) - \int_X M_m(f_n)\mathrm{d}\mu\right\|_\infty$$

$$\leqslant \frac{1}{2^{n+1}} + \|A_{r_m}M_m(g) - A_{r_m}M_m(f_{n-1})\|_\infty$$

$$+ \left\|A_{r_m}M_m(f_{n-1}) - \int_X M_m(f_n)\mathrm{d}\mu\right\|_\infty$$

$$< \frac{1}{2^{n+1}} + \frac{1}{2^{n+1}} + \left\|A_{r_m}M_m(f_{n-1}) - \int_X M_m(f_{n-1})\mathrm{d}\mu\right\|_\infty$$

$$+ \left|\int_X (f_{n-1} - f_n)\mathrm{d}\mu\right|$$

$$< \frac{1}{2^{n+1}} + \frac{1}{2^{n+1}} + 2\left(\frac{1}{2^m} + \frac{1}{2^{m+1}} + \ldots + \frac{1}{2^{n-1}}\right) + \frac{1}{2^n}$$

$$= 2\left(\frac{1}{2^m} + \frac{1}{2^{m+1}} + \ldots + \frac{1}{2^{n-1}} + \frac{1}{2^n}\right),$$

by induction.

4.4. Topological representation

Thus the f_n and r_n satisfying (1) are defined for all n, and f_n converges a.e. to a total function f satisfying (2).

Given any monomial $M(f)$ in $f, fT, \ldots, fT^{k-1}$, M appears in our list as an M_m for infinitely many values of m. Therefore

$$\left\| \int_X M(f)\,d\mu - A_{r_m} M(f) \right\|_\infty \leq 4 \cdot \frac{1}{2^m}$$

for infinitely many m. This implies that $M(f)$ is uniform. For given $\varepsilon > 0$, choose m so that $2^{m-2} < \varepsilon/2$ and R_0 so that $2r_m \|M(f)\|_\infty / R_0 < \varepsilon/2$. Then for $R \geq R_0$ write $R = nr_m + q$ and

$$A_R M(f) = \frac{1}{R} \sum_{j=0}^{n-1} \sum_{k=0}^{r_m - 1} M(f) T^k T^{jr_m} + \frac{1}{R} \sum_{k=0}^{q-1} M(f) T^k T^{nr_m}$$

to see that

$$\left\| \int_X M(f)\,d\mu - A_R M(f) \right\|_\infty < \varepsilon.$$

E. Conclusion of the argument

It only remains now to prove the statement to which the Jewett–Krieger Theorem was reduced in Remark 4.6: given an ergodic system $(X, \mathcal{B}, \mu, T)$, there is a compact totally disconnected $D \subset [0, 1]$ and a total function $h: X \to D$ such that the algebra generated by $\{hT^n : n \in \mathbb{Z}\}$ consists entirely of uniform functions.

Choose a total function $f: X \to (0, 1]$ such that the algebra generated by $\{fT^n : n \in \mathbb{Z}\}$ consists entirely of uniform functions (Lemma 4.13). Avoiding the at most countably many atoms of the measure μf^{-1}, choose a countable dense set $B \subset \mathbb{R}$ such that $\mu(f^{-1}B) = 0$. Use an old trick to construct a Cantor-type function: if $B = \{b_1, b_2, \ldots\}$, let

$$\phi(t) = \sum_{b_k < t} \frac{1}{2^k}.$$

Then $D = \overline{\phi(\mathbb{R})} \subset [0, 1]$ is totally disconnected: B is dense and at b_k the strictly increasing function ϕ has a jump of $1/2^k$.

Now $h = \phi \circ f$ is total, since f is total and ϕ is strictly increasing (Remark 4.5(5)). Choose continuous ϕ_n, ψ_n with

$$0 \leq \phi_n \nearrow \phi \quad \text{and} \quad 1 \geq \psi_n \searrow \phi \text{ on } B^c.$$

Let $h_n = \phi_n \circ f$ and $i_n = \psi_n \circ f$. Then h_n and i_n are in the supremum-norm closure of the algebra generated by f and the constants (by the Stone–Weierstrass Theorem), and hence h_n and i_n are uniform (by Remark 4.4(1)

208   4. More about recurrence

concerning uniform functions). Then Remark 4.4(2) about uniform functions implies that h is uniform.

In the same way, any monomial

$$hT^{m_1} \cdot hT^{m_2} \ldots hT^{m_r}$$

is the increasing limit as $n \to \infty$ of monomials

$$\phi_n f T^{m_1} \cdot \phi_n f T^{m_2} \ldots \phi_n f T^{m_r}$$

and decreasing limit as $n \to \infty$ of monomials

$$\psi_n f T^{m_1} \cdot \psi_n f T^{m_2} \ldots \psi_n f T^{m_r},$$

each of which is uniformly approximable by a polynomial in $fT^{m_1}, \ldots, fT^{m_r}$. Thus

$$hT^{m_1} \cdot hT^{m_2} \ldots hT^{m_r} \in \text{Alg}_T(f)$$

and hence is uniform.

Exercises

1. Construct a strictly ergodic subsystem of the space of bilateral sequences of 0s and 1s with the shift transformation which is metrically isomorphic to an irrational rotation of the unit circle.
2. Consider the three following m.p.t.s.
   (a) The von Neumann–Kakutani adding machine transformation T is defined on the unit interval as follows. Starting with the unit interval, at each stage cut in half vertically and stack the right half on top of the left half.

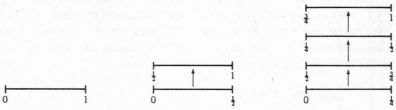

T maps each interval isometrically to the one above it and at each stage is left momentarily undefined on the topmost interval. Here is another picture of the action of T.

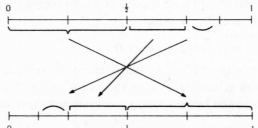

(b) $\mathbb{Z}_2^{\mathbb{N}}$ is a compact topological group under 'addition with carry to the right':

$$
\begin{array}{r}
1\ 0\ 1\ 1\ 0\ 1\ \ldots \\
+0\ 1\ 1\ 0\ 1\ 0\ \ldots \\
\hline
1\ 1\ 0\ 0\ 0\ 0
\end{array}
$$

Let $S\omega = \omega + (1, 0, 0, \ldots)$ for $\omega \in \mathbb{Z}_2^{\mathbb{N}}$.

(c) Define a 'Toeplitz sequence' $\tau \in \{0, 1\}^{\mathbb{Z}}$ by entering a 0 at all even places, then entering a 1 at every other one of the remaining places (i.e., write 1, blank, 1, blank, ...), then a 0 at every other one, etc. Let X be the orbit closure in $\{0, 1\}^{\mathbb{Z}}$ of τ under the shift σ.

Show that $(\mathbb{Z}_2^{\mathbb{N}}, S)$ and (X, σ) are strictly ergodic and metrically isomorphic to $([0, 1], T)$.

3. Show that the set of limit points of the forward orbit of the Morse sequence ... 0 1 10 1001 10010110 ... (at each stage write down the 'dual' of what is at hand, where 0 and 1 are the duals of each other) under the shift is strictly ergodic.

4. Note that the Bellow–Furstenberg Theorem is clear for weakly mixing transformations.

5. (a) Show by example that if (X, T) and (Y, S) are uniquely ergodic topological systems (i.e. compact metric spaces with homeomorphisms), then $(X \times Y, T \times S)$ need not be uniquely ergodic.

(b) If $\phi:(X, T) \to (Y, S)$ is a homomorphism, does unique ergodicity of (X, T) imply that of (Y, S)? Conversely?

(c) Let $T:X \to X$ be an ergodic m.p.t. on $(X, \mathcal{B}, \mu)$. Do the uniform functions in $L^\infty(X, \mathcal{B}, \mu)$ form either an algebra or a lattice? If not, what can be said about the algebra and the lattice that they generate?

4.5. Two examples

Although in the sense of Baire category almost every m.p.t. is weakly mixing but not strongly mixing, concrete examples of such transformations were not easy to find. We will look at a modification of one such example constructed by Kakutani (1973). Our version (adapted from Petersen and Shapiro 1973 – see Petersen 1973b) is a derivative rather than a primitive induced transformation of a group rotation and has the advantage that it can easily be shown not even to be topologically strongly mixing, so that it also provides an example of a minimal, uniquely ergodic

4. More about recurrence

cascade that is topologically weakly mixing but not topologically strongly mixing.

Recall that for strictly ergodic cascades, the following implications hold:

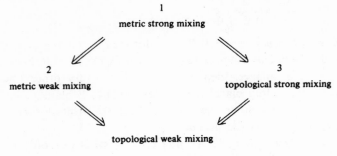

This example shows that $2 \not\Rightarrow 1$, $4 \not\Rightarrow 3$, and $2 \not\Rightarrow 3$. Other examples showing that $2 \not\Rightarrow 1$ are due to Chacon (1969), Katok and Stepin (1967), and Dekking and Keane (1978). That $3 \not\Rightarrow 1$ is seen in examples of Petersen (1970a) and, more recently, Dekking (1980), and that $4 \not\Rightarrow 2$ in one by Kolmogorov (1953) (see Parry 1981). Finally, it seems that topological strong mixing can hold without metric weak mixing for uniquely ergodic cascades. The following way to see this emerged in discussions among myself, Brian Marcus, and Marina Ratner. Every continuous reparametrization of a horocycle flow $\{T_t\}$ is topologically strongly mixing (Marcus 1975). On the other hand, since it is loosely Bernoulli (Ratner 1978), the horocycle flow has a measurable reparametrization which is an irrational flow on a torus, hence not metrically weakly mixing. By a result of Ornstein and Smorodinsky (1978), a system metrically isomorphic to this one can be arrived at by a *continuous* reparametrization of the horocycle flow. Again by a result of Marcus (1975), T_t is uniquely ergodic for a.e. t. For such a t, $T = T_t$ will have the desired properties. A different example has been found by Oren (personal communication). In fact Lehrer (not yet published) proved that the topological model in the Jewett–Krieger Theorem can always be made topologically strongly mixing.

The second example, due to Chacon (1969), is also uniquely ergodic and metrically weakly mixing but not metrically strongly mixing. Recently del Junco (1978) showed that this transformation is actually *prime* – it has no proper factors, or, equivalently, the σ-algebra of all measurable subsets contains no proper invariant sub-σ-algebras. The transformation first shown to be prime was constructed by Ornstein (1967). More recent work on prime transformations is in Rudolph (1979), Fieldsteel, del Junco,

4.5. Two examples

Rahe and Swanson (1980) and del Junco (1981). For the topological versions, see Petersen (1969), Furstenberg, Keynes and Shapiro (1973), Keynes and Newton (1976).

A. Metric weak mixing without topological strong mixing

The basis of this example is the von Neumann–Kakutani 'adding machine' transformation $\phi:[0,1) \to [0,1)$. For $n = 0, 1, 2, \ldots$, let $I_n = [1 - 1/2^n, 1 - 1/2^{n+1})$:

We define
$$\phi(x) = x - \left(1 - \frac{1}{2^n} - \frac{1}{2^{n+1}}\right) \text{ on } I_n,$$
so that ϕ slides I_n to a symmetric position on the other side of 1/2. Then ϕ is an ergodic m.p.t. with discrete spectrum, having eigenvalues $e^{2\pi i k \lambda}$ with λ any dyadic rational. The map ϕ is isomorphic to translation by a generator on the compact group of 2-adic integers. (See Exercise 2, 4.4.)

Proposition 5.1 For any $f \in L^2$, $f \circ \phi^{2^n} \to f$ in L^2.

Proof: For each dyadic rational λ, let f_λ be the corresponding eigenfunction of modulus 1. Given $f \in L^2$, we have
$$f = \sum_\lambda a_\lambda f_\lambda \text{ in } L^2,$$
where $a_\lambda = (f, f_\lambda)$. Then
$$f(\phi^{2^n} x) = \sum a_\lambda e^{2\pi i 2^n \lambda} f_\lambda(x),$$
so that
$$\|f \circ \phi^{2^n} - f\|_2^2 = \sum_\lambda |a_\lambda|^2 |e^{2\pi i 2^n \lambda} - 1|^2.$$
Now $2^n \lambda = 0 \mod 1$ if $\lambda = p/2^k$ for $k \leq n$ and p odd, i.e. if λ has *rank* no larger than n. Since $|e^{2\pi i 2^n \lambda} - 1| \leq 2$ for all n, we see that
$$\|f \circ \phi^{2^n} - f\|_2^2 \leq 4 \sum_{\text{rank}\lambda > n} |a_\lambda|^2 \to 0,$$
since the series $\sum |a_\lambda|^2$ converges.

Now let $A = \bigcup_{n=0}^{\infty} I_{2n}$.

212 4. More about recurrence

Clearly if $x \in A$ then either $\phi x \in A$ or $\phi^2 x \in A$, so that A has the first-return decomposition $A = A_1 \cup A_2$, where

$$A_1 = (I_2 \cup I_4 \cup \ldots) \cup (I_0 \cap \phi^{-1}(I_2 \cup I_4 \cup \ldots))$$
$$A_2 = I_0 \cap \phi^{-1}(I_1 \cup I_3 \cup I_5 \cup \ldots).$$

(Note that writing the intersections with I_0 here is actually redundant.) Let $\phi_A : A \to A$ be the induced (derivative) transformation: $\phi_A = \phi^i$ on A_i, $i = 1, 2$.

Proposition 5.2 ϕ_A is weakly mixing.

Proof: Suppose that $f : A \to \{z \in \mathbb{C} : |z| = 1\}$ is an eigenfunction of ϕ_A with eigenvalue ξ^2 (for convenience). Extend f to $[0, 1)$ by letting $f(x) = \xi f(\phi^{-1} x)$ for $x \in A^c$. Then

$$\frac{f(\phi x)}{f(x)} = \begin{cases} \xi & \text{on } A_2 \cup \phi A_2 \\ \xi^2 & \text{on } A_1. \end{cases}$$

Thus if $\xi = e^{2\pi i \lambda}$, we have $f(\phi x) = e^{2\pi i u(x) \lambda} f(x)$, where

$$u(x) = \begin{cases} 1 & \text{if } x \in A_2 \cup \phi A_2 \\ 2 & \text{if } x \in A_1. \end{cases}$$

If $u_n(x) = u(x) + u(\phi x) + \ldots + u(\phi^{n-1} x)$ for $n \geq 1$, then

$$f(\phi^n x) = e^{2\pi i u_n(x) \lambda} f(x).$$

Consider the case when $n = 2^p$ and $p = 2q$ is even. Decompose $[0, 1)$ into 2^p equal subintervals. The map ϕ permutes these intervals cyclically, mapping all but the rightmost one by translation. Thus if $x \in [0, 1)$ is not a dyadic rational, the points $x, \phi x, \ldots, \phi^{2^p - 1} x$ are distributed one to each of these subintervals. We can tell on which of these intervals $u = 2$; we are uncertain only of $u(x') = u(x'')$, where x' and x'' are those points of $\{x, \phi x, \ldots, \phi^{2^p - 1} x\}$ which are in the rightmost subintervals of $(0, 1/2)$ and $(1/2, 1)$.

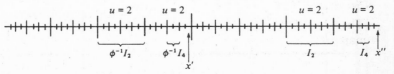

4.5. Two examples

Beginning at 0 and 1/2 and moving to the right, we encounter 2^{p-1} subintervals (of length $1/2^p$) on which $u = 1$, then 2^{p-2} on which $u = 2$, 2^{p-3} on which $u = 1, \ldots, 2$ on which $u = 1$, and finally 2 on which u is not constant. Thus

$$u_{2^p}(x) = 2^{p-1} + 2\cdot 2^{p-2} + 2^{p-3} + 2\cdot 2^{p-4} + \ldots + 2\cdot 2^2 + 2^1 + 2u(x').$$

Hence

$$u_{2^p}(x) = M_q + 2 \quad \text{or} \quad M_q + 4,$$

where

$$M_q = 2^{2q-1} + 2\cdot 2^{2q-2} + \ldots + 2\cdot 2^2 + 2^1.$$

Now

$$\mu\{x : u_{2^{2q}}(x) = M_q + 2\} = \mu\{x : u(x'') = 1\}$$
$$= \mu\{x : x'' \in A_2 \cup \phi A_2\} = \mu\{x : x'' \in \phi A_2\}$$
$$= \mu\{x : x'' \in I_1 \cup I_3 \cup I_5 \cup \ldots\}.$$

The rightmost subinterval of our partition looks like this, where the desirable region for x'' is darkened:

```
          I_{2q}              I_{2q-1}     I_{2q-2}
    |--------------------|------------|------|--|-|
  1 - 1/2^p
```

In $[1 - 1/2^p, 1)$, the darkened region has relative measure

$$\tfrac{1}{4} + \tfrac{1}{16} + \ldots = \tfrac{1}{3},$$

so also

$$\mu\{x : u_{2^{2q}}(x) = M_q + 2\} = \tfrac{1}{3}$$

(sliding x slides x'').

By the preceding Proposition,

$$0 = \lim_{q \to \infty} \| f \circ \phi^{2^p} - f \|_2^2 = \lim_{q \to \infty} \| e^{2\pi i \lambda u_{2^p}(x)} - 1 \|_2^2$$
$$= \lim_{q \to \infty} \int |e^{2\pi i \lambda u_{2^p}(x)} - 1|^2 \, dx$$
$$= \lim_{q \to \infty} (\tfrac{1}{3} |e^{2\pi i \lambda (M_q + 2)} - 1|^2 + \tfrac{2}{3} |e^{2\pi i \lambda (M_q + 4)} - 1|^2).$$

Therefore $\lambda(M_q + 2) \to 0$ and $\lambda(M_q + 4) \to 0 \pmod{1}$ as $q \to \infty$. Subtracting, $2\lambda \to 0 \bmod 1$. Hence $\xi^2 = 1$ and ϕ_A is metrically weakly mixing.

The maps ϕ and ϕ_A can be realized as uniquely ergodic cascades. Define a map ω from $[0, 1)$ to $\prod_{-\infty}^{\infty} \{0, 1\}$ by

$$\omega_n(x) = \chi_{A^c}(\phi^n x).$$

214   4. *More about recurrence*

Then ϕ on $[0, 1)$ is carried over to the shift σ on $X = \overline{\mathcal{O}(\omega(0))}$. (X, σ) is uniquely ergodic since any invariant measure must be carried to Lebesgue measure by this isomorphism. (See Exercise 2, 4.4)

Let $X_0 = \{\omega \in X : \omega_0 = 0\}$. Then ϕ_A is isomorphic to $\sigma_0 : X_0 \to X_0$, where σ_0 is the 'shift to the next 0'. (X_0, σ_0) is again uniquely ergodic.

Proposition 5.3 (X_0, σ_0) is metrically weakly mixing but not topologically strongly mixing.

Proof: Since ϕ_A is metrically weakly mixing, so is σ_0. Let B_r be the initial 2^r-block of

$$\downarrow \text{center}$$
$$\omega(0) = \ldots 010001010100010 \ldots$$

For (X_0, σ_0) to be topologically strongly mixing, it must be the case that if $\mathcal{N}_r$ is the set of places in $\omega(0)$ at which B_r appears, then the difference set $\mathcal{N}_r - \mathcal{N}_r$ contains all integers from some point on. Let p_r denote the number of 0s in B_r. We will prove that if r is odd and $n > r \geq 3$, then there do not exist two appearances of B_r in $\omega(0)$ separated by $p_n - 1$ places.

Denote by B^* the block which agrees with B everywhere except in the last entry, which is 'dualized' – 0 is changed to 1 and 1 to 0. Notice that $B_{n+1} = B_n B_n^*$ for all n. We establish a list of simple observations.

(1) B_n appears at the $k \cdot 2^{n+1}$st place in $\omega(0)$ for all $k \in \mathbb{Z}$.

Proof: The 2^{n+1}-block appearing at this place is either B_{n+1} or B_{n+1}^*.

(2) If B_n appears at the mth place in $\omega(0)$, Then $m = k \cdot 2^n$ for some $k \in \mathbb{Z}$.

Proof: By induction. The statement is clearly true if $n = 0$. If $n \geq 1$, since $B_n = B_{n-1} B_{n-1}^*$, we have $m = k \cdot 2^{n-1}$ for some k. If k were odd, by (1) we would have both B_{n-1} and B_{n-1}^* appearing at the $(k+1)2^{n-1}$st place.

(3) (a) If n is odd, $p_{n+1} = 2p_n + 1 = 1 + \sum_{i=0}^{n} p_i$.

  (b) If n is even, $p_{n+1} = 2p_n - 1 = \sum_{i=0}^{n} p_i$.

  (c) For r odd and $n > r \geq 3$, $p_n - \sum_{i=r}^{n-1} p_i = \begin{cases} p_r & \text{if } n \text{ is odd} \\ p_r + 1 & \text{if } n \text{ is even} \end{cases}$

Proof: (a) and (b) are proved easily by induction, and then (c) follows.

(4) $\mathcal{N}_r - \mathcal{N}_r$ (the difference set of the appearances of B_r) is contained in the collection of all numbers of the form

$$a = \varepsilon_r p_r + \ldots + \varepsilon_m p_m,$$

where $m \geq r$, each ε_i is 0, 1, or -1, and $\varepsilon_m = 1$.

4.5. Two examples

Proof: Because of (2), it is enough to show that each initial $k \cdot 2^r$-block of $\omega(0)$, for any $k = 1, 2, \ldots 2^n - 1$, contains

$$\delta_r p_r + \ldots + \delta_{r+n-1} p_{r+n-1}$$

0s, for some choice of the δ_is each equal to 0 or 1.

We use induction on n. The statement is clear for $n = 1$. Suppose now that it holds for n. If $k = 2^n$, there are $p_{n+r} = 1 \cdot p_{r+(n+1)-1}$ 0s in the block in question. If k is one $2^n + 1, \ldots, 2^{n+1} - 1$, then the initial $k \cdot 2^r$-block of $\omega(0)$ is as shown below.

```
       B_{n+r}                        B
  _____/              _____
        ↓                              ↑
    p_{n+r}  0s                  initial (k-2^n)2^r-block of ω(0)
```

By induction the block B contains $\delta_r p_r + \ldots + \delta_{r+n-1} p_{r+n-1}$ 0s for some choice of $\delta_r, \ldots, \delta_{r+n-1}$. Thus all told there are $\delta_r p_r + \ldots + \delta_{r+n-1} p_{r+n-1} + p_{n+r}$ 0s.

5. If r is odd and $n > r \geqslant 3$, then $p_n - 1$ is *not* a number of the form

$$a = \varepsilon_r p_r + \ldots + \varepsilon_m p_m,$$

where $m \geqslant r$, each ε_i is 0, 1, or -1, and $\varepsilon_m = 1$.

Proof: Note that if $m < n$, then

$$a \leqslant \sum_{i=r}^{n-1} p_r = \left\{ \begin{array}{c} p_n - p_r \\ \text{or} \\ p_n - p_r - 1 \end{array} \right\} < p_n - 1.$$

Thus we may assume that $m \geqslant n$.

The smallest number a of the above form is

$$p_m - \sum_{i=r}^{m-1} p_i = \left\{ \begin{array}{c} p_r \\ \text{or} \\ p_r + 1 \end{array} \right..$$

If for some k with $n \leqslant k \leqslant m-1$ we change ε_k from -1 to 0, we will get $p_r + p_k$ or $p_r + p_k + 1$, each of which is greater than $p_n - 1$. Thus in order to achieve $a = p_n - 1$, we must keep all these $\varepsilon_k = -1$:

$$a = p_m - \sum_{i=n}^{m-1} p_i + \varepsilon_{n-1} p_{n-1} + \ldots + \varepsilon_r p_r$$

$$= \varepsilon_{n-1} p_{n-1} + \ldots + \varepsilon_r p_r + \left\{ \begin{array}{ll} p_n & (m \text{ odd}) \\ \text{or} & \\ p_n + 1 & (m \text{ even}). \end{array} \right.$$

It is enough, then, to show that
$$b = \varepsilon_{n-1} p_{n-1} + \ldots + \varepsilon_r p_r$$
can never be -1 or -2. Let ε_q be the first nonzero coefficient, so that
$$b = \varepsilon_q p_q + \ldots + \varepsilon_r p_r,$$
and $q \leq n-1$. If $\varepsilon_q = 1$, then $b > 0$. Hence $\varepsilon_q = -1$. But then the largest possible b is
$$-p_q + \sum_{i=r}^{q-1} p_i = \begin{cases} -p_r \\ \text{or} \\ -1 - p_r \end{cases} \leq -5.$$

B. A prime transformation

Like the Kakutani example, Chacon's can be constructed either as a strictly ergodic symbolic cascade or as a cutting and sliding around of subintervals of $[0, 1]$ which preserves Lebesgue measure.

In the symbolic version of the construction, we define a sequence of blocks by

$$A_0 = 0$$
$$A_1 = 0010$$
$$A_2 = 0010\ 0010\ 1\ 0010$$
$$\vdots$$
$$A_{n+1} = A_n A_n 1 A_n$$
$$\vdots$$

The length of A_n is $l(A_n) = l_n = \frac{1}{2}(3^{n+1} - 1)$. Let ω_1 be a point of $\{0,1\}^{\mathbb{Z}}$ whose initial l_n-block is A_n for all n, and let Ω be a minimal subset of $\text{cl}\{\sigma^n \omega_1 : n = 0, 1, 2, \ldots\}$, where σ is the shift. Then Ω consists of all those 0, 1 sequences which contain only those blocks that appear in the A_n for some n. Thus there is a point $\omega \in \Omega$ whose initial l_n-block, for each n, is A_n.

We will show that (Ω, σ) is strictly ergodic, metrically weakly mixing, not metrically strongly mixing, and prime.

For symbolic cascades, unique ergodicity is easy to determine by counting frequencies of blocks. If A and B are blocks on the symbols 0 and 1, i.e. $A = a_0 a_1 \ldots a_n$ and $B = b_0 b_1 \ldots b_m$ where all the a_i and b_i are 0 or 1, denote by $N(A, B)$ the number of appearances of A in B, i.e.

$$N(A, B) = \text{card }\{j : 0 \leq j \leq m,\ b_j b_{j+1} \ldots b_{j+n} = A\}.$$

Lemma 5.4 $(\overline{\mathcal{O}(x)}, \sigma)$ *is uniquely ergodic if and only if there is a*

4.5. Two examples

sequence of integers $k_n \nearrow \infty$ and a sequence of blocks A_m which appear in x with $l(A_m) \nearrow \infty$ such that the frequency of appearance of each A_m in any k_n-block in x tends uniformly to a constant as $n \to \infty$: For each m there is a constant μ_m such that given $\varepsilon > 0$, there is an n_0 such that if $n \geq n_0$ and B is any k_n-block which appears in x, then

$$\left| \frac{N(A_m, B)}{k_n} - \mu_m \right| < \varepsilon.$$

Proof: Suppose that $(\overline{\mathcal{O}(x)}, \sigma)$ is uniquely ergodic with invariant measure μ. For each block A, the *cylinder set*

$$[A] = \{ y : y_0 y_1 \ldots y_{l(A)-1} = A \}$$

is open and closed, hence its characteristic function is continuous. By 2.8,

$$\frac{1}{n} \sum_{k=0}^{n-1} \chi_{[A]}(\sigma^k y) \to \mu[A]$$

uniformly for $y \in \overline{\mathcal{O}(x)}$. The expression on the left-hand side is the frequency of A in the initial n-block of y. Clearly this implies the condition in the statement of the Lemma.

Suppose now that the condition is satisfied, and that ν and μ are different ergodic invariant Borel probability measures for $(\overline{\mathcal{O}(x)}, \sigma)$. Since $\{\sigma^j[A_m] : j \in \mathbb{Z}, m = 1, 2, \ldots\}$ generates the σ-algebra of all measurable sets, we must have $\nu[A_m] \neq \mu[A_m]$ for some m, which we now fix. Then for ν-almost all $y \in \overline{\mathcal{O}(x)}$,

$$\frac{N(A_m, y_0 y_1 \ldots y_{k_n - 1})}{k_n} \to \nu[A_m],$$

by the Ergodic Theorem. Since $y_0 y_1 \ldots y_{k_n - 1}$ is a k_n-block which appears in x, though, $N(A_m, y_0 y_1 \ldots y_{k_n - 1})/k_n$ should be close to μ_m for large n. Hence $\nu[A_m] = \mu_m$, and similarly $\mu[A_m] = \mu_m$.

The following two simple observations will be very useful.

Lemma 5.5 (1) Every block in ω of length $l_n + 1$ contains the initial point of an appearance of A_n in ω.

(2) A_n appears in ω only where we expect it (i.e. only at those places where it is explicitly and consciously written down during the above construction).

Proof: (1) is clear. As for (2), the statement is obvious for $A_0 = 0$, and we can check visually that A_1 appears only twice in

$$A_1 A_1 = 0010\,0010$$

and in

$$A_1 1 A_1 = 0010\ 1\ 0010.$$

Then by induction A_{n+1} appears only twice in $A_{n+1}A_{n+1}$ and in $A_{n+1}1A_{n+1}$, since otherwise A_n would appear at an 'unexpected' place in one of these blocks.

Proposition 5.6 (Ω, σ) is uniquely ergodic.

Proof: Every l_{n+m}-block in ω appears either in $A_{n+m}A_{n+m}$ or $A_{n+m}1A_{n+m}$, and every l_{n+m}-sub-block of each of these two blocks contains between 3^n and 3^n-3 appears of A_m. Therefore, if B is any l_{n+m}-block which appears in ω, we have

$$\frac{N(A_m, B)}{l_{n+m}} \sim \frac{3^n}{\frac{1}{2}(3^{n+m+1} - 1)} \to \frac{2}{3}\frac{1}{3^m} \quad \text{as } n \to \infty.$$

Since the convergence is uniform in B, an application of the Lemma is sufficient to conclude the proof.

Remark 5.7 Let μ be the unique invariant probability measure on (Ω, σ). Then

$$\mu[A_m] = \frac{2}{3}\frac{1}{3^m} \quad \text{for } m = 0, 1, 2, \ldots$$

A system isomorphic to (Ω, σ, μ) can be constructed by 'cutting and stacking'. Divide $[0, 1]$ into $[0, 2/3)$ and $[2/3, 1)$. The interval $[0, 2/3)$ forms the 'initial stack', and $[2/3, 1)$ forms the 'reservoir'.

At each stage we cut the stack on the left into thirds, which we place above one another (bottom to top corresponding to left to right), except that we interpose a subinterval of the reservoir, of the appropriate length, between the second and third thirds of the stack. Thus the next step is:

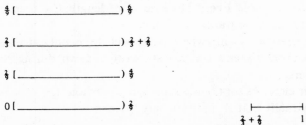

Continue in this manner.

4.5. Two examples

The transformation T is defined by mapping each interval in a stack linearly to the one above it. Eventually T is defined on all of $[0, 1]$ in this way.

The isomorphism with (Ω, σ, μ) arises from corresponding $[0, 2/3)$ with $[0]$ and $[2/3, 1]$ with $[1]$: to $x \in [0, 1]$ we assign the sequence $\{\chi_{[2/3, 1]}(T^k x)\}$.

The ideas of the Kakutani and Chacon constructions are similar: in each case a delay is inserted into a fairly regular (discrete spectrum) system. This is sufficient to produce weak mixing but not strong mixing.

Proposition 5.8 (Ω, σ, μ) is weakly mixing.

Proof: Suppose that $f : \Omega \to \mathbb{C}$ is an eigenfunction with eigenvalue $e^{2\pi i \lambda}$. Then for large n, f is approximately constant on each cylinder set

$$[A_n] = \{x \in \Omega : x_0 x_1 \ldots x_{l_n-1} = A_n\}.$$

If f actually *were* constant on such a set, we could find

$$x = -\cdot A_n A_n 1 A_n \ldots \in [A_n],$$

in which case we would have

$$\sigma^{l_n} x = -\cdot A_n 1 A_n \ldots \in [A_n]$$

and

$$\sigma^{2l_n+1} x = -\cdot A_n \ldots \in [A_n].$$

Then $f(\sigma^{l_n} x) = e^{2\pi i l_n \lambda} f(x)$ and $f(\sigma^{2l_n+1} x) = e^{2\pi i (2l_n+1)\lambda} f(x)$, but these are both equal to $f(x)$, so we could conclude that

$$(e^{2\pi i \lambda})^{l_n} = (e^{2\pi i \lambda})^{2l_n+1} = 1,$$

and

$$e^{2\pi i \lambda} = 1.$$

The actual argument applies Lusin's Theorem to find, given $\varepsilon > 0$, a set F with $\mu(F) > 1 - \varepsilon$ such that if $f(x) = e^{2\pi i \Theta(x)}$, then $\Theta(x)$ is uniformly continuous on F. Find a cylinder set E of the form

$$E = \sigma^k [A_n]$$

such that $\mu(E \cap F) > (1 - \varepsilon) \mu(E)$ and the oscillation of Θ on $E \cap F$ is less than ε. Then

$$\Theta(\sigma^{l_n} x) = l_n \lambda + \Theta(x), \quad \Theta(\sigma^{2l_n+1} x) = (l_n + 1)\lambda + \Theta(\sigma^{l_n} x) \pmod{2\pi},$$

so that

$$\Theta(\sigma^{2l_n+1} x) - \Theta(\sigma^{l_n} x) = \lambda + \Theta(\sigma^{l_n} x) - \Theta(x).$$

For small ε, there is an $x \in (E \cap F) \cap \sigma^{-l_n}(E \cap F) \cap \sigma^{-(2l_n+1)}(E \cap F)$ (say, $x \in \sigma^k [A_{n+1}]$); then both the left side and $\Theta(\sigma^{l_n} x) - \Theta(x)$ are small, so we must have $\lambda = 0$.

220   4. *More about recurrence*

Proposition 5.9 (Ω, σ, μ) is not strongly mixing.

Proof: For any $k = 0, 1, 2\ldots$, $\sigma^{l_k}[A_k] \cap [A_k]$, which consists of all points of the form—. $A_k A_k$—, has measure greater than or equal to that of $[A_{k+1}]$, namely $\frac{2}{3}(1/3^{k+1}) = \frac{1}{3}\mu[A_k]$.

For each k, the sets in $\mathscr{F}_k = \{[A_k], \sigma[A_k], \ldots, \sigma^{l_k-1}[A_k]\}$ are pairwise disjoint, and together with $\sigma^{l_k}[A_k]$ they cover Ω. Moreover, for each n, $\bigcup_{k \geq n} \mathscr{F}_k$ generates the full σ-algebra of measurable sets. Therefore, given any measurable set E, any $\varepsilon > 0$, and any n, we can find a $k \geq n$ and finitely many integers $i_j = 0, 1, \ldots, l_k - 1$ such that

$$\mu(E \Delta \bigcup_j \sigma^{i_j}[A_k]) < \varepsilon.$$

Then

$$\mu(\sigma^{l_k} E \cap E) \geq \mu\left[\left(\sigma^{l_k} \bigcup_j \sigma^{i_j}[A_k]\right) \cap \left(\bigcup_j \sigma^{i_j}[A_k]\right)\right] - 2\varepsilon$$

$$\geq \sum_j \mu(\sigma^{l_k}\sigma^{i_j}[A_k] \cap \sigma^{i_j}[A_k]) - 2\varepsilon = \sum_j \mu(\sigma^{l_k}[A_k] \cap [A_k]) - 2\varepsilon$$

$$\geq \frac{1}{3}\sum_j \mu[A_k] - 2\varepsilon \geq \frac{1}{3}[\mu(E) - \varepsilon] - 2\varepsilon.$$

This is inconsistent with strong mixing if $\mu(E)$ and ε are small enough.

Proposition 5.10 (*del Junco* 1978) (Ω, σ, μ) is *metrically prime*: if $\phi: \Omega \to Y$ is a measurable map to a system (Y, S, ν) such that $\mu\phi^{-1} = \nu$ and $\phi\sigma = S\phi$ a.e., then either ϕ is one-to-one a.e. or ϕ is constant a.e.

Thus the system (Ω, σ, μ) has *no proper factors*; equivalently, the σ-algebra of measurable subsets of Ω admits *no proper invariant sub-σ-algebras*.

Some preliminary discussion is in order before we make the combinatorial observations about ω that constitute the actual proof. We may assume that $Y \subset \{0, 1\}^{\mathbb{Z}}$. For if we choose a set $E \subset Y$ with $0 < \mu(E) < 1$ and define

$$\psi(x)_k = \chi_E(\phi(\sigma^k x)) \quad \text{for} \quad x \in \Omega$$

and

$$\lambda(G) = \mu(\psi^{-1} G)$$

for measurable $G \subset \{0, 1\}^{\mathbb{Z}}$, then $\psi: (\Omega, \sigma, \mu) \to (\{0, 1\}^{\mathbb{Z}}, \sigma, \lambda)$ will be a nontrivial homomorphism if ϕ is.

Second, we will employ the $\bar{d}$ *distance* between sequences of 0s and 1s, which is defined as follows: if $x, y \in \{0, 1\}^{\mathbb{Z}}$, then

$$\bar{d}_i^j(x, y) = \frac{1}{j - i + 1} \quad \text{card } \{k : i \leq k \leq j, x_k \neq y_k\}$$

4.5. Two examples

and
$$\bar{d}(x, y) = \limsup_{i,j \to \infty} \bar{d}^j_{-i}(x, y).$$

Thus $\bar{d}(x, y)$ measures the average frequency of the disagreements between x and y. We have
$$\bar{d}(\sigma x, \sigma y) = \bar{d}(x, y)$$
and
$$\bar{d}(x, z) \leq \bar{d}(x, y) + \bar{d}(x, z),$$
so that $\bar{d}$ is an invariant pseudometric.

A map $\psi: \{0, 1\}^{\mathbb{Z}} \to \{0, 1\}^{\mathbb{Z}}$ is called a *finite code* if it commutes with the shift and there is an n such that $\psi(x)_0$ depends only on x_k for $-n \leq k \leq n$. Thus the finite codes are exactly the shift-commuting continuous maps – or *endomorphisms* – of $\{0, 1\}^{\mathbb{Z}}$. (There is quite a lot known about such maps – see Hedlund 1969). The smallest such n is denoted by $|\psi|$.

Given our measurable map $\phi: \{0, 1\}^{\mathbb{Z}} \to \{0, 1\}^{\mathbb{Z}}$ and $\varepsilon > 0$, we can find a finite code ψ such that $\bar{d}(\phi(x), \psi(x)) < \varepsilon$ a.e. For let $\mathscr{B}^j_i$ denote the σ-algebra generated by the sets $\sigma^k[0]$, $i \leq k \leq j$. We can choose an n and a (cylinder) set $B \in \mathscr{B}^n_{-n}$ such that if $\mathcal{O} = \{x : \phi(x)_0 = 0\}$, then
$$\mu(B \Delta \mathcal{O}) < \varepsilon.$$
Define ψ by
$$\psi(x)_k = \chi_{B^c}(\sigma^k x).$$
Then $\phi(x)_k \neq \psi(x)_k$ if and only if $\sigma^k x \in B \Delta \mathcal{O}$, so that
$$\bar{d}(\phi(x), \psi(x)) = \limsup_{i,j \to \infty} \frac{1}{j+i+1} \sum_{k=-i}^{j} \chi_{B \Delta \mathcal{O}}(\sigma^k x) = \mu(B \Delta \mathcal{O}) < \varepsilon \text{ a.e.},$$
by the Ergodic Theorem and ergodicity of (Ω, σ, μ).

Our goal is to prove that *if $\phi: \Omega \to Y$ is the given homeomorphism, and if ϕ is not one-to-one a.e., then $\bar{d}(\phi(x), \phi(\sigma x)) = 0$ a.e.* For then, as above, we will have for a.e. x that
$$0 = \bar{d}(\phi(x), \phi(\sigma x))$$
$$= \limsup_{i,j \to \infty} \frac{1}{j+i+1} \text{ card } \{k : -i \leq k \leq j, \phi(x)_k \neq \phi(x)_{k+1}\}$$
$$= \limsup_{i,j \to \infty} \frac{1}{j+i+1} \sum_{k=-i}^{j} \chi_D(\sigma^k \phi(x)),$$
where $D = \{z \in \{0, 1\}^{\mathbb{Z}} : z_0 \neq z_1\}$. Choosing x so that $\phi(x)$ is ν-generic, we see that $\nu(D) = 0$. Thus $\mu(\phi^{-1} D) = 0$, $\phi(x)_0 = \phi(x)_1$ a.e. $d\mu$, and ϕ is constant a.e. on Ω.

4. More about recurrence

Proof of the Proposition: Suppose that $\phi: \Omega \to Y \subset \{0,1\}^Z$ is not one-to-one a.e. Let $\varepsilon > 0$. Choose a finite code ψ such that $\bar{d}(\phi(x), \psi(x)) < \varepsilon$ a.e. Then

$$\bar{d}(\phi(x), \phi(\sigma x)) \leq \bar{d}(\phi(x), \psi(x)) + \bar{d}(\psi(x), \psi(\sigma x)) + \bar{d}(\psi(\sigma x), \phi(\sigma x))$$
$$< \bar{d}(\psi(x), \psi(\sigma x)) + 2\varepsilon,$$

and if $\phi(x) = \phi(y)$ then $\bar{d}(\psi(x), \psi(y)) < 2\varepsilon$.

Let us define ψ on finite blocks B by agreeing that $\psi(B)$ is the initial $l(B)$-block of the image under ψ of the point $...000.B0000... \in \Omega$. For a block A, let $\bar{A}$ and $\underline{A}$ denote the block A with its first (respectively last) element deleted. We will show that $\psi(\bar{A}_n)$ and $\psi(\underline{A}_n)$ differ in only a small percentage of places for infinitely many n. In fact, if q_n is the number of places in which $\psi(\bar{A}_n)$ and $\psi(\underline{A}_n)$ differ, we will show that, for infinitely many n, $q_{n-1}/l_{n-1} < 36\varepsilon$. For this purpose, it is enough to show that if $\phi(x) = \phi(y)$, then for infinitely many n

$$\bar{d}(\psi(x), \psi(y)) \geq \frac{1}{18} \frac{q_{n-1}}{l_{n-1}}.$$

Suppose that $\phi(x) = \phi(y)$. We may assume that $y \notin \mathcal{O}(x)$, since for each k the invariant set $\{x: \phi(x) = \phi(\sigma^k x)\}$ must have measure 0 – otherwise $\phi = \phi\sigma^k$ a.e. for some k, and ϕ would have a non-constant eigenfunction.

We may also assume that x and y are not in the orbit of any point which is of the form $—.A_n—$ for every n or $—A_n.—$ for every n, since the sets

$$\bigcap_{n=0}^{\infty} [A_n] \quad \text{and} \quad \bigcap_{n=0}^{\infty} \sigma^{l_n}[A_n]$$

both have measure 0.

Notice that a point which is in $[A_n]$ for infinitely many n is actually in $[A_n]$ for all n. Thus for large n the central coordinates of x and y are each contained in blocks A_{n+1}. Note that because of (2) of the Lemma (which says that A_{n+1} does not overlap itself in x and y), for each n x and y can be resolved uniquely into sequences on the two symbols A_{n+1} and 1. Let us call the appearance of A_{n+1} in x which contains x_0 the central A_{n+1} in x, even though it may not be exactly centered. Define $k_n(x) = 1, 2,$ or 3 according to whether x_0 is contained in the first, second, or third A_n which comprises A_{n+1}.

Let $E_n = \{z: k_n(z) = 1 \text{ or } 3\}$. The cutting and stacking construction allows us to compute $\mu(E_n \cap E_{n+1} \cap ... \cap E_{n+r})$. For consider the stack present at the $(n+1)$th stage.

4.5. Two examples

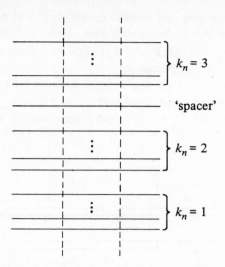

When this stack is cut into thirds along the dotted lines and restacked, with a spacer between the second and third chunks, exactly 2/3 of the part where $k_n = 1$ or 3 falls into the new first and third chunks, i.e. the set where $k_{n+1} = 1$ or 3. Continuing to cut into thirds and stack (with spacers), we see that

$$\mu(E_n \cap E_{n+1} \cap \ldots \cap E_{n+r}) = (2/3)^r \mu(E_n).$$

Therefore, for each n

$$\mu\left(\bigcap_{k \geq n} E_k\right) = 0,$$

and it follows that with probability 1 $k_n = 2$ for infinitely many n. We assume that our x and y are elements of this set of full measure.

We will show now that *there are infinitely many j_0 such that $k_{j_0}(x) \neq k_{j_0}(y)$ and the central A_{j_0}s of x and y overlap at least l_{j_0-1} places.* (See del Junco, Rahe and Swanson 1980.) The idea is that the spacers (1s) in the central A_{j_0+1}'s of x and y should be far enough apart that, using a sort of differential delay in x and y, we will be able to find A_{j_0-1} and $\bar{A}_{j_0-1}$ in one sequence facing the same block B in the other.

Fix an n_0 such that $k_{n_0}(x) = 2$. There is a $j > n_0$ such that $k_j(x) \neq k_j(y)$. For otherwise, since $k_{n_0+1}(x) = k_{n_0+1}(y)$, there is an i with $|i| \leq l_{n_0+1}$ such that $\sigma^i x$ and y agree on a block of length l_{n_0+1} which includes the center of y. Then $k_{n_0+2}(x) = k_{n_0+2}(y)$ implies that $\sigma^i x$ and y agree on a block of length l_{n_0+2} which includes the center of y, and so on. Hence either $\sigma^i x$ and y are different and positively or negatively asymptotic – a possibility that we have excluded by discarding the orbits of the sets $\bigcap [A_n]$ and $\bigcap \sigma^{l_n}[A_n]$ – or else $\sigma^i x = y$, which is also not allowed.

Let j_0 be the smallest $j > n_0$ such that $k_j(x) \neq k_j(y)$. I claim that the central A_{j_0}'s of x and y overlap at least $l_{j_0 - 1}$ places.

Suppose first that $j_0 = n_0 + 1$. Then the central position falls in the second A_{n_0} in the central A_{n_0+1} in x and in either the first or third A_{n_0} in the central A_{n_0+1} in y, so the central A_{n_0+1}'s in x and y must overlap at least $l_{n_0} = l_{j_0 - 1}$ places:

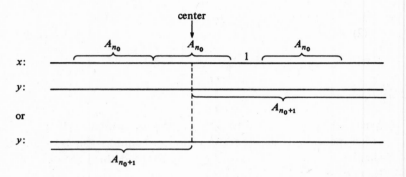

The other possibility is that $j_0 > n_0 + 1$. Then $k_{j_0-1}(x) = k_{j_0-1}(y)$, so that the central A_{j_0}'s in x and y overlap at least $2l_{j_0-1}$ places.

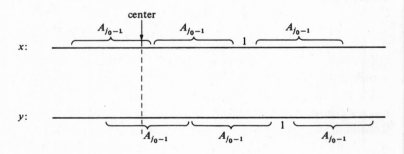

Fix a j_0 such that $k_{j_0}(x) \neq k_{j_0}(y)$ and the central A_{j_0}'s of x and y overlap at least l_{j_0-1} places. We will show now that for such j_0, we can find within the set of places occupied by the central A_{j_0+1} of x the blocks $\bar{A}_{j_0-1}$ and $\underline{A}_{j_0-1}$ (in one of x or y) facing the same block B (in the other). Because of symmetry, there are only three cases to consider.

(1) Suppose $k_{j_0}(x) = 1$ and $k_{j_0}(y) = 2$.

4.5. Two examples

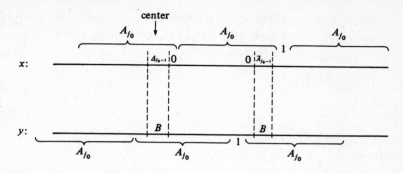

(2) Suppose $k_{j_0}(x) = 2$ and $k_{j_0}(y) = 3$.

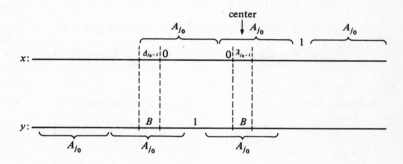

(3) Suppose $k_{j_0}(x) = 1$ and $k_{j_0}(y) = 3$.

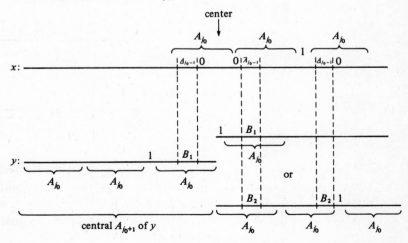

The central A_{j_0+1} of y is followed either by $1A_{j_0+1}$ or A_{j_0+1}. If it is $1A_{j_0+1}$, use B_1; if it is A_{j_0+1}, use B_2.

4. More about recurrence

Thus if $\psi(\underline{A}_{j_0-1})$ and $\psi(\bar{A}_{j_0-1})$ differ in q_{j_0-1} places, then (allowing for edge effects) $\psi(x)$ and $\psi(y)$ differ in at least $q_{j_0-1} - 4|\psi|$ places in $[-l_{j_0+1}, l_{j_0+1} - 1]$. Therefore

$$2\varepsilon > \bar{d}(\psi(x), \psi(y)) \geq \limsup_{j_0} \frac{q_{j_0-1}}{2l_{j_0+1}} = \limsup_{j_0} \frac{1}{2} \frac{q_{j_0-1}}{9l_{j_0-1} + 4}$$

$$= \frac{1}{18} \limsup_{j_0} \frac{q_{j_0-1}}{l_{j_0-1}},$$

so that $q_{j_0-1}/l_{j_0-1} < 36\varepsilon$ for infinitely many j_0.

Apply this observation to central A_j's of a typical $z \in \Omega$, for which the limit used to compute $\bar{d}(\phi(z), \phi(\sigma z))$ exists, to see that

$$\bar{d}(\phi(z), \phi(\sigma z)) = \lim_{i,j \to \infty} \bar{d}^j_{-i}(\phi(z), \phi(\sigma z))$$

$$\leq \liminf_{i,j \to \infty} [\bar{d}^j_{-i}(\phi(z), \psi(z)) + \bar{d}^j_{-i}(\psi(z), \psi(\sigma z)) + \bar{d}^j_{-i}(\psi(\sigma z), \phi(\sigma z))]$$

$$\leq 2 \limsup_{i,j \to \infty} \bar{d}^j_{-i}(\phi(z), \psi(z)) + \liminf_{i,j \to \infty} \bar{d}^j_{-i}(\psi(z), \psi(\sigma z))$$

$$< 2\varepsilon + 36\varepsilon = 38\varepsilon.$$

Therefore $\bar{d}(\phi(z), \phi(\sigma z)) = 0$ a.e., and the proof is complete.

5
Entropy

The concept of entropy was invented by Clausius in 1854; Shannon carried it over to information theory in 1948 and Kolmogorov to ergodic theory in 1958. In each setting entropy is a measure of randomness or disorder.

The importance of entropy in ergodic theory arises from its usefulness in connection with the isomorphism problem for measure-preserving transformations. There are two major theorems concerning ergodic-theoretic entropy. The first, due to Kolmogorov and Sinai, says that the full entropy of a transformation can be computed by finding its entropy with respect to a generator. This makes possible the actual computation of entropy in a variety of cases and leads to the conclusion that Bernoulli schemes of different entropies are not isomorphic; in this way Kolmogorov and Sinai settled in the negative the old question of whether or not $\mathscr{B}(\frac{1}{2}, \frac{1}{2})$ and $\mathscr{B}(\frac{1}{3}, \frac{1}{3}, \frac{1}{3})$ are isomorphic. The second, due to Ornstein in 1970, says that for Bernoulli schemes entropy is a *complete* invariant: two Bernoulli schemes are isomorphic if and only if they have the same entropy. We will give a brief introduction to this subject in Sections 6.4 and 6.5.

Entropy is a sort of categorical concept, versions of it existing in group theory, graph theory, and other areas. We begin with a brief look, for background, at the idea of entropy in the three fields in which it plays especially important roles.

5.1. Entropy in physics, information theory, and ergodic theory

A. Physics

Rudolf Clausius created the thermodynamical concept of entropy in 1854. He coined the actual term 'entropy' fourteen years later, from the Greek roots for 'transformation content'. When a system changes from state Σ_1 to state Σ_2, the change in entropy is given by the curve integral

$$S = \int_{\Sigma_1}^{\Sigma_2} \frac{dQ}{T}$$

228   5. Entropy

in state space, where Q is the heat content of the system and T is its temperature. Later Boltzmann showed that (after normalization) the entropy S of a system is proportional to the logarithm of the relative probability of its state:

$$S = k \log P,$$

where k is Boltzmann's constant. By choosing the state so that S (or, equivalently, P) is maximized, one can derive, for example, the Maxwell–Boltzmann velocity distribution for an ideal gas.

Let us consider a simple illustration. Suppose that an isolated system consists of N identical molecules, each of which can occupy any one of a number of states $\Sigma_1, \Sigma_2, \ldots$ in phase (position-momentum) space. Consider a state of the system in which there are N_i molecules in state Σ_i ($i = 1, 2, \ldots, k$). The number of ways to achieve such a distribution is

$$\binom{N}{N_1}\binom{N - N_1}{N_2} \cdots \binom{N - (N_1 + N_2 + \cdots + N_{k-1})}{N_k}$$
$$= \frac{N!}{N_1! N_2! \ldots N_k!}.$$

This number is proportional to the relative probability of the state, and so for large N the entropy of the state approximately is proportional to (using Stirling's estimate $\log n! \sim n \log n - n$)

$$\log N! - \sum \log N_i! \sim N \log N - \sum N_i \log N_i.$$

If $p_i = N_i/N$ is the probability of a given molecule being in state Σ_i, then

$$S \sim N \log N - \sum N_i \log N_i = N \log N - N \sum p_i (\log p_i + \log N)$$
$$= N \sum p_i \log p_i,$$

and the entropy per particle is

(1) $-\sum p_i \log p_i.$

Suppose that the state Σ_i has energy E_i associated with it. Then maximizing S under the constraints

$$\sum p_i E_i = E \text{ ' (fixed total energy)},$$
$$\sum p_i = 1,$$

yields the Maxwell–Boltzmann distribution, in which

$$p_i = \frac{e^{-E_i/kT}}{\sum_j e^{-E_j/kT}}.$$

The high-probability, high-entropy states are pictured as ones which achieve a high degree of disorder and randomness. They also exhibit a sort of statistical attractiveness for other states; see our discussion of the Recurrence Theorem. In fact, according to the Second Law of Thermodynamics the entropy of a developing isolated system does not decrease. If we start in a relatively low probability state, say with a warm brick next to a cold one, the two-brick system will move to the maximal-entropy state in which both bricks have the same temperature. (Of course there will always be very slight fluctuations of the temperatures of the two bricks about their exact mean; and there is a positive probability of a fluctuation large enough to recreate the original state. Note that laws of physics, however, talk about what *does* happen and not about what *may* happen.) When the bricks reach thermal equilibrium, the system is somehow more disorganized than at first, and there has been some loss of information. Thus there is a connection between the concepts of entropy in physics and in information theory.

B. Information theory

There is concrete justification for the feeling that increased entropy, hence increased disorder and randomness, is associated with a decrease in available information and an increase in uncertainty. Suppose a gas molecule moves freely in a box which is divided into two equal halves by a partition;

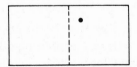

then at any time we know which half of the box the molecule is in. Suppose now that we allow the partition to slide under the impacts of the molecule, as the temperature of the molecule is kept constant by contact with a heat reservoir. Since the molecule does work in pushing back the wall, it must draw heat from the reservoir. The heat content of the system increases and so does its entropy:

$$\Delta S = \frac{\Delta Q}{T}.$$

The increase in entropy corresponds to a loss of information, since we no longer have as precise an idea of where the molecule is at any moment.

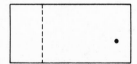

There has been an increase in uncertainty together with the increase in disorder.

In 1948 Claude Shannon initiated the mathematical study of information transmission. The (initial) key ideas are that (1) the amount of information transmitted can be measured by the amount of uncertainty removed, and (2) the uncertainty regarding a transmission is proportional to the expected value of the logarithm of the probability of the symbol received. Let us be specific.

Consider a *source*, which is thought of as a ticker-tape or teletype emitting a string of symbols $\ldots x_0 x_1 x_2 \ldots$, where each x_i is an element of a finite alphabet $A = \{a_1, a_2, \ldots, a_n\}$. Suppose the probability of receiving a_i is p_i for each $i = 1, \ldots, n$. If each symbol is transmitted independently of what has come before, we can take as a measure of the *amount of information transmitted per symbol on the average* (i.e., perhaps over many transmissions) the quantity

(2) $$H = -\sum_{i=1}^{n} p_i \log_2 p_i.$$

(We use logarithms to the base 2 because usually information is measured in terms of *binary bits*, counting the number of yes-no questions that must be answered in order to convey it.) Four remarks suggest the appropriateness of the quantity H.

(1) Compare with formula (1) above.

(2) If some $p_{i_0} = 1$ while all other $p_i = 0$, then there is no uncertainty at all about what the source will do, and so there is no information being conveyed. In this case, under the convention $0 \log 0 = \lim_{x \to 0^+} x \log_2 x = 0$, we also have $H = 0$.

(3) H is a maximum when all $p_i = 1/n$. To see this, consider the function f defined on $[0, 1]$ by

$$f(t) = \begin{cases} 0 & \text{if } t = 0 \\ -t \log_2 t & \text{if } 0 < t \leq 1. \end{cases}$$

The function f is continuous, nonnegative, and concave downward:

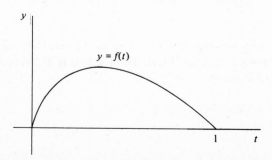

5.1. Entropy in physics, information, and ergodic theory

From Jensen's Inequality it follows that for any positive $\lambda_1, \lambda_2, \ldots, \lambda_n \in [0, 1]$,

$$f\left(\frac{1}{n}\sum_{i=1}^{n} \lambda_i\right) \geq \frac{1}{n}\sum_{i=1}^{n} f(\lambda_i).$$

Thus for any probability vector $(\lambda_1, \lambda_2, \ldots, \lambda_n)$,

$$\frac{1}{n}\sum_{i=1}^{n} f(\lambda_i) \leq f\left(\frac{1}{n}\sum_{i=1}^{n} \lambda_i\right) = f\left(\frac{1}{n}\right) = -\frac{1}{n}\log_2 \frac{1}{n} = \frac{1}{n}\log_2 n,$$

so $H(\lambda) \leq \log_2 n$. But clearly $H(1/n, 1/n, \ldots, 1/n) = \log_2 n$. This case, when all symbols are equally likely, is the one in which there is the most uncertainty about the output of the source and hence the one in which the source as it functions conveys the maximum amount of information.

(4) H is the average of the information conveyed by the receipt of each symbol, if we take the information content of an event E to be $-\log_2 P(E)$. This is a reasonable choice since the only measure of information of the form $f(P(E))$, where $f: \mathbb{R} \to [0, \infty)$ is continuous, which is additive for independent events $-f(P(E_1 \cap E_2)) = f(P(E_1)P(E_2)) = f(P(E_1)) + f(P(E_2))$ – is $-\log_r P(E)$ for some $r > 0$. For another justification of the definition of H, see Khintchine (1957).

For a general source, the probability of a given symbol's being received may depend on what has been received beforehand. For example, if the ticker tape is putting out English prose, the probability of receiving a 'u' will rise dramatically each time a 'q' comes through. Such effects can be taken into account in the calculation of average information per symbol by grouping into blocks. For each $k = 1, 2, \ldots$, let $\mathscr{C}_k$ be the family of all blocks of length k on the symbols $a_1, a_2, \ldots, a_n$. Each block C of $\mathscr{C}_k$ has a certain probability $P(C)$ (possibly 0) of being received. The average information *per symbol* in a transmission of length k is then

(3) $$H_k = -\frac{1}{k} \sum_{C \in \mathscr{C}_k} P(C) \log_2 P(C).$$

We define the *entropy of the source* itself to be

(4) $$h = \lim_{k \to \infty} -\frac{1}{k} \sum_{C \in \mathscr{C}_k} P(C) \log_2 P(C).$$

The limit can be proved to exist (Proposition 2.10), and it represents the average amount of information which the source conveys with every symbol it prints.

Equivalently, h measures the average degree of uncertainty the reader of the ticker tape has about what the next symbol will be, given whatever he or she has received so far. By letting people guess the next letter in samples of English prose, having seen 0 or 1 or 2 or ... immediately preceding letters, Shannon estimated the entropy of typical English prose to be about 1. (cf. Proposition 2.12). This is rather low compared to the $\log_2 26$ entropy of a source printing out the 26 letters independently. The reason for this is that English prose is extremely *redundant*. The rules of word formation, grammar and logic all restrict the number of combinations that actually occur and thereby decrease our uncertainty about what is coming next. Of course this redundancy is essential for the practical functioning of language, since it allows for the correction of errors, lapses in attention, etc. Sources putting out poetry or mathematical proofs, for example, will have higher entropy but will also suffer more disastrously if an occasional misprint occurs.

C. Ergodic theory

The source discussed above was nothing but a finite-state stationary process. Outputs of the source are infinite strings $\ldots x_0 x_1 x_2 \ldots$ of symbols of an alphabet $A = \{a_1 a_2, \ldots, a_n\}$, that is to say points of the sequence space A^Z. If we assume that the source is indeed stationary, so that the probability of receiving any string does not change with time, then there is a measure on A^Z which is preserved by the shift transformation $\sigma: A^Z \to A^Z$. If μ is the product measure generated by the weights p_i for a_i, then we are dealing with the Bernoulli shift $\mathscr{B}(p_1, p_2, \ldots, p_n)$; this is the case when the symbols are transmitted independently of what has gone before. In any case, in calculating the entropy of the source according to formula (4), we have found the 'entropy' of a measure-preserving system $(A^Z, \mathscr{B}, \mu, \sigma)$.

Given any measure-preserving system $(X, \mathscr{B}, \mu, T)$, it is easy to associate many finite-state stationary processes. Just take a *partition*

$$\alpha = \{A_1, A_2, \ldots, A_n\}$$

of X into finitely many measurable sets (the A_i are pairwise disjoint, have positive measure, and cover X, all up to sets of measure 0). Each point $x \in X$ produces a single output of the ticker-tape as follows: at time j the tape prints out symbol A_i if $T^j x \in A_i$. Let us define the *entropy of a partition* $\alpha = \{A_1, A_2, \ldots, A_n\}$ to be

$$H(\alpha) = - \sum_{i=1}^{n} \mu(A_i) \log_2 \mu(A_i).$$

(cf. (2)). The appearance at the jth place on the tape of a block of symbols $A_{i_1} A_{i_2} \ldots A_{i_k}$ corresponds to the entry of $T^j x$ into the set

5.1. Entropy in physics, information, and ergodic theory

$A_{i_1} \cap T^{-1} A_{i_2} \cap \ldots \cap T^{-k+1} A_{i_k}$. This set is an element of the partition $\alpha_0^{k-1} = \alpha \vee T^{-1} \alpha \vee \ldots \vee T^{-k+1} \alpha$ which is the *least common refinement* of the partitions $\alpha, T^{-1}\alpha, \ldots, T^{-k+1}\alpha$.

Definitions: If $\alpha = \{A_1, \ldots, A_n\}$ and $\beta = \{B_1, \ldots, B_m\}$, then

$T^{-1}\alpha$ is the partition $\{T^{-1}A_1, \ldots, T^{-1}A_n\}$,
$\alpha \vee \beta$ is the partition $\{A_i \cap B_j : i = 1, \ldots, n; j = 1, \ldots, m\}$, and
β is a *refinement* of α, written $\beta \geqslant \alpha$, if each B_j is, up to a set of measure 0, a subset of some A_i.

Thus the quantity H_k in (3) is

$$\frac{1}{k} H(\alpha \vee T^{-1}\alpha \vee \ldots \vee T^{-k+1}\alpha),$$

and the entropy of the source defined by (4) is

$$h(\alpha, T) = h_\mu(\alpha, T) = \lim_{k \to \infty} \frac{1}{k} H(\alpha \vee T^{-1}\alpha \vee \ldots \vee T^{-k+1}\alpha).$$

The quantity $h(\alpha, T)$ (which we will soon prove exists) is called the *entropy of the transformation T with respect to the partition* α.

The number $h(\alpha, T)$ is a measure of the average uncertainty per unit time we have about which element of the partition α the point x will enter next (as it is moved by T), given its preceding history. Of course if α is foolishly chosen, there may not be much uncertainty – telling which cells of α the points $T^j x$ actually land in may not convey much information that could not have been guessed *a priori*. Therefore the *entropy of the transformation T* is defined to be the maximal uncertainty over all the finite-state processes associated with T:

$$h(T) = h_\mu(T) = \sup_\alpha h(\alpha, T).$$

The number $h(T)$ is a measure of our average uncertainty about where T moves the points of X. The size of $h(T)$ reflects the 'randomness' of T and the degree to which T disorganizes the space. *Clearly $h(T)$ is an isomorphism invariant of T.* More generally, if S is a factor of T, then $h(S) \leqslant h(T)$ (because each partition downstairs determines one of equal entropy upstairs, but not necessarily conversely). Thus in fact entropy is an invariant of *weak isomorphism*, where we say that S and T are *weakly isomorphic* if each is a factor of the other. (Only recently Steve Polit (1974) gave an example of two weakly isomorphic transformations that are not isomorphic. Another example has been given by Rudolph (1979).)

We may also think of a partition α as the collection of possible outcomes of an experiment (e.g. reading the most recent symbol printed out by a ticker-tape or rolling dice). The number $H(\alpha)$ is a measure of our (expected) uncertainty about the outcome of the experiment or equivalently of the amount of information that is gained by performing the experiment. The common refinement (or *join*) $\alpha \vee \beta$ represents the compound experiment formed by performing the experiments α and β simultaneously. Thus

$$\frac{1}{k}H(\alpha \vee T^{-1}\alpha \vee \ldots \vee T^{-k+1}\alpha)$$

is the average amount of information per repetition obtained from k repetitions of the experiment α, $h(\alpha, T)$ is the time average of the information content of the experiment α, and $h(T)$ is the maximum information per repetition that can be obtained from any experiment, so long as T is used to advance the time (i.e., to develop the system).

Is it possible to design an experiment so cleverly that it extracts all the information that is to be had, i.e., so that

$$h(\alpha, T) = h(T)?$$

It seems that if the partitions $\alpha \vee T^{-1}\alpha \vee \ldots \vee T^{-k+1}\alpha$ generate the full σ-algebra $\mathcal{B}$, which is to say that α is a *generator*, this might just be the case, since then the outcomes of any experiment could be described to any desired degree of accuracy in terms of some finite-length compound of α. Kolmogorov (1958) defined the entropy of T to be $h(\alpha, T)$ if T has a generator α and ∞ otherwise; the preceding definition is due to Sinai (1959a); the equivalence of the two is what is usually called the Kolmogorov–Sinai Theorem, which we will prove below. It provides the chief tool for the computation of entropy and permits us to show that, for example, $\mathcal{B}(\frac{1}{2}, \frac{1}{2})$ and $\mathcal{B}(\frac{1}{3}, \frac{1}{3}, \frac{1}{3})$ are not isomorphic. Before proving this theorem we need to establish a few properties of entropy and conditional entropy and to tie up the loose end concerning the existence of the limit

$$h(\alpha, T) = \lim_{k \to \infty} \frac{1}{k} H(\alpha \vee T^{-1}\alpha \vee \ldots \vee T^{-k+1}\alpha).$$

5.2. Information and conditioning

In the preceding section we saw that $-\log_2 \mu(E)$ is a reasonable measure of the information content of the announcement of the occurrence of an event E. Let $\alpha = \{A_1, A_2, \ldots\}$ be a *countable* partition of X into measurable sets. (For convenience we will deal sometimes with countable partitions, sometimes with finite ones.) For each $x \in X$, denote by $\alpha(x)$ the

5.2. Information and conditioning

element of α to which x belongs. Then the *information function* associated to α is defined to be

$$I_\alpha(x) = -\log_2 \mu(\alpha(x)) = -\sum_{A \in \alpha} \log_2 \mu(A) \chi_A(x),$$

so that $I_\alpha(x)$ takes the constant value $-\log_2 \mu(A)$ on the cell A of α. $I_\alpha(x)$ measures the gain in information when we learn to which element of α the point x belongs. Clearly

$$H(\alpha) = -\sum_{A \in \alpha} \mu(A) \log_2 \mu(A) = \int_X I_\alpha(x) d\mu(x),$$

A point x can be thought of as a particular complete world history. $I_\alpha(x)$ tells how much we learn by performing the experiment α in the case of a particular history, and $H(\alpha)$ is the average information gain over all possible histories.

It is useful to consider conditional information and entropy, which take into account information that may already be in hand. Let $\mathscr{F} \subset \mathscr{B}$ be a sub-σ-algebra of the family of measurable subsets of X. Recall that for $\phi \in L^1(X)$, the *conditional expectation* $E(\phi|\mathscr{F})$ of ϕ given $\mathscr{F}$ is an $\mathscr{F}$-measurable function on X which satisfies

$$\int_F E(\phi|\mathscr{F}) d\mu = \int_F \phi \, d\mu$$

for all $F \in \mathscr{F}$; $E(\phi|\mathscr{F})(x)$ represents our expected value for ϕ if we are *given the foreknowledge* $\mathscr{F}$, i.e., told exactly which sets of $\mathscr{F}$ the point x belongs to. The *conditional probability* of a set $A \in \mathscr{B}$ given $\mathscr{F}$ is

$$\mu(A|\mathscr{F}) = E(\chi_A|\mathscr{F});$$

it represents our revised estimate of the likelihood of the occurrence of A once we know the information in $\mathscr{F}$. Now if receipt of a symbol A from a ticker-tape source conveys an amount $-\log_2 \mu(A)$ of information, then if the information in $\mathscr{F}$ is already known receipt of the symbol A will have information content $-\log_2 \mu(A|\mathscr{F})$. Thus we define the *conditional information function* of a countable partition α given a σ-algebra $\mathscr{F} \subset \mathscr{B}$ to be

$$I_{\alpha|\mathscr{F}}(x) = -\sum_{A \in \alpha} \log_2 \mu(A|\mathscr{F})(x) \chi_A(x).$$

The *conditional entropy* of α given $\mathscr{F}$ is defined by

$$H(\alpha|\mathscr{F}) = \int_X I_{\alpha|\mathscr{F}}(x) d\mu(x).$$

Thus $I_{\alpha|\mathscr{F}}(x)$ measures our uncertainty about the outcome of the experiment α once we know to which elements of $\mathscr{F}$ the point x belongs, and $H(\alpha|\mathscr{F})$ is the average of these uncertainties over all possible x. Of course these quantities can be infinite; for finite partitions they are always finite.

Remarks 2.1
(i) $I_{\alpha|\mathscr{F}} \geq 0$ a.e., so that $H(\alpha|\mathscr{F}) \geq 0$.
(ii) $I_{\alpha|\{\varnothing,X\}} = I_\alpha$ a.e. and $H(\alpha|\{\varnothing, X\}) = H(\alpha)$.
(iii) $I_{\alpha|\mathscr{B}} = 0$ a.e., so that $H(\alpha|\mathscr{B}) = 0$.

The proof of these statements is left as an exercise.

Remark 2.2 Two alternative definitions of $I_{\alpha|\mathscr{F}}(x)$ may suggest themselves to the reader: $E(I_\alpha|\mathscr{F})$ and

$$-\sum_{A \in \alpha} \log_2 \mu(A|\mathscr{F}) \cdot \mu(A|\mathscr{F}).$$

The first is not as interesting as $I_{\alpha|\mathscr{F}}$, since its integral is just $H(\alpha)$. The second also has integral $H(\alpha|\mathscr{F})$, since for each $A \in \alpha$

$$E(\log_2 \mu(A|\mathscr{F}) \cdot \chi_A | \mathscr{F}) = \log_2 \mu(A|\mathscr{F}) \cdot \mu(A|\mathscr{F}),$$

so that

$$H(\alpha|\mathscr{F}) = E(I_{\alpha|\mathscr{F}}) = E(E(I_{\alpha|\mathscr{F}}|\mathscr{F})) = E\left(-\sum_{A \in \alpha} \log_2 \mu(A|\mathscr{F}) \cdot \mu(A|\mathscr{F})\right);$$

however, it is $\mathscr{F}$-measurable and so may not be sufficiently sensitive to α.

It is easy to describe $I_{\alpha|\mathscr{F}}$ and $H(\alpha|\mathscr{F})$ in case $\mathscr{F}$ is the (finite) σ-algebra $\mathscr{B}(\beta)$ generated by a finite partition $\beta = \{B_1, B_2, \ldots, B_m\}$. Then on each set $B \in \mathscr{B}(\beta)$ we have $\mu(A|\mathscr{B}(\beta)) = \mu(A|B) = \mu(A \cap B)/\mu(B)$, so that

$$I_{\alpha|\mathscr{B}(\beta)} = -\log_2 \mu(A|\mathscr{B}(\beta)) = -\log_2 \frac{\mu(A \cap B)}{\mu(B)} \quad \text{on } A \cap B$$

and

$$H(\alpha|\mathscr{B}(\beta)) = \sum_{A,B} -\log_2 \mu(A|B) \mu(A \cap B)$$

$$= \sum_{B \in \beta} \left(-\sum_{A \in \alpha} \log_2 \mu(A|B) \mu(A|B)\right) \mu(B)$$

$$= -\int_X \log_2 \mu(\alpha(x)|\beta(x)) \, d\mu(x).$$

In this case we will abbreviate $I_{\alpha|\mathscr{B}(\beta)}$ to $I_{\alpha|\beta}$ and $H(\alpha|\mathscr{B}(\beta))$ to $H(\alpha|\beta)$. Notice that $I_{\alpha|\beta}$ is, on a cell B of β, just the ordinary information function of the partition of the restricted probability space $(B, \mathscr{B} \cap B, \mu_B)$ (where

5.2. Information and conditioning

$\mu_B(E) = \mu(E)/\mu(B)$) determined by α. $I_{\alpha|\beta}(x)$ measures the amount of information gained when we are told which cell of α the point x is in, if we already know which cell of β it's in. $H(\alpha|\beta)$ measures our (average) uncertainty about the outcome of the experiment α if we already know the outcome of another experiment β.

We establish some useful computational properties of the conditional information and entropy. The reader should supply philosophical support for these formulas by interpreting them in terms of ticker-tapes and experiments. Recall that if $\mathscr{F}_1$ and $\mathscr{F}_2$ are σ-algebras, then $\mathscr{F}_1 \vee \mathscr{F}_2$ denotes the smallest σ-algebra containing $\mathscr{F}_1 \cup \mathscr{F}_2$. Similarly, we define $\bigvee_{n=1}^{\infty} \mathscr{F}_n = \mathscr{B}(\bigcup_{n=1}^{\infty} \mathscr{F}_n)$; also, by the join of infinitely many partitions we mean the σ-algebra they generate: $\bigvee_{n=1}^{\infty} \alpha_n = \mathscr{B}(\bigcup_{n=1}^{\infty} \mathscr{B}(\alpha_n)) = \bigvee_{n=1}^{\infty} \mathscr{B}(\alpha_n)$.

Proposition 2.3 If α and β are countable measurable partitions of X and $\mathscr{F}$ is a sub-σ-algebra of $\mathscr{B}$, then

$$I_{\alpha \vee \beta | \mathscr{F}} = I_{\alpha|\mathscr{F}} + I_{\beta|\mathscr{B}(\alpha) \vee \mathscr{F}} \quad \text{a.e.}$$

Proof: For each $B \in \beta$,

$$\mu(B|\mathscr{B}(\alpha) \vee \mathscr{F}) = \sum_{A \in \alpha} \frac{\mu(B \cap A|\mathscr{F})}{\mu(A|\mathscr{F})} \chi_A \quad \text{a.e.,}$$

where we take $\frac{0}{0} = 0$. This is so because the right-hand side is measurable with respect to $\mathscr{B}(\alpha) \vee \mathscr{F}$; and for a typical generating member $A' \cap F$ (where $A' \in \alpha$ and $F \in \mathscr{F}$) of $\mathscr{B}(\alpha) \vee \mathscr{F}$ we have

$$\int_{A' \cap F} \left(\sum_{A \in \alpha} \frac{\mu(B \cap A|\mathscr{F})}{\mu(A|\mathscr{F})} \chi_A \right) d\mu = \int_F \chi_{A'} \frac{\mu(B \cap A'|\mathscr{F})}{\mu(A'|\mathscr{F})} d\mu$$

$$= \int_F E\left(\chi_{A'} \frac{\mu(B \cap A'|\mathscr{F})}{\mu(A'|\mathscr{F})} \Big| \mathscr{F} \right) d\mu = \int_F E(\chi_{A'}|\mathscr{F}) \frac{\mu(B \cap A'|\mathscr{F})}{\mu(A'|\mathscr{F})} d\mu$$

$$= \int_F \mu(B \cap A'|\mathscr{F}) d\mu = \mu(B \cap A' \cap F) = \int_{A' \cap F} \chi_B d\mu.$$

Therefore $\sum_{A \in \alpha} [\log_2 \mu(A \cap B|\mathscr{F})/\mu(A|\mathscr{F})] \chi_A = \log_2 \mu(B|\mathscr{B}(\alpha) \vee \mathscr{F})$ a.e., and

$$\begin{aligned}
I_{\alpha \vee \beta | \mathscr{F}} &= -\sum_{\substack{A \in \alpha \\ B \in \beta}} \log_2 \mu(A \cap B|\mathscr{F}) \chi_{A \cap B} \\
&= -\sum_{\substack{A \in \alpha \\ B \in \beta}} \log_2 \mu(A|\mathscr{F}) \chi_A \chi_B - \sum_{B \in \beta} \log_2 \mu(B|\mathscr{B}(\alpha) \vee \mathscr{F}) \chi_B \\
&= -\sum_{A \in \alpha} \log_2 \mu(A|\mathscr{F}) \chi_A - \sum_{B \in \beta} \log_2 \mu(B|\mathscr{B}(\alpha) \vee \mathscr{F}) \chi_B \\
&= I_{\alpha|\mathscr{F}} + I_{\beta|\mathscr{B}(\alpha) \vee \mathscr{F}}.
\end{aligned}$$

5. Entropy

Corollary 2.4 For countable measurable partitions α and β,
$$I_{\alpha \vee \beta} = I_\alpha + I_{\beta|\alpha} \text{ a.e.}$$
Proof: Take $\mathscr{F} = \{\emptyset, X\}$ in the Proposition.

Proposition 2.5 Let α and β be countable measurable partitions of X and $\mathscr{F}, \mathscr{F}_1$, and $\mathscr{F}_2$ sub-σ-algebras of $\mathscr{B}$.

(1) $H(\alpha \vee \beta | \mathscr{F}) = H(\alpha | \mathscr{F}) + H(\beta | \mathscr{B}(\alpha) \vee \mathscr{F})$.
(2) If $\mathscr{F}_1 \supset \mathscr{F}_2$, then $H(\alpha | \mathscr{F}_1) \leq H(\alpha | \mathscr{F}_2)$.
(3) $H(T^{-1}\alpha | T^{-1}\mathscr{F}) = H(\alpha | \mathscr{F})$.
(4) If $\alpha \leq \beta$, then $H(\alpha | \mathscr{F}) \leq H(\beta | \mathscr{F})$.
(5) $H(\alpha \vee \beta | \mathscr{F}) \leq H(\alpha | \mathscr{F}) + H(\beta | \mathscr{F})$.
  (1') $H(\alpha \vee \beta) = H(\alpha) + H(\beta | \alpha)$.
  (3') $H(T^{-1}\alpha) = H(\alpha)$.
  (4') If $\alpha \leq \beta$, then $H(\alpha) \leq H(\beta)$.
  (5') $H(\alpha \vee \beta) \leq H(\alpha) + H(\beta)$.

Proof: The primed statements follow from the corresponding unprimed ones by taking $\mathscr{F} = \{\emptyset, X\}$.

(1) This follows directly from Proposition 2.3 by integrating.

(2) Apply Jensen's Inequality (1.4.8) to the concave function $f(t) = -t \log_2 t$: for each $A \in \alpha$,
$$E(f \circ \mu(A | \mathscr{F}_1) | \mathscr{F}_2) \leq f \circ E(\mu(A | \mathscr{F}_1) | \mathscr{F}_2) = f \circ \mu(A | \mathscr{F}_2).$$
Integrating gives
$$\int_X f \circ \mu(A | \mathscr{F}_1) d\mu \leq \int_X f \circ \mu(A | \mathscr{F}_2) d\mu,$$
and (2) follows by summing over all $A \in \alpha$, bearing in mind Remark 2.2.

(3) It is easily checked that $\mu(T^{-1}A | T^{-1}\mathscr{F})(x) = \mu(A | \mathscr{F})(Tx)$ a.e. Therefore
$$H(T^{-1}\alpha | T^{-1}\mathscr{F}) = -\int \sum_{A \in \alpha} \log_2 \mu(T^{-1}A | T^{-1}\mathscr{F})(x) \chi_{T^{-1}A}(x) d\mu(x)$$
$$= -\int \sum_{A \in \alpha} \log_2 \mu(A | \mathscr{F})(Tx) \chi_A(Tx) d\mu(x)$$
$$= -\int \sum_{A \in \alpha} \log_2 \mu(A | \mathscr{F})(x) \chi_A(x) d\mu(x) = H(\alpha | \mathscr{F}).$$

(4) Since $H(\beta | \mathscr{B}(\alpha) \vee \mathscr{F}) \geq 0$ and $\alpha \vee \beta = \beta$ when $\alpha \leq \beta$, by (1)
$$H(\beta | \mathscr{F}) = H(\alpha \vee \beta | \mathscr{F}) = H(\alpha | \mathscr{F}) + H(\beta | \mathscr{B}(\alpha) \vee \mathscr{F}) \geq H(\alpha | \mathscr{F}).$$

(5) Since $\mathscr{F} \subset \mathscr{B}(\alpha) \vee \mathscr{F}$, this follows from (1) and (2).

5.2. Information and conditioning

Corollary 2.6 $H(\alpha|\mathscr{F}) \leq H(\alpha)$.
Proof: Take $\mathscr{F}_2 = \{\emptyset, X\}$ in (2) above.

Proposition 2.7 $H(\alpha|\mathscr{F}) = 0$ if and only if $\alpha \subset \mathscr{F}$ up to sets of measure 0.

Proof: If $\alpha \subset \mathscr{F}$, then $\mu(A|\mathscr{F}) = \chi_A$ a.e. for each $A \in \alpha$, so that $I_{\alpha|\mathscr{F}} = 0$ a.e.
Conversely, suppose that $H(\alpha|\mathscr{F}) = 0$. Then (by Remark 2.2) for each $A \in \alpha$ we must have $\log_2 \mu(A|\mathscr{F}) \cdot \mu(A|\mathscr{F}) = 0$ a.e. This implies that $\mu(A|\mathscr{F}) = 0$ or 1 a.e., and hence $\mu(A|\mathscr{F})$ is a.e. equal to χ_F for some $F \in \mathscr{F}$. Clearly $\mu(F) = \int \mu(A|\mathscr{F}) d\mu = \mu(A)$. On the other hand, $\mu(A \cap F) = \int_F \chi_A d\mu = \int_F \mu(A|\mathscr{F}) d\mu = \int_F \chi_F d\mu = \mu(F)$. Therefore $A = F$ up to a set of measure 0.

Corollary 2.8 $H(\alpha|\beta) = 0$ if and only if $\alpha \leq \beta$.

We say that two measurable partitions α and β are *independent*, and write $\alpha \perp \beta$, in case $\mu(A \cap B) = \mu(A)\mu(B)$ for each $A \in \alpha$ and $B \in \beta$. It is reasonable that the information gained from independent experiments should be the sum of the amounts produced by each.

Proposition 2.9 The following statements about two countable measurable partitions α and β of finite entropy are equivalent:
(1) $\alpha \perp \beta$
(2) $H(\alpha \vee \beta) = H(\alpha) + H(\beta)$.
(3) $H(\alpha|\beta) = H(\alpha)$.

Proof: The equivalence of (2) and (3) is clear in view of Proposition 2.5 (1').
If $\alpha \perp \beta$, then for each $A \in \alpha$ we have $\mu(A|\mathscr{B}(\beta)) = \mu(A)$ a.e.; hence $I_{\alpha|\beta} = I_\alpha$ a.e. and $H(\alpha|\beta) = H(\alpha)$.
Conversely, suppose that $H(\alpha|\beta) = H(\alpha)$. Then, with $f(t) = -t \log_2 t$,

$$\sum_{A \in \alpha} \left[f(\mu(A)) - \int_X f(\mu(A|\mathscr{B}(\beta))) d\mu \right] = 0.$$

In the first step of the proof of Proposition 2.5 (2) take $\mathscr{F}_2 = \{\emptyset, X\}$ to see that each term in this sum is nonnegative. Therefore

$$f(\mu(A)) = \int_X f(\mu(A|\mathscr{B}(\beta))) d\mu \quad \text{for each } A \in \alpha,$$

i.e.,

$$-\mu(A) \log_2 \mu(A) = \sum_{B \in \beta} -\log_2 \frac{\mu(A \cap B)}{\mu(B)} \frac{\mu(A \cap B)}{\mu(B)} \mu(B).$$

Suppose that $\beta = \{B_1, B_2, \ldots\}$; let $\lambda_k = \mu(B_k)$ and $x_k = \mu(A|B_k) = \mu(A \cap B_k)/\mu(B_k)$ for each k. We have

$$(\mu(A), f(\mu(A))) = \sum \lambda_k (x_k, f(x_k)).$$

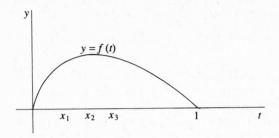

Now the region below the graph of f is convex, and we have written an extreme point of this region as a convex combination of other extreme points. This is impossible unless all the x_ks are the same, that is to say $\mu(A \cap B_k)/\mu(B_k) = \mu(A)$ for all k. Therefore $\alpha \perp \beta$.

Proposition 2.10 For each countable measurable partition α of X, $h(\alpha, T) = \lim_{n \to \infty} H(\alpha \vee T^{-1}\alpha \vee \ldots \vee T^{-n+1}\alpha)/n$ exists (it is possibly $+\infty$).

Proof: Suppose that $H(\alpha) < \infty$, since otherwise clearly $h(\alpha, T) = \infty$. For each $n = 1, 2, \ldots$ let $H_n = H(\alpha \vee T^{-1}\alpha \vee \ldots \vee T^{-n+1}\alpha)$. By Proposition 2.5 (4'), $\{H_n\}$ is increasing. Notice that this sequence is *subadditive*:

$$H_{n+m}$$
$$= H(\alpha \vee T^{-1}\alpha \vee \ldots \vee T^{-n+1}\alpha \vee T^{-n}(\alpha \vee T^{-1}\alpha \vee \ldots \vee T^{-m+1}\alpha))$$
$$\leq H(\alpha \vee T^{-1}\alpha \vee \ldots \vee T^{-n+1}\alpha) + H(\alpha \vee T^{-1}\alpha \vee \ldots \vee T^{-m+1}\alpha)$$
$$= H_n + H_m.$$

Now clearly $H_{jn} \leq jH_n$ and $H_j \leq jH_1$, so that $\{H_j/j : j = 1, 2, \ldots\}$ is bounded; hence

$$L = \liminf_{j \to \infty} \frac{H_j}{j} < \infty.$$

Given $\varepsilon > 0$, choose a large j such that $H_j/j < L + \varepsilon$. For each $n > j$, choose $i(n)$ such that

$$[i(n) - 1]j < n \leq i(n)j.$$

Then $H_n \leq H_{i(n)j}$, and hence

$$\frac{H_n}{n} \leq \frac{H_{i(n)j}}{n} \leq \frac{H_{i(n)j}}{[i(n)-1]j} \leq \frac{i(n)H_j}{[i(n)-1]j} < \frac{i(n)}{i(n)-1}(L + \varepsilon).$$

Thus if n is large enough, we will have

$$L - \varepsilon < \frac{H_n}{n} < L + 2\varepsilon.$$

5.2. Information and conditioning

Proposition 2.11 If $\mathscr{B}_1 \subset \mathscr{B}_2 \subset \ldots$ are sub-σ-algebras of $\mathscr{B}$ and $\mathscr{B}_\infty = \bigvee_{n=1}^\infty \mathscr{B}_n$, and α is a *finite* measurable partition, then

$$\lim_{n\to\infty} H(\alpha|\mathscr{B}_n) = H(\alpha|\mathscr{B}_\infty).$$

Proof: By the Martingale Convergence Theorem (see Section 3.4, noting that $E(\mu(A|\mathscr{B}_n)) = \mu(A)$ for all n)

$$\mu(A|\mathscr{B}_n) \to \mu(A|\mathscr{B}_\infty) \quad \text{a.e.} \quad \text{for each } A \in \alpha.$$

Composing with the bounded continuous function $f(t) = -t\log_2 t$, we have

$$f(\mu(A|\mathscr{B}_n)) \to f(\mu(A|\mathscr{B}_\infty)) \quad \text{a.e. for each } A \in \alpha.$$

By the Bounded Convergence Theorem,

$$\int_X f(\mu(A|\mathscr{B}_n))\,d\mu \to \int_X f(\mu(A|\mathscr{B}_\infty))\,d\mu \quad \text{for each } A \in \alpha,$$

and the result follows by summing over the *finitely* many $A \in \alpha$.

Proposition 2.12 If α is a *finite* partition, then

$$h(\alpha, T) = \lim_{n\to\infty} H\!\left(\alpha \,\Big|\, \bigvee_{k=1}^n T^{-k}\alpha\right) = H\!\left(\alpha \,\Big|\, \bigvee_{k=1}^\infty T^{-k}\alpha\right).$$

Proof: By Proposition 2.5 (1′),

$$H\!\left(\alpha \,\Big|\, \bigvee_{k=1}^j T^{-k}\alpha\right) = H(\alpha \vee T^{-1}\alpha \vee \ldots \vee T^{-j}\alpha) - H\!\left(\bigvee_{k=1}^j T^{-k}\alpha\right)$$

$$= H\!\left(\bigvee_{k=0}^j T^{-k}\alpha\right) - H\!\left(\bigvee_{k=0}^{j-1} T^{-k}\alpha\right).$$

Sum over $j = 1, \ldots, n$ to get

$$\sum_{j=1}^n H\!\left(\alpha \,\Big|\, \bigvee_{k=1}^j T^{-k}\alpha\right) = H\!\left(\bigvee_{k=0}^n T^{-k}\alpha\right) - H(\alpha).$$

Notice that $H(\alpha|\bigvee_{k=1}^n T^{-k}\alpha)$ is nonnegative and decreasing, so that its limit exists as $n \to \infty$. Therefore the limit of its Cesaro averages exists as well. Thus dividing the preceding equation through by $n+1$ and taking limits gives

$$\lim_{n\to\infty} H\!\left(\alpha \,\Big|\, \bigvee_{k=1}^n T^{-k}\alpha\right) = \lim_{n\to\infty} \frac{1}{n+1} H\!\left(\bigvee_{k=0}^n T^{-k}\alpha\right) = h(\alpha, T).$$

(This also provides another proof of Proposition 2.10.)

For the remaining equality, apply Proposition 2.11 with

$$\mathscr{B}_n = \mathscr{B}\!\left(\bigvee_{k=1}^n T^{-k}\alpha\right) \quad \text{and} \quad \mathscr{B}_\infty = \mathscr{B}\!\left(\bigvee_{k=1}^\infty T^{-k}\alpha\right).$$

This formula justifies our speaking earlier of the entropy of a source as the average uncertainty about what symbol it will print next, given its output so far. Of course mathematically it is difficult to distinguish the past from the future: if T is the ordinary left shift and α is the partition according to the entries observed at time 0, then $\bigvee_{k=1}^{\infty} T^{-k}\alpha$ actually represents the readings made in the future rather than the past. We could redefine our terms, transformations, or directions – or note that it doesn't really make any difference which way we go; see Exercise 4.

Remark Both 2.11 and 2.12 remain true if α is assumed to a *countable* partition of *finite entropy*. For the proof, we need Corollary 6.2.2 below.

Several more formulas will be needed for the computations of the next section. We prove one and leave several others for exercises.

Proposition 2.13 For any countable measurable partitions α and β,
$$h(\beta, T) \leq h(\alpha, T) + H(\beta|\alpha).$$

Proof: Let $\beta_0^{m-1} = \beta \vee T^{-1}\beta \vee \ldots \vee T^{-m+1}\beta$ and $\alpha_0^{m-1} = \alpha \vee T^{-1}\alpha \vee \ldots \vee T^{-m+1}\alpha$ for $m = 1, 2, \ldots$. Then for each m,

$$\begin{aligned} H(\beta_0^{m-1}|\alpha_0^{m-1}) &\leq H(\beta|\alpha_0^{m-1}) + H(T^{-1}\beta|\alpha_0^{m-1}) + \ldots + H(T^{-m+1}\beta|\alpha_0^{m-1}) \\ &\leq H(\beta|\alpha) + H(T^{-1}\beta|T^{-1}\alpha) + \ldots + H(T^{-m+1}\beta|T^{-m+1}\alpha) \\ &= mH(\beta|\alpha). \end{aligned}$$

Therefore
$$\begin{aligned} H(\beta_0^{m-1}) &\leq H(\beta_0^{m-1} \vee \alpha_0^{m-1}) = H(\alpha_0^{m-1}) + H(\beta_0^{m-1}|\alpha_0^{m-1}) \\ &\leq H(\alpha_0^{m-1}) + mH(\beta|\alpha), \end{aligned}$$

and
$$h(\beta, T) = \lim_{m \to \infty} \frac{1}{m} H(\beta_0^{m-1}) \leq \lim_{m \to \infty} \frac{1}{m} H(\alpha_0^{m-1}) + H(\beta|\alpha)$$
$$= h(\alpha, T) + H(\beta|\alpha).$$

Exercises
1. Show that $\alpha \leq \beta$ implies $I_{\alpha|\mathscr{F}} \leq I_{\beta|\mathscr{F}}$ a.e., but it is not even always true that $I_{\alpha|\mathscr{F}} \leq I_\alpha$ a.e.
2. Extend Proposition 2.9 by proving that $H(\alpha|\mathscr{F}) = H(\alpha)$ if and only if the two σ-algebras $\mathscr{B}(\alpha)$ and $\mathscr{F}$ are independent, in that $\mu(E \cap F) = \mu(E)\mu(F)$ for all $E \in \mathscr{B}(\alpha)$ and all $F \in \mathscr{F}$.

3. Supply the proofs of Remarks 2.1.
4. Show that
   (a) $h(\alpha, T) = \lim_{n\to\infty} H(T^{-n}\alpha | \bigvee_{k=0}^{n-1} T^{-k}\alpha)$.
   (b) $h(\alpha, T) = \lim_{n\to\infty} H(\alpha | \bigvee_{k=1}^{n-1} T^k\alpha) = H(\alpha | \bigvee_{k=1}^{\infty} T^k\alpha)$.
   (c) $h(\alpha, T) = h(\alpha, T^{-1})$.

   Interpret these formulas in terms of ticker-tapes.
5. Define $I_{\mathscr{F}|\mathscr{G}}$ for $\mathscr{F}$ and $\mathscr{G}$ sub-σ-algebras of $\mathscr{B}$. Prove that
$$I_{\mathscr{F}_1 \vee \mathscr{F}_2 | \mathscr{G}} = I_{\mathscr{F}_1 | \mathscr{G}} + I_{\mathscr{F}_2 | \mathscr{F}_1 \vee \mathscr{G}}.$$
6. Show that $\alpha \leq \beta$ implies $h(\alpha, T) \leq h(\beta, T)$.
7. Show that $h(\bigvee_{k=m}^{n} T^{-k}\alpha, T) = h(\alpha, T)$.
8. Show that $h(T^k) = |k| h(T)$.
   (Hint: For any partition α, consider the partition $\alpha^k_{-k} = \sigma^{-k}\alpha \vee \cdots \vee \sigma^k \alpha$ and estimate $h(\alpha^k_{-k}, T^k)$.)
9. Show that $I_{T^{-1}\alpha} = I_\alpha \circ T$.
10. Show that $h_\mu(\alpha, T)$ and $h_\mu(T)$ are affine functions of μ on the space $\mathscr{M}_T$ of T-invariant probability measures on $(X, \mathscr{B})$.
11. Show that $h(T) = \sup\{h(\alpha, T); \alpha$ is a countable measurable partition with $H(\alpha) < \infty\}$.
12. Prove *Pinsker's Formula*:
$$H(\alpha \vee \beta | (\alpha \vee \beta)_1^\infty) = H(\beta | \beta_1^\infty) + H(\alpha | \alpha_1^\infty \vee \beta_{-\infty}^\infty).$$

(Hint: Note that
$$H(\alpha | \alpha_1^\infty \vee \beta_{-\infty}^\infty) = \lim_{n\to\infty} H(\alpha_{-n}^0 | \alpha_1^\infty \vee \beta_1^\infty \vee \beta_{-n}^0)$$
and
$$H(\beta | \beta_1^\infty) = \lim_{n\to\infty} \frac{1}{n} H(\beta_{-n}^0 | \alpha_1^\infty \vee \beta_1^\infty).)$$

5.3. Generators and the Kolmogorov–Sinai Theorem

The computation of $h(T)$ from its definition as $\sup_\alpha h(\alpha, T)$ is seldom feasible. However, if there should exist a partition α for which $h(\alpha, T) = h(T)$, then of course $h(T)$ can be found. This is the case when T has a partition which generates the full σ-algebra $\mathscr{B}$.

In this section we will deal again mainly with *finite* measurable partitions of X. For any partition α we use the notations $\alpha_m^n = \bigvee_{k=m}^{n} T^{-k}\alpha$; in case $m = -\infty$ or $n = \infty$, recall that we intend by this notation the σ-algebra generated by all the partitions involved:

$$\alpha_1^\infty = \bigvee_{k=1}^\infty \mathscr{B}(T^{-k}\alpha),$$

$$\alpha_{-\infty}^{-1} = \bigvee_{k=1}^\infty \mathscr{B}(T^k\alpha),$$

$$\alpha_{-\infty}^\infty = \bigvee_{k=-\infty}^\infty \mathscr{B}(T^k\alpha).$$

These σ-algebras may be thought of as representing all past (or future, depending on how you look at it) performances of the experiment α, all future (or past) performances of α, and all performances of α, respectively.

A finite partition α is called a *generator* with respect to T in case $\alpha_{-\infty}^\infty = \mathscr{B}$ up to sets of measure 0. We now have the tools to prove the following important result rather easily. Sinai's interesting proof, based on the *entropy metric*, is sketched in the Exercises.

Theorem 3.1 *Kolmogorov–Sinai Theorem* (Kolmogorov 1958, Sinai 1959b) If α is a generator with respect to T, then

$$h(T) = h(\alpha, T).$$

Proof: It is enough to show that for each finite partition β we have $h(\beta, T) \leq h(\alpha, T)$. Now for any $n = 1, 2, \ldots$, by Proposition 2.13 and Exercise 2.7

$$h(\beta, T) \leq h(\alpha_{-n}^n, T) + H(\beta|\alpha_{-n}^n) = h(\alpha, T) + H(\beta|\alpha_{-n}^n).$$

But since $\beta \subset \alpha_{-\infty}^\infty$ up to sets of measure 0, by Propositions 2.7 and 2.11 we have

$$0 = H(\beta|\alpha_{-\infty}^\infty) = \lim_{n \to \infty} H(\beta|\alpha_{-n}^n).$$

Thus the result follows by letting $n \to \infty$ in the first formula.

According to the following important theorem of Krieger, entropy can be computed in the most interesting cases. For the proof, see Denker, Grillenberger and Sigmund (1976).

Theorem 3.2 *Krieger Generator Theorem* (1970) If T is an ergodic m.p.t. on a Lebesgue space with $h(T) < \infty$, then T has a finite generator.

Examples of the computation of entropy

Example 3.3 *Zero entropy* Suppose that T has a *one-sided generator*, i.e. that there is a finite partition α such that $\alpha_1^\infty = \mathscr{B}$ up to sets of

5.3. Generators and the Kolmogorov–Sinai Theorem

measure 0. The existence of such a partition α means that in some sense the present and future of the system $(X, \mathscr{B}, \mu, T)$ are completely determined by its past. Then of course α is a generator, so we have

$$h(T) = h(\alpha, T) = H(\alpha|\alpha_1^\infty) = H(\alpha|\mathscr{B}) = 0,$$

as might be expected. For example, *every ergodic rotation of the circle has entropy zero*: if α is a partition of the circle into two disjoint semi-circles, then $\alpha_1^\infty = \mathscr{B}$ up to sets of measure 0.

Another way to see this is to notice that, with α and T as above, the partition $\alpha \vee T^{-1}\alpha \vee \ldots \vee T^{-m+1}\alpha$ consists of $2m$ intervals on the circle. Thus $H(\alpha \vee T^{-1}\alpha \vee \ldots \vee T^{-m+1}\alpha) \leq \log_2(2m)$ (the entropy is always a maximum for a partition into pieces of equal measures), and $H(\alpha \vee T^{-1}\alpha \vee \ldots \vee T^{-m+1}\alpha)/m \to 0$. This argument shows that $h(\alpha, T)$ depends on the asymptotic numbers and relative sizes of the cells of $\alpha \vee T^{-1}\alpha \vee \ldots \vee T^{-m+1}\alpha$. We will have more to say about this later (Shannon–McMillan–Breiman Theorem).

More generally *every discrete spectrum system has entropy zero*. For, by Exercise 10 of Section 2.4, if T has discrete spectrum then there is a sequence of integers $n_k \to \infty$ with $T^{n_k}f \to f$ in L^2 for each $f \in L^2$. Let $\alpha = \{A_1, A_2, \ldots, A_n\}$ be a partition of X and $f(x) = i$ if $x \in A_i$. Then $T^{n_k}f \to f$ in L^2 implies that, up to sets of measure 0, $\alpha \subset \alpha_1^\infty$, so that $h(\alpha, T) = H(\alpha|\alpha_1^\infty) = 0$.

Example 3.4 *Entropy of Bernoulli schemes* Consider the Bernoulli scheme $\mathscr{B}(p_1, p_2, \ldots, p_n)$ on the alphabet $\{a_1, a_2, \ldots, a_n\}$. The time-zero cylinder sets

$$A_i = \{x : x_0 = a_i\}, \quad i = 1, 2, \ldots, n,$$

form a measurable partition α. Since $\sigma^{-m}A_i = \{x : x_m = a_i\}$, $\bigvee_{-\infty}^{\infty} \mathscr{B}(\sigma^{-m}\alpha)$ contains all cylinder sets and hence equals $\mathscr{B}$ up to sets of measure 0. Thus α is a generator and

$$h(\sigma) = h(\alpha, \sigma) = \lim_{m \to \infty} \frac{1}{m} H(\alpha \vee \sigma^{-1}\alpha \vee \ldots \vee \sigma^{-m+1}\alpha).$$

Now $\alpha \vee \sigma^{-1}\alpha \vee \ldots \vee \sigma^{-m+1}\alpha$ is the partition by m-blocks: its elements are sets of the form

$$A_{i_1} \cap \sigma^{-1}A_{i_2} \cap \ldots \cap \sigma^{-m+1}A_{i_m}.$$

Because the measure is product measure, the partitions $\alpha, \sigma^{-1}\alpha, \ldots, \sigma^{-m+1}\alpha$ are independent. By Proposition 2.9,

$$H(\alpha \vee \sigma^{-1}\alpha \vee \ldots \vee \sigma^{-m+1}\alpha) = mH(\alpha) = -m\sum_{i=1}^{n} p_i \log_2 p_i,$$

246   5. Entropy

and therefore the average uncertainty about what will appear next in any sequence is

$$h(\sigma) = - \sum_{i=1}^{n} p_i \log_2 p_i.$$

Thus $\mathscr{B}(\frac{1}{2},\frac{1}{2})$ has entropy $\log_2 2$ and $\mathscr{B}(\frac{1}{3},\frac{1}{3},\frac{1}{3})$ has entropy $\log_2 3$, so that these two systems cannot be isomorphic. In this way Kolmogorov settled a question that had been embarrassing ergodic theorists for two decades.

Example 3.5 *Entropy of Markov shifts* Let $A = (a_{ij})$ be an $n \times n$ stochastic matrix with fixed (row) probability vector p $(pA = p)$. Again let α be the partition into time-zero cylinder sets $\{x: x_0 = a_i\}$, $i = 1, 2, \ldots, n$. This time, a typical element $\{x: x_0 = i_0, x_1 = i_1, \ldots, x_m = i_m\}$ of $\alpha \vee \sigma^{-1}\alpha \vee \ldots \vee \sigma^{-m}\alpha$ has measure $p_{i_0} a_{i_0 i_1} \ldots a_{i_{m-1} i_m}$, so that

$$\begin{aligned}
H(\alpha \vee \sigma^{-1}\alpha \vee \ldots \vee \sigma^{-m}\alpha) &= - \sum_{i_0,\ldots,i_m} f(p_{i_0} a_{i_0 i_1} \ldots a_{i_{m-1} i_m}) \\
&= - \sum_{i_0,\ldots,i_m} p_{i_0} a_{i_0 i_1} \ldots a_{i_{m-2} i_{m-1}} (a_{i_{m-1} i_m} \log_2 p_{i_0} a_{i_0 i_1} \ldots a_{i_{m-2} i_{m-1}} \\
&\quad + a_{i_{m-1} i_m} \log_2 a_{i_{m-1} i_m}) \\
&= - \sum_{i_0,\ldots,i_{m-1}} \left(\sum_{i_m} a_{i_{m-1} i_m} \right) f(p_{i_0} a_{i_0 i_1} \ldots a_{i_{m-2} i_{m-1}}) \\
&\quad - \sum_{i_{m-1},i_m} \left(\sum_{i_0,\ldots i_{m-2}} p_{i_0} a_{i_0 i_1} \ldots a_{i_{m-2} i_{m-1}} \right) f(a_{i_{m-1} i_m}) \\
&= - \sum_{i_0,\ldots,i_{m-1}} f(p_{i_0} a_{i_0 i_1} \ldots a_{i_{m-2} i_{m-1}}) - \sum_{i_{m-1},i_m} p_{i_{m-1}} f(a_{i_{m-1} i_m}) \\
&= \ldots = - \sum_i p_i \log_2 p_i - m \sum_{i,j} p_i a_{ij} \log_2 a_{ij}.
\end{aligned}$$

It follows that the Markov shift determined by A has entropy

$$h(\sigma) = - \sum_{i,j} p_i a_{ij} \log_2 a_{ij}.$$

If we read sequences one entry at a time from right to left, this number can be interpreted as the average uncertainty about what symbol will appear next, given the one we see at present. In fact, in this case $H(\alpha | \bigvee_{k=1}^{\infty} T^{-k}\alpha) = H(\alpha | T^{-1}\alpha)$; this is consistent with the Markov property, according to which the probability of what happens depends only on the immediately preceding situation rather than on the full past.

The following fact is often useful when computing entropy.

Proposition 3.6 If $\alpha_1 \leq \alpha_2 \leq \ldots$ is an increasing sequence of finite

5.3. Generators and the Kolmogorov–Sinai Theorem

partitions and $\bigvee_{n=1}^{\infty} \mathscr{B}(\alpha_n) = \mathscr{B}$ up to sets of measure 0, then

$$h(T) = \lim_{n \to \infty} h(\alpha_n, T).$$

Proof: Let β be any finite measurable partition of X. According to Proposition 2.13, for each $n = 1, 2, \ldots$ we have

$$h(\beta, T) \leq h(\alpha_n, T) + H(\beta|\alpha_n).$$

By Proposition 2.11, $H(\beta|\alpha_n) \to H(\beta|\mathscr{B}) = 0$. Therefore

$$h(\beta, T) \leq \lim_{n \to \infty} \uparrow h(\alpha_n, T).$$

It follows that $h(T) = \lim_{n \to \infty} h(\alpha_n, T)$.

Example 3.7 *Entropy of a product transformation* Let $T_1 : X_1 \to X_1$ and $T_2 : X_2 \to X_2$ be m.p.t.s on probability spaces $(X_1, \mathscr{B}_1, \mu_1)$ and $(X_2, \mathscr{B}_2, \mu_2)$. Recall that the *product transformation* $T = T_1 \times T_2$ on $X_1 \times X_2$ is defined by $T(x_1, x_2) = (T_1 x_1, T_2 x_2)$. If α_1 and α_2 are finite partitions of X_1 and X_2, respectively, define partitions $\bar{\alpha}_1$ and $\bar{\alpha}_2$ of $X_1 \times X_2$ by

$$\bar{\alpha}_1 = \{A \times X_2 : A \in \alpha_1\}$$
$$\bar{\alpha}_2 = \{X_1 \times A : A \in \alpha_2\}.$$

Then $\bar{\alpha}_1$ and $\bar{\alpha}_2$ are independent, as are $T^{-j}\bar{\alpha}_1$ and $T^{-j}\bar{\alpha}_2$ for all j. Let $\alpha = \bar{\alpha}_1 \vee \bar{\alpha}_2$. Because $T^{-j}\alpha = T^{-j}\bar{\alpha}_1 \vee T^{-j}\bar{\alpha}_2$, for each $n = 1, 2, \ldots$

$$H(\alpha_0^{n-1}) = H((\bar{\alpha}_1)_0^{n-1}) + H((\bar{\alpha}_2)_0^{n-1}) = H((\alpha_1)_0^{n-1}) + H((\alpha_2)_0^{n-1})$$

and consequently

$$h(\alpha, T) = h(\alpha_1, T_1) + h(\alpha_2, T_2).$$

Now if we take increasing sequences of partitions $\{\alpha_1^{(n)}\}$ of X_1 and $\{\alpha_2^{(n)}\}$ of X_2 with $\bigvee_{n=1}^{\infty} \alpha_1^{(n)} = \mathscr{B}_1$ and $\bigvee_{n=1}^{\infty} \alpha_2^{(n)} = \mathscr{B}_2$ up to sets of measure 0, then $\bigvee_{n=1}^{\infty} \alpha_1^{(n)} \times \alpha_2^{(n)}$ is the σ-algebra of $X_1 \times X_2$ up to sets of measure 0. By Proposition 3.6,

$$h(T_1 \times T_2) = h(T_1) + h(T_2).$$

Exercises

1. Calculate the entropy of a Markov shift by using the formula $h(\alpha, T) = H(\alpha|\alpha_1^{\infty})$. Obtain the entropy of a Bernoulli shift as a corollary.
2. What can you say about the entropy of an *infinite product transformation*? What about an inverse limit of m.p.t.s?

3. The entropy of a stationary stochastic process $\{f_n: -\infty < n < \infty\}$ is 0 if and only if f_0 is a measurable function of $f_{-1}, f_{-2}, f_{-3}, \ldots$.
4. Given any $r > 0$, there is a Bernoulli shift of entropy r.
5. Show that the set Φ of equivalence classes (mod 0) of finite measurable partitions of $(X, \mathcal{B}, \mu)$ is a metric space under the entropy metric $d(\alpha, \beta) = H(\alpha | \beta) + H(\beta | \alpha)$.
6. Show that $|h(\alpha, T) - h(\beta, T)| \leq d(\alpha, \beta)$ for $\alpha, \beta \in \Phi$, so that $h(\alpha, T)$ is a continuous function of α.
7. Show that if $\alpha_1 \leq \alpha_2 \leq \ldots$ are finite partitions with $\bigvee_{k=1}^{\infty} \alpha_k = \mathcal{B}$ up to sets of measure 0, then $\{\beta \in \Phi: \beta \leq \alpha_n \text{ for some } n\}$ is dense in Φ (with respect to d). (Hint: Given $\alpha \in \Phi$, approximate all the elements of α well by members of some $\mathcal{B}(\alpha_n)$, and take β to be the partition the latter generate. Then use Exercise 6.)
8. Use Exercises 5–7 to give another (namely Sinai's) proof of the Kolmogorov–Sinai Theorem. (Hint: Let α be a generator and $\alpha_n = \alpha_{-n}^n$. Show that if $\beta \leq \alpha_n$ for some n, then $2h(\beta, T) \leq 2h(\alpha, T)$.)
9. Show that if $p_i \leq q_j$ for all i and j, then the entropy of the Bernoulli shift $\mathcal{B}(p_1, p_2, \ldots, p_n)$ is no less than that of $\mathcal{B}(q_1, q_2, \ldots, q_m)$.
10. Show that if α is a *countable* generator for T with $H(\alpha) < \infty$, then $h(T) = h(\alpha, T)$. What can you say if you find a countable generator α for T with $H(\alpha) = \infty$?

6
More about entropy

This chapter treats several topics in entropy theory that are somewhat beyond the basics. We begin by computing the entropies of automorphisms of the torus, skew products, and induced transformations. The following sections discuss convergence of the information per unit time (Shannon–McMillan–Breiman Theorem) and the topological version of entropy for cascades. We give an introduction to the Ornstein Isomorphism Theorem, which says that two Bernoulli schemes are isomorphic if and only if they have the same entropy. Ornstein's associated theory of sufficient conditions for m.p.t.s to be isomorphic to Bernoulli shifts has produced a surprising list of examples, including classical ones like geodesic maps and automorphisms of the torus, that are metrically indistinguishable from repeated independent random experiments. In the final section we present the Keane–Smorodinsky construction of the isomorphism whose existence is implied by Ornstein's theorem. Their work actually strengthens Ornstein's result, since they are able to construct the isomorphism explicitly, and the map is *finitary*: each coordinate of the image of a point can be calculated from knowledge of only a finite piece of the history of that point. (Alternatively, the map is a homeomorphism once a set of measure 0 has been deleted). This means that in principle such a coding can actually be carried out mechanically.

6.1. More examples of the computation of entropy

A. Entropy of an automorphism of the torus

Let $\mathbb{K}^n = \mathbb{R}^n/\mathbb{Z}^n$ be the n-dimensional torus and $T: \mathbb{K}^n \to \mathbb{K}^n$ the Haar-measure-preserving map determined by an $n \times n$ integer matrix A with determinant ± 1. Assume that T is ergodic, so that among the eigenvalues $\lambda_1, \lambda_2, \ldots, \lambda_n$ of A there are no roots of unity. (Then in fact T is even strongly mixing–see Section 2.5.) The formula

$$h(T) = \sum_{|\lambda_i| > 1} \log_2 |\lambda_i|$$

for the entropy of T was found by Sinai (1959b). We sketch, for $n = 2$, the proof given by Berg (1968).

250   6. *More about entropy*

The idea is to construct a sequence of partitions $\alpha_1 \leq \alpha_2 \leq \ldots$ with $\bigvee_{n=1}^{\infty} \alpha_n = \mathscr{B}$ up to sets of measure 0 and such that the entropy of T with respect to each α_n is the same. Then the result will follow from Proposition 5.3.6.

Consider now the integer matrix $A = \binom{a\ b}{c\ d}$ with determinant ± 1 which determines $T: \mathbb{K}^2 \to \mathbb{K}^2$ by $T\binom{x}{y} = A\binom{x}{y}$ mod 1. A has characteristic polynomial $\chi_A(x) = x^2 - (\text{trace } A)x + \det A$. I claim that the *eigenvalues* λ_1 and λ_2 *of A are real, with, say* $|\lambda_1| > 1$ *and* $|\lambda_2| < 1$. For if A had a complex eigenvalue λ, then its other eigenvalue would be the complex conjugate $\bar{\lambda}$. Then $\lambda + \bar{\lambda} = 2 \text{ Re } \lambda = \text{trace } A$ and $\lambda\bar{\lambda} = |\lambda|^2 = \det A$, so that $\det A = 1$ and $|\lambda| = 1$. Since trace A is an integer, we would have Re $\lambda = 0, 1$, or $\frac{1}{2}$. The corresponding λs would be $\pm 1, \pm i$, and $\pm \frac{1}{2} \pm i\sqrt{3}/2$, all roots of unity, which we have ruled out.

The eigenspaces V_1 and V_2 corresponding to the eigenvalues λ_1 and λ_2, when A is considered as a linear transformation of the plane, are two lines through the origin, each with *irrational slope*. (This is so because if $m = $ trace A, then $\lambda = (m \pm \sqrt{m^2 \pm 4})/2$, and $m^2 \pm 4$ is never a perfect square for nonzero integer m. If $m = 0$, then $\lambda = 1$, which is impossible.) Remembering the Kronecker–Weyl Theorem, the projection of each of these lines to the torus $\mathbb{K}^2 = \mathbb{R}^2/\mathbb{Z}^2$ is dense. Assume without loss of generality that $\lambda_1, \lambda_2 > 0$ and for the sake of the pictures that both eigenvectors are in the first quadrant. One of our partitions α is constructed as follows. Beginning at the origin, take a long segment of V_1 and project it to the torus. Call the projection P_1.

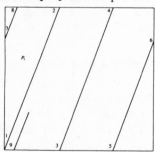

Do the same for V_2 to get P_2.

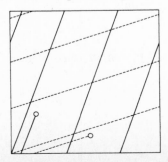

6.1. More examples of the computation of entropy

Extend any 'hanging ends' like those indicated by the circles so as just to touch the projection of the other half-line.

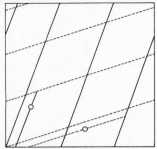

Finally, beginning at whichever of $(1,0)$, $(1,1)$, or $(0,1)$ is possible, retreat along V_1, V_2, or both until just coming into contact with the already existing dividing lines; this guarantees that there will not be a 'hanging edge' at $(0,0) \in \mathbb{K}^2$.

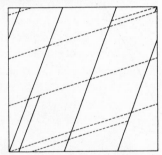

We obtain a partition α of $\mathbb{K}^2$ into finitely many parallelograms with disjoint interiors whose boundaries are made up of segments P_1 and P_2 of the projections of V_1 and V_2 into the torus.

Now the action of A on $\mathbb{K}^2$ is to expand vectors in the direction of V_1 by a factor of λ_1 and to contract those in the V_2 direction by a factor λ_2. Thus one of these small parallelograms behaves under A as shown:

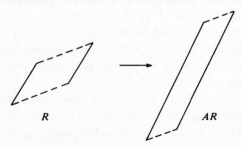

(after which it is reduced mod 1). Because V_2 is an invariant subspace for A, we have $AP_2 \subseteq P_2$. This means that the dotted (contracting) boundaries of the above parallelogram must again reach the dotted boundary, and they can't stop short as one end does here:

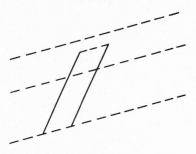

Similarly, $A^{-1}P_1 \subset P_1$. This means that the image AR of R contains no part of the solid (expanding) boundary P_1 in its interior. Therefore, A carries R into something that sits nicely with respect to the partition α:

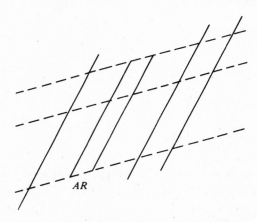

In the expanding direction, AR extends fully across each element α whose interior it hits; and AR contains no part of the expanding boundary in its interior. These are the key properties of a *Markov partition*. Such partitions were discovered by Adler and Weiss (1967) and have been studied extensively by Sinai (1968) and Bowen (1970a), (1970b), (1978). The existence of such a partition greatly facilitates the representation of a transformation by symbolic dynamics and, what is of most immediate interest for us, the calculation of entropy.

Fix $k = 1, 2, \ldots$. The partition $\alpha \vee T\alpha \vee \ldots \vee T^k\alpha$ arises by subdividing each parallelogram of α into long thin parallelograms by subdividing the contracting boundary of α only. In the figure, the large parallelogram is an element of α, the small ones belong to $\alpha \vee T\alpha \vee \ldots \vee T^k\alpha$.

6.1. More examples of the computation of entropy

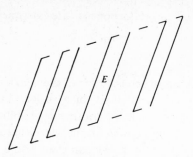

The contracting boundary of each element E of $\alpha \vee T\alpha \vee \ldots \vee T^{k-1}\alpha$ is between λ_1^{-k} times the lengths of the shortest and longest contracting boundaries of the elements of α, and its expanding boundary has the same length as an expanding boundary of one of the elements of α; therefore

$$\frac{1}{c} \leq \lambda_1^k \mu(E) \leq c$$

for some constant $c > 0$, for all $E \in \alpha \vee T\alpha \vee \ldots \vee T^k\alpha$ and all $k = 1, 2, \ldots$
Now we can write

$$H(\alpha \vee T\alpha \vee \ldots \vee T^k\alpha) = -\sum_E \mu(E)\log_2 \mu(E)$$

$$= -\sum_E \delta_E \lambda_1^{-k} \log_2(\delta_E \lambda_1^{-k}) \text{ (where } \delta_E = \lambda_1^k \mu(E),$$

$$\text{so that } \frac{1}{c} \leq \delta_E \leq c \text{ for all } E)$$

$$= -\sum_E \mu(E)\log_2 \delta_E + k \log_2 \lambda_1 \sum_E \mu(E),$$

and make the estimate

$$\left| H(\alpha \vee T\alpha \vee \ldots \vee T^k\alpha) - k \log_2 \lambda_1 \right| \leq \sum_E \mu(E) |\log_2 \delta_E| \leq |\log_2 c|.$$

Then clearly

$$h(\alpha, T) = \lim_{k \to \infty} \frac{1}{k+1} H(\alpha \vee T\alpha \vee \ldots \vee T^k\alpha) = \log_2 \lambda_1.$$

By beginning with longer and longer segments of the eigenspaces V_1 and V_2, we can produce a sequence of partitions $\alpha_1, \alpha_2, \ldots$ for which $\bigvee_{n=1}^\infty \alpha_n = \mathscr{B}$ up to sets of measure 0. That $h(T) = \log_2 \lambda_1$ then follows from Proposition 5.3.6.

Adler and Weiss (1970) used these Markov partitions to show that entropy is a *complete* invariant for automorphisms of the two-dimensional torus. This result was subsumed in the later work of Katznelson (1971),

254   6. *More about entropy*

who showed that ergodic toral automorphisms are isomorphic to Bernoulli shifts.

B. Entropy of a skew product

Let us consider a slightly more general type of skew product than we discussed in 2.4. Let $T: X \to X$ be a m.p.t. of a probability space $(X, \mathcal{B}, \mu)$, and $\{S_x : x \in X\}$ a family of m.p.t.s of another probability space $(Y, \mathcal{C}, \nu)$. Suppose that $\{S_x : x \in X\}$ is *measurable* in the sense that the map $(x, y) \to S_x y$ is measurable from $X \times Y$ to Y. We can then define a transformation $\tau = T \times \{S_x\}$ on $X \times Y$ by

$$\tau(x, y) = (Tx, S_x y).$$

As before, τ is easily verified to be measurable and measure-preserving.

The entropy of τ was computed by Abramov and Rokhlin (1962). For a finite partition β of Y and $n = 1, 2, \ldots$ define

$$\beta_1^n(x) = S_x^{-1}\beta \vee S_x^{-1} S_{Tx}^{-1} \beta \vee \ldots \vee S_x^{-1} S_{Tx}^{-1} \ldots S_{T^{n-1}x}^{-1} \beta$$

and $\beta_1^\infty(x) = \vee_{n=1}^\infty \beta_1^n(x)$. Then $H(\beta | \beta_1^\infty(x))$ can be seen to be a measurable function of x. We let

$$h_T(\beta, S) = \int_X H(\beta | \beta_1^\infty(x)) d\mu(x)$$

and define the *fiber entropy* to be

$$h_T(S) = \sup\{h_T(\beta, S) : \beta \text{ a finite measurable partition of } Y\}.$$

We will show that

$$h(T \times \{S_x\}) = h(T) + h_T(S).$$

First a couple of preliminary results. Let $\beta_0^n(x) = \beta \vee \beta_1^n(x)$.

Proposition 1.1 $h_T(\beta, S) = \lim_{n \to \infty} (1/n) \int_X H(\beta_0^n(x)) d\mu(x)$.

Proof: For each $x \in X$,

$$H(\beta_0^n(x)) = H(\beta \vee S_x^{-1} \beta_0^{n-1}(Tx))$$
$$= H(\beta_0^{n-1}(Tx)) + H(\beta | S_x^{-1} \beta_0^{n-1}(Tx)).$$

Since $\int_X H(\beta_0^k(Tx)) d\mu = \int_X H(\beta_0^k(x)) d\mu$ for all k, we may repeat this calculation to find that

$$\int_X H(\beta_0^n(x)) d\mu = \int_X [H(\beta | S_x^{-1} \beta_0^{n-1}(Tx)) + H(\beta | S_x^{-1} \beta_0^{n-2}(Tx))$$
$$+ \ldots + H(\beta | S_x^{-1} \beta_0^1(Tx)) + H(\beta | S_x^{-1} \beta_0^0(Tx))] d\mu,$$

6.1. More examples of the computation of entropy

so that

$$\frac{1}{n}\int_X H(\beta_0^n(x))\mathrm{d}\mu(x) = \frac{1}{n}\sum_{k=0}^{n-1}\int_X H(\beta|S_x^{-1}\beta_0^k(Tx))\mathrm{d}\mu(x).$$

Now $H(\beta|S_x^{-1}\beta_0^k(Tx)) = H(\beta|\beta_1^k(x))$ decreases a.e. to

$$H(\beta|\beta_1^\infty(x)) \quad \text{(Proposition 5.2.11)};$$

the integrals of these functions converge to

$$\int_X H(\beta|\beta_1^\infty(x))\mathrm{d}\mu = h_T(\beta, S)$$

by the Bounded Convergence Theorem. Since the Cesaro means of a convergent sequence have the same limit as the sequence itself,

$$\lim_{n\to\infty}\frac{1}{n}\int_X H(\beta_0^n(x))\mathrm{d}\mu(x) = \lim_{n\to\infty}\frac{1}{n}\sum_{k=0}^{n-1}\int_X H(\beta|\beta_1^k(x))\mathrm{d}\mu(x)$$
$$= h_T(\beta, S).$$

(The similarity of this part of the argument to the proof of Proposition 5.2.12 is more than coincidental.)

Let α be a finite partition of X. Extend α to a partition $\bar{\alpha} = \{A \times Y : A \in \alpha\}$ of $X \times Y$. Similarly, Let $\bar{\mathcal{B}} = \{A \times Y : A \in \mathcal{B}\}$. For a partition ξ of $X \times Y$ and $x \in X$, let ξ_x denote the partition of Y determined by the x-sections of members of ξ: the member of ξ_x determined by a set $Z \in \xi$ is $Z_x = \{y \in Y : (x, y) \in Z\}$.

Proposition 1.2 For a finite partition ξ of $X \times Y$,

$$H(\xi|\bar{\mathcal{B}}) = \int_X H(\xi_x)\mathrm{d}\mu(x).$$

Proof: Note first that for any $Z \in \xi$,

$$\mu \times \nu(Z|\bar{\mathcal{B}}) = \nu(Z_x) \text{ a.e. } d(\mu \times \nu).$$

For $\nu(Z_x)$, being a function of x alone, is $\bar{\mathcal{B}}$-measurable on $X \times Y$; and its integral over a typical set $A \times Y \in \bar{\mathcal{B}}$ is

$$\iint_{A \times Y} \nu(Z_x)\mathrm{d}\mu(x)\mathrm{d}\nu(y) = \int_A \nu(Z_x)\mathrm{d}\mu(x)$$
$$= \int_A \nu\{y \in Y : (x, y) \in Z\}\mathrm{d}\mu(x) = \int_A \int_Y \chi_Z(x, y)\mathrm{d}\nu(y)\mathrm{d}\mu(x)$$
$$= \iint_{A \times Y} \chi_Z(x, y)\mathrm{d}\nu\mathrm{d}\mu.$$

256   6. *More about entropy*

Therefore on each $Z \in \xi$ we have
$$-\log_2(\mu \times \nu)(Z|\bar{\mathscr{B}}) = -\log_2 \nu(Z_x) \quad \text{a.e.,}$$
and hence
$$I_{\xi|\bar{\mathscr{B}}}(x, y) = I_{\xi_x}(y) \text{ a.e.}$$
Then clearly
$$H(\xi|\bar{\mathscr{B}}) = \int\!\!\int_{X \times Y} I_{\xi|\bar{\mathscr{B}}}(x, y) \,d\nu\,d\mu = \int_X\!\!\int_Y I_{\xi_x}(y)\,d\nu(y)\,d\mu(x)$$
$$= \int_X H(\xi_x)\,d\mu(x).$$

Proposition 1.3 $h(T \times \{S_x\}) = h(T) + h_T(S)$.

Proof: As in the computation of the entropy of a direct product, it is enough to show that the supremum of $h(\xi, \tau)$ over all finite partitions ξ of $X \times Y$ which are of the form $\xi = \alpha \times \beta$, where α and β are finite partitions of X and Y, respectively, is $h(T) + h_T(S)$. Now for such a ξ and $n = 1, 2, \ldots$,

$$H(\xi_0^n) = H(\bar{\alpha}_0^n \vee \xi_0^n) = H(\bar{\alpha}_0^n) + H(\xi_0^n|\bar{\alpha}_0^n) = H(\alpha_0^n) + H(\xi_0^n|\bar{\alpha}_0^n)$$
$$\geq H(\alpha_0^n) + H(\xi_0^n|\bar{\mathscr{B}}) = H(\alpha_0^n) + \int_X H((\xi_0^n)_x)\,d\mu(x)$$

(where $\bar{\alpha}_0^n = \bar{\alpha} \vee \tau^{-1}\bar{\alpha} \vee \ldots \vee \tau^{-n}\bar{\alpha}$ and $\alpha_0^n = \alpha \vee T^{-1}\alpha \vee \ldots \vee T^{-n}\alpha$). Because $\xi = \alpha \times \beta$, each $\xi_x = \beta$ and $(\xi_0^n)_x = \beta_0^n(x)$. Using Proposition 1.1, then,

$$h(\xi, \tau) = \lim_{n \to \infty} \frac{1}{n+1} H(\xi_0^n) \geq \lim_{n \to \infty} \frac{1}{n+1} H(\alpha_0^n)$$
$$+ \lim_{n \to \infty} \frac{1}{n+1} \int_X H(\beta_0^n(x))\,d\mu(x) = h(\alpha, T) + h_T(\beta, S).$$

It follows by taking suprema over α and β that
$$h(\tau) \geq h(T) + h_T(S).$$

For the reverse inequality, begin by fixing finite partitions ξ of $X \times Y$ and α of X. We have, for $n, m = 1, 2, \ldots$,

$$\frac{H(\xi_0^{nm})}{nm} \leq \frac{H(\bar{\alpha}_0^{nm} \vee \xi_0^{nm})}{nm} = \frac{H(\bar{\alpha}_0^{nm})}{nm} + \frac{H(\xi_0^{nm}|\bar{\alpha}_0^{nm})}{nm}$$
$$= \frac{H(\alpha_0^{nm})}{nm} + \frac{H(\xi_0^{nm}|\bar{\alpha}_0^{nm})}{nm}.$$

Because

$$H(\zeta_0^{nm}|\bar\alpha_0^{nm}) = H((\zeta \vee \tau^{-1}\zeta \vee \ldots \vee \tau^{-m+1}\zeta)$$
$$\vee (\tau^{-m}\zeta \vee \ldots \vee \tau^{-2m+1}\zeta) \vee \ldots |\bar\alpha_0^{nm})$$
$$= H(\zeta_0^m \vee \tau^{-m}\zeta_0^m \vee \ldots \vee \tau^{-(n-1)m}\zeta_0^m|\bar\alpha_0^{nm})$$
$$\leq \sum_{k=0}^{n-1} H(\tau^{-km}\zeta_0^m|\bar\alpha_0^{nm}) \leq \sum_{k=0}^{n-1} H(\tau^{-km}\zeta_0^m|\tau^{-km}\bar\alpha)$$
$$= nH(\zeta_0^m|\bar\alpha),$$

this says that

$$\frac{H(\zeta_0^{nm})}{nm} \leq \frac{H(\alpha_0^{nm})}{nm} + \frac{H(\zeta_0^m|\bar\alpha)}{m}.$$

Taking limits as $n \to \infty$ gives

$$h(\zeta, \tau) \leq h(\alpha, T) + \frac{H(\zeta_0^m|\bar\alpha)}{m}.$$

Now choose α from a sequence $\alpha_1 \leq \alpha_2 \leq \ldots$ for which $\vee_{k=1}^\infty \mathcal{B}(\alpha_k) = \mathcal{B}$ up to sets of measure 0. We can conclude that

$$h(\zeta, \tau) \leq h(T) + \frac{H(\zeta_0^m|\bar{\mathcal{B}})}{m},$$

which in a case when $\zeta = \alpha \times \beta$ (a different α from the one that just disappeared) says that

$$h(\zeta, \tau) \leq h(T) + \frac{1}{m}\int_X H(\beta_0^m(x))\,d\mu(x).$$

Letting $m \to \infty$ gives

$$h(\zeta, \tau) \leq h(T) + h_T(\beta, S),$$

and taking suprema over α and β shows that

$$h(\tau) \leq h(T) + h_T(S).$$

C. Entropy of an induced transformation

Let $T: X \to X$ be a m.p.t. on $(X, \mathcal{B}, \mu)$ and $E \in \mathcal{B}$ a set of positive measure. Recall that the *induced* or *derivative transformation* $T_E: E \to E$ on $(E, \mathcal{B} \cap E, \mu_E)$ is defined by

$$T_E x = T^{n_E(x)} x,$$

where

$$n_E(x) = \inf\{n \geq 1 : T^n x \in E\}.$$

The formula

$$h(T_E) = \frac{h(T)}{\mu(E)}$$

was found by Abramov (1959); we give here the simple proof due to H. Scheller (see Krengel 1970, Denker, Grillenberger and Sigmund 1976, Neveu 1969). Notice that in general $h(T_E) > h(T)$: our uncertainty about the action of T_E is greater, since points have some chance to wander around while they're in E^c.

Let α be a countable measurable partition of E with $H(E) < \infty$; we will also consider α to be a partition of X, by adjoining E^c. We may assume also that all the sets $E_n = \{x \in E : n_E(x) = n\}$ are in $T_E^{-1}\alpha$. To see this, it is necessary to verify that the partition $\{E_n\}$ of E has finite entropy. If the measures of the E_n are $\epsilon_{n_1} \geq \epsilon_{n_2} \ldots$, then the non-ergodic version of Kac's formula (2.4.6) gives

$$1 \geq \sum_{k=1}^{\infty} n_k \epsilon_{n_k} \geq \left(\sum_{k=1}^{n} n_k\right) \epsilon_n \geq \frac{n(n+1)}{2} \epsilon_n,$$

so that $\epsilon_n \leq 2/n^2$ for all n. Since $-x \log_2 x \leq x^{2/3}$ for x sufficiently close to 0,

$$-\epsilon_n \log_2 \epsilon_n \leq \epsilon_n^{2/3} \leq 2^{2/3} n^{-4/3} \qquad \text{for large } n,$$

so that

$$H\{E_n\} = -\sum \epsilon_k \log_2 \epsilon_k < \infty.$$

Since it is possible to find an increasing sequence $\{\alpha_k\}$ of such partitions α for which $\mathscr{B}([\alpha_k]_{-k}^k)$ increases to $\mathscr{B}$ (up to sets of measure 0), it follows from Proposition 5.3.6 and Exercises 5.2.7 and 5.2.11 that $h(T)$ is the supremum of $h(\alpha, T)$ over all such α.

For each $n = 1, 2, \ldots, E_n = T^{-n}(T_E E_n)$, since $T_E | E_n = T^n$, so that each $E_n \in \alpha_1^\infty$ and hence also $E \in \alpha_1^\infty$.

We claim that the following equation holds relating two sub-σ-algebras of $\mathscr{B} \cap E$:

$$\bigvee_{k=1}^{\infty} T_E^{-k} \alpha = \alpha_1^\infty \cap E.$$

First notice that if $A \subset E$, then

$$T_E^{-1} A = \bigcup_{k=1}^{\infty} T_E^{-1}(T_E E_k \cap A) = \bigcup_{k=1}^{\infty} E_k \cap T^{-k} A.$$

Since each $E_k \in \alpha_1^\infty \cap E$, if $A \in \alpha$ then each set in the above union is in $\alpha_1^\infty \cap E$ also, and hence $T_E^{-1} A \in \alpha_1^\infty \cap E$. If $A \in \alpha \vee T_E^{-1}\alpha$, again $T_E^{-1} A \in \alpha_1^\infty \cap E$. Continuing in this way, we see that

$$\bigvee_{k=1}^{\infty} T_E^{-k} \alpha \subset \alpha_1^\infty \cap E.$$

6.2. The Shannon–McMillan–Breiman Theorem

For the reverse inclusion, notice that if $A \in \alpha$ is a subset of E, then

$$T^{-n}A \cap E = (T_E^{-1}A \cap E_n) \cup \left(\bigcup_{j_1=1}^{n-1} T_E^{-2} A \cap E_{j_1} \cap T_E^{-1} E_{n-j_1} \right) \cup \ldots$$

$$\cup (T_E^{-n}A \cap E_1 \cap T_E^{-1}E_1 \cap \ldots \cap T_E^{-(n-1)}E_1) \in \bigvee_{k=1}^{\infty} T_E^{-k}\alpha,$$

because each $E_j \in T_E^{-1}\alpha$.

For a sub-σ-algebra $\mathscr{F} \subset \mathscr{B}$ and $A \subset E \in \mathscr{F}$,

$$\mu(A|\mathscr{F}) = \mu(A|E \cap \mathscr{F}) \quad \text{a.e. on } E;$$

and if $A \in E \cap \mathscr{F}$, then of course $\mu(A|E \cap \mathscr{F}) = \chi_A$ a.e. Therefore $E^c \in \alpha_1^\infty$ and $\alpha(x) = E^c$ on E^c imply that $I_{\alpha|\alpha_1^\infty}(x) = 0$ a.e. on E^c, and

$$h(\alpha, T) = H(\alpha|\alpha_1^\infty) = \int_X I_{\alpha|\alpha_1^\infty} \, d\mu = \int_E I_{\alpha|\alpha_1^\infty} \, d\mu + \int_{E^c} I_{\alpha|\alpha_1^\infty} \, d\mu$$

$$= \int_E I_{\alpha|\alpha_1^\infty} \, d\mu = \int_E I_{\alpha|\alpha_1^\infty \cap E} \, d\mu = \int_E I_{\alpha|\bigvee_{k=1}^{\infty} T_E^{-k}\alpha} \, d\mu$$

$$= \mu(E) \int_E I_{\alpha|\bigvee_{k=1}^{\infty} T_E^{-k}\alpha} \, d\mu_E = \mu(E) h(\alpha, T_E).$$

Abramov's formula follows by taking the supremum over all α of the type under consideration.

Exercises

1. Show that if Y is a compact group, ν is Haar measure on Y, $f: X \to Y$ is measurable, and $\tau(x, y) = (Tx, y + f(x))$, then $h(\tau) = h(T)$.
2. Show that given any $r \in (0, \infty)$, there is a Bernoulli scheme of entropy r. Is the same true of toral automorphisms?
3. Give examples of m.p.t.s of infinite entropy.
4. Derive the formula for the entropy of a primitive (skyscraper) transformation built over a base $T: X \to X$.

6.2. The Shannon–McMillan–Breiman Theorem

Consider again a ticker-tape-type source like in our discussion of information theory. At each time, measure the amount of information per symbol received that has been conveyed up to that time. It is plausible that in the long run this average amount of information per symbol should converge to the entropy of the source, at least provided that the message that we are receiving accurately reflects the statistics of the source (which happens almost surely if the source is *ergodic*).

Let $T: X \to X$ be an ergodic m.p.t. on $(X, \mathcal{B}, \mu)$ and α a countable partition of X with $H(\alpha) < \infty$. Recall that $I_{\alpha_0^{n-1}}(x) = I_{\alpha \vee T^{-1}\alpha \vee \ldots \vee T^{-n+1}\alpha}(x)$ measures the amount of information conveyed when the first n symbols of the message x are received (that is, when we know to which elements of α the points $x, Tx, \ldots, T^{n-1}x$ belong). We know (Proposition 5.2.10)

$$\frac{1}{n+1} H(\alpha_0^n) = \frac{1}{n+1} \int_X I_{\alpha_0^n}(x) \, d\mu \to h(\alpha, T);$$

the preceding paragraph proposes that in fact the integrand

$$\frac{1}{n+1} I_{\alpha_0^n}(x)$$

already converges a.e. to $h(\alpha, T)$. That this does in fact occur when T is ergodic is the content of the theorem proved in increasing degrees of generality by C. Shannon (1948), B. McMillan (1953), L. Carleson (1958), Leo Breiman (1957), (1960), A. Ionescu Tulcea (1960), and K. L. Chung (1961).

This theorem can also be interpreted as providing an estimate of the speed with which the process of moving by T and intersecting chops up the atoms of α. For a typical $A_n \in \alpha_0^n$ and large n, we have

$$-\frac{1}{n+1} \log_2 \mu(A_n) \approx h(\alpha, T),$$

so that

$$\mu(A_n) \approx 2^{-(n+1)h(\alpha, T)}.$$

Thus the sizes of the atoms decrease exponentially at a rate determined by the entropy. We will formulate this version of the theorem more precisely below (see Corollary 2.5).

The proof of the Shannon–McMillan–Breiman Theorem again depends on a maximal inequality, which in this case is stronger than the usual weak-type $(1, 1)$ estimate. Continue to let $T: X \to X$ denote an ergodic m.p.t. on $(X, \mathcal{B}, \mu)$ and α a countable measurable partition of X with *finite entropy*.

Lemma 2.1 For each $n = 1, 2, \ldots$ let $f_n = I_{\alpha | T^{-1}\alpha \vee \ldots \vee T^{-n}\alpha} = I_{\alpha | \alpha_1^n}$ and $f^* = \sup_{n \geq 1} f_n$. Then for each $\lambda \geq 0$ and each $A \in \alpha$,

$$\mu\{x \in A : f^*(x) > \lambda\} \leq 2^{-\lambda}.$$

6.2. The Shannon–McMillan–Breiman Theorem

Proof: For each $A \in \alpha$ and $n = 1, 2, \ldots$, let

$$f_n^A = I_{A|\alpha_1^n} = -\log_2 \mu(A \mid T^{-1}\alpha \vee \ldots \vee T^{-n}\alpha)$$

and

$$B_n^A = \{x : f_1^A(x), \ldots, f_{n-1}^A(x) \leq \lambda, f_n^A(x) > \lambda\}.$$

Of course $f_n^A = f_n$ on A.

Since $B_k^A \in \mathcal{B}(T^{-1}\alpha \vee \ldots \vee T^{-k}\alpha)$,

$$\mu(B_k^A \cap A) = \int_{B_k^A} \chi_A \, d\mu = \int_{B_k^A} \mu(A \mid T^{-1}\alpha \vee \ldots \vee T^{-k}\alpha) \, d\mu$$

$$= \int_{B_k^A} 2^{-f_k^A} \, d\mu \leq 2^{-\lambda} \mu(B_k^A).$$

Therefore

$$\mu\{x \in A : f^*(x) > \lambda\} = \sum_{k=1}^{\infty} \mu(B_k^A \cap A) \leq 2^{-\lambda} \sum_{k=1}^{\infty} \mu(B_k^A) \leq 2^{-\lambda}.$$

Corollary 2.2 $f^* \in L^1$.

Proof: Of course $\mu\{x \in A : f^*(x) > \lambda\} \leq \mu(A)$, so $\mu\{x \in A : f^*(x) > \lambda\} \leq \min\{\mu(A), 2^{-\lambda}\}$. As in the proof of Theorem 3.1.16,

$$\int_X f^* \, d\mu = \int_0^\infty \mu\{f^* > \lambda\} \, d\lambda = \sum_{A \in \alpha} \int_0^\infty \mu\{x \in A : f^*(x) > \lambda\} \, d\lambda$$

$$\leq \sum_{A \in \alpha} \int_0^\infty \min\{\mu(A), 2^{-\lambda}\} \, d\lambda$$

$$= \sum_{A \in \alpha} \left[\int_0^{-\log_2 \mu(A)} \mu(A) \, d\lambda + \int_{-\log_2 \mu(A)}^\infty 2^{-\lambda} \, d\lambda \right]$$

$$= \sum_{A \in \alpha} \left[-\mu(A) \log_2 \mu(A) + \frac{\mu(A)}{\log 2} \right] = H(\alpha) + \frac{1}{\log 2}.$$

Theorem 2.3 *Shannon–McMillan–Breiman Theorem*   Let $T: X \to X$ be an *ergodic* m.p.t. on $(X, \mathcal{B}, \mu)$ and α a countable measurable partition of X with $H(\alpha) < \infty$. Then

$$\lim_{n \to \infty} \frac{1}{n+1} I_{\alpha \vee T^{-1}\alpha \vee \ldots \vee T^{-n}\alpha}(x) = h(\alpha, T) \text{ a.e.}$$

Proof: Let $f_n = I_{\alpha \mid T^{-1}\alpha \vee \ldots \vee T^{-n}\alpha}$ for each $n = 1, 2, \ldots$, as before. By making

6. More about entropy

repeated use of the formulas $I_{\beta \vee \gamma} = I_\beta + I_{\gamma|\beta}$ and $I_{T^{-1}\alpha} = I_\alpha \circ T$ (see Proposition 5.2.3 and Exercise 5.2.9), one can see that

$$I_{\alpha_0^n} = I_{\alpha \vee T^{-1}\alpha \vee \ldots \vee T^{-n}\alpha} = I_{T^{-1}\alpha \vee \ldots \vee T^{-n}\alpha} + I_{\alpha|T^{-1}\alpha \vee \ldots \vee T^{-n}\alpha}$$
$$= I_{\alpha \vee \ldots \vee T^{-n+1}\alpha} \circ T + I_{\alpha|T^{-1}\alpha \vee \ldots \vee T^{-n}\alpha}$$
$$= f_n + f_{n-1} T + \ldots + f_1 T^{n-1} + I_{T^{-1}\alpha} \circ T^n = \sum_{k=0}^{n} f_{n-k} T^k,$$

where we have defined $f_0 = I_{T^{-1}\alpha}$.

Let $f = I_{\alpha|\alpha_1^\infty}$; By Corollary 2.2 and the Martingale Convergence Theorem (see 3.4), $f = \lim_{n\to\infty} f_n$ both pointwise a.e. and in L^1. Then we may write

$$\frac{1}{n+1} I_{\alpha_0^n} = \frac{1}{n+1} \sum_{k=0}^{n} f_{n-k} T^k = \frac{1}{n+1} \sum_{k=0}^{n} f T^k$$
$$+ \frac{1}{n+1} \sum_{k=0}^{n} (f_{n-k} - f) T^k.$$

By the Ergodic Theorem,

$$\lim_{n\to\infty} \frac{1}{n+1} \sum_{k=0}^{n} f T^k = \int_X f \, d\mu = \int_X I_{\alpha|\alpha_1^\infty} \, d\mu = H(\alpha|\alpha_1^\infty)$$
$$= h(\alpha, T) \text{ a.e.};$$

therefore, in order to prove the Theorem it suffices to show that

(1) $$\lim_{n\to\infty} \frac{1}{n+1} \sum_{k=0}^{n} |f_{n-k} - f| T^k = 0 \text{ a.e.}$$

For each $N = 1, 2, \ldots$, let $F_N = \sup_{k \geq N} |f_k - f|$. Then

$$\frac{1}{n+1} \sum_{k=0}^{n} |f_{n-k} - f| T^k = \frac{1}{n+1} \sum_{k=0}^{n-N} |f_{n-k} - f| T^k$$
$$+ \frac{1}{n+1} \sum_{k=n-N+1}^{n} |f_{n-k} - f| T^k$$
$$\leq \frac{1}{n+1} \sum_{k=0}^{n-N} F_N T^k + \frac{1}{n+1} \sum_{k=n-N+1}^{n} |f_{n-k} - f| T^k.$$

Fix N and let $n \to \infty$. Because $|f_{n-k} - f| \leq f^* + f \in L^1$ for all k, the second term above tends to zero a.e. Similarly, $0 \leq F_N \leq f^* + f \in L^1$, so we may apply the Ergodic Theorem to the first term:

$$\lim_{n\to\infty} \frac{1}{n+1} \sum_{k=0}^{n-N} |f_{n-k} - f| T^k \leq \lim_{n\to\infty} \frac{1}{n+1} \sum_{k=0}^{n-N} F_N T^k = \int_X F_N \, d\mu.$$

6.2. The Shannon–McMillan–Breiman Theorem

The Dominated Convergence Theorem and the observation that $F_N \to 0$ a.e. then yield (1).

Remark 2.4 Because the convergence in the Ergodic Theorem also takes place in L^1, and the F_N above are dominated by an integrable function, also

$$\frac{1}{n+1} I_{\alpha_0^n} \to h(\alpha, T) \text{ in } L^1.$$

The following consequence of the Shannon–McMillan–Breiman Theorem is sometimes called the *entropy equipartition property*.

Corollary 2.5 Let $T: X \to X$ be ergodic and α a countable measurable partition of X with $H(\alpha) < \infty$. Given $\varepsilon > 0$ there is an n_0 such that for $n \geq n_0$ the elements of α_0^n can be divided into two classes, $\mathscr{g}$ and $\mathscr{b}$ (the 'good' and 'bad' atoms), such that:

(i) $\mu \bigcup_{B \in \mathscr{b}} B < \varepsilon$

(ii) For each $G \in \mathscr{g}$,

$$2^{-(n+1)(h(\alpha, T) + \varepsilon)} < \mu(G) < 2^{-(n+1)(h(\alpha, T) - \varepsilon)}.$$

Proof: Since $I_{\alpha_0^n}/(n+1) \to h(\alpha, T)$ a.e., the convergence also takes place in measure: for each $\varepsilon > 0$ there is an n_0 such that if $n \geq n_0$ then

$$\mu\left\{ x : \left| \frac{1}{n+1} I_{\alpha_0^n}(x) - h(\alpha, T) \right| \geq \varepsilon \right\} < \varepsilon.$$

Of course $I_{\alpha_0^n}(x)$ is constant on each atom of α_0^n. Thus we need only to let $\mathscr{g}$ be the collection of all those atoms G of α_0^n on which

$$\left| \frac{1}{n+1} I_{\alpha_0^n}(x) - h(\alpha, T) \right| < \varepsilon,$$

and let $\mathscr{b}$ be all those atoms that are left over.

Exercises
1. Formulate and prove a version of the Shannon–McMillan–Breiman Theorem for nonergodic T.
2. Compare the sizes of $\mathscr{g} = \mathscr{g}_n$ and r^n in case $\alpha = \{A_1, A_2, \ldots A_r\}$ is a finite partition with $h(\alpha, T) < \log_2 r$. Interpret in terms of teletypes.
3. Let $\mathscr{B}_1 \subset \mathscr{B}_2 \subset \ldots$ be an increasing sequence of sub-σ-algebras of $\mathscr{B}$, $\mathscr{B}_\infty = \bigvee_{n=1}^\infty \mathscr{B}_n$, and α a countable measurable partition of X with $H(\alpha) < \infty$. Then $I_{\alpha|\mathscr{B}_n} \to I_{\alpha|\mathscr{B}_\infty}$ a.e. and in L^1.

4. Use the Shannon–McMillan–Breiman Theorem to compute the entropy of an ergodic Markov shift.

6.3. Topological entropy

A topological analogue of entropy for a cascade (X, T) (X compact Hausdorff, $T: X \to X$ a homeomorphism) was introduced by Adler, Konheim, and McAndrew (1965). Later, Bowen (1971) gave a different definition for (possibly noncompact) metric or uniform spaces. A similar idea appears in Dinaburg (1970). The relationship between topological entropy and metric entropy was found by Goodwyn (1969, 1972), Dinaburg (1970), and Goodman (1971): the topological entropy of a homeomorphism is the supremum of its metric entropies over all ergodic invariant Borel probability measures. Given this 'variational principle', many of the properties of topological entropy follow immediately from the corresponding statements for metric entropy.

In the case of a homeomorphism $T: X \to X$ of a compact Hausdorff space X, we deal with open covers $\mathscr{U}$ of X rather than with measurable partitions. We say that $\mathscr{V}$ is a *refinement* of $\mathscr{U}$, and write $\mathscr{V} \geq \mathscr{U}$, if every $V \in \mathscr{V}$ is a subset of some $U \in \mathscr{U}$. The *least common refinement* or *join*, $\mathscr{U} \vee \mathscr{V}$, of $\mathscr{U}$ and $\mathscr{V}$ consists of the nonempty members of $\{U \cap V : U \in \mathscr{U}, V \in \mathscr{V}\}$.

For each open cover $\mathscr{U}$ of X, let

$N(\mathscr{U})$ = the minimum of the cardinality of the subcovers of $\mathscr{U}$

and

$$H(\mathscr{U}) = \log_2 N(\mathscr{U}).$$

Proposition 3.1 $H(\mathscr{U} \vee \mathscr{V}) \leq H(\mathscr{U}) + H(\mathscr{V})$.
Proof: $N(\mathscr{U} \vee \mathscr{V}) \leq N(\mathscr{U}) N(\mathscr{V})$.

For $-\infty < m \leq n < \infty$, define

$$\mathscr{U}_m^n = \bigvee_{k=m}^{n} T^{-k} \mathscr{U} = T^{-m} \mathscr{U} \vee T^{-(m+1)} \mathscr{U} \vee \dots \vee T^{-n} \mathscr{U}.$$

Proposition 3.2 $h(\mathscr{U}, T) = \lim_{n \to \infty} (1/n) H(\mathscr{U}_0^{n-1})$ exists, and $h(\mathscr{U}, T) \leq H(\mathscr{U})$.
Proof: If $H_n = H(\mathscr{U}_0^{n-1})$, then

$$H_{n+m} = H(\mathscr{U}_0^{n-1} \vee T^{-n} \mathscr{U}_0^{m-1}) \leq H(\mathscr{U}_0^{n-1}) + H(T^{-n} \mathscr{U}_0^{m-1})$$
$$= H_n + H_m.$$

Then the proof of Proposition 5.2.10 applies to show that $\{H_n/n\}$ is bounded by $H_1 = H(\mathscr{U})$ and has a limit as $n \to \infty$.

6.3. Topological entropy

Definition 3.3 The *topological entropy* of a cascade (X, T) is defined to be
$$h_{\text{top}}(T) = \sup_{\mathcal{U}} h(\mathcal{U}, T).$$

Notice that $\mathcal{V} \geq \mathcal{U}$ implies $h(\mathcal{V}, T) \geq h(\mathcal{U}, T)$. Thus $h_{\text{top}}(T)$ is the increasing limit of the net $h(\mathcal{U}, T)$, where the index set is the family of all open covers of X. The following fact, which is useful for the computation of topological entropy, is apparent.

Proposition 3.4 If $\{\mathcal{U}_n : n = 1, 2, \ldots\}$ is a *refining sequence* of open covers in that $\mathcal{U}_1 \leq \mathcal{U}_2 \leq \ldots$ and for each finite open cover $\mathcal{V}$ of X we have $\mathcal{V} \leq \mathcal{U}_n$ for some n, then
$$h_{\text{top}}(T) = \lim_{n \to \infty} h(\mathcal{U}_n, T).$$

Remark 3.5 If $\mathcal{U}_1 \leq \mathcal{U}_2 \leq \ldots$ and
$$\operatorname{diam} \mathcal{U}_n = \sup_{U \in \mathcal{U}_n} \operatorname{diameter}(U) \to 0,$$
then $\{\mathcal{U}_n\}$ is a refining sequence. For given any open cover $\mathcal{V}$, we can choose n large enough that all the members of $\mathcal{U}_n$ have diameter smaller than the Lebesgue number of $\mathcal{V}$. This forces each member of $\mathcal{U}_n$ to be contained in a member of $\mathcal{V}$.

Example 3.6 Let σ be the shift transformation on $\{0, 1\}^{\mathbb{Z}}$ and X a closed shift-invariant subset. The covers consisting of the cylinder sets of rank n form a natural refining sequence. Here $\mathcal{U}_n$ consists of all non-empty cylinder sets of the form
$$\{x \in X : x_{-n} = i_{-n}, x_{-n+1} = i_{-n+1}, \ldots, x_0 = i_0, \ldots, x_n = i_n\},$$
for some choice of $i_{-n}, i_{-n+1}, \ldots, i_n \in \{0, 1\}$. Note that, writing $\mathcal{U} = \mathcal{U}_0$,
$$\mathcal{U}_n = (\mathcal{U})_{-n}^n.$$
As in Exercise 5.2.7, $h(\mathcal{U}_n, T) = h(\mathcal{U}, T)$ for all n, because
$$h(\mathcal{U}, T) \leq h(\mathcal{U}_n, T) = \lim_{k \to \infty} \frac{1}{k} H((\mathcal{U}_n)_0^{k-1})$$
$$= \lim_{k \to \infty} \frac{1}{k} H(\mathcal{U}_{-n}^{n+k-1}) = \lim_{k \to \infty} \frac{1}{k} H(\mathcal{U}_n^{n+k-1} \vee \mathcal{U}_n^{n-1})$$
$$\leq \lim_{k \to \infty} \frac{1}{k} [H(\mathcal{U}_n^{n+k-1}) + H(\mathcal{U}_n^{n-1})] = \lim_{k \to \infty} \frac{1}{k} H(\mathcal{U}_0^{k-1})$$
$$= h(\mathcal{U}, T).$$

Thus
$$h_{\text{top}}(T) = \lim_{n \to \infty} h(\mathcal{U}_n, T) = \lim_{n \to \infty} h(\mathcal{U}, T) = h(\mathcal{U}, T) = \lim_{m \to \infty} \frac{H(\mathcal{U}_0^{m-1})}{m}.$$

The number of (nonempty) elements in $\mathscr{U}_0^{m-1}$ is just the number of m-blocks that can appear in a sequence in X. Therefore we have the important result that *the topological entropy of a symbolic cascade is*

$$\lim_{m\to\infty} \frac{1}{m} \log_2 (\text{number of } m\text{-blocks}).$$

The topological entropy of the full two-shift is 1. If X consists of a single periodic orbit, then the number N_n of n-blocks is bounded, so that in this case $h(\sigma|_X) = 0$. If X consists of all sequences containing only even-length maximal strings of 0's and of 1's, then $N_{2k} = 2^k$, so that $h_{\text{top}}(\sigma|_X) = \frac{1}{2}$. (This is reasonable because in this case $(\sigma|_X)^2$ is isomorphic to the full two-shift.) Exercise 4 asks for the (less trivial) computation of the topological entropy of an arbitrary subshift of finite type.

This example shows that the topological entropy is indeed a measure of the absolute information content of a transformation T, or, equivalently, of the extent to which T scatters points around the space X. It is a bit of a miracle that this definition in terms of open covers has proved to be so right and fruitful, as convincingly shown by Theorem 3.10.

We turn now to the definition of entropy given by Bowen. Let $T: X \to X$ be a homeomorphism of a compact metric space X. In analogy with the preceding example, we wish to count the number of different orbit-blocks of length n that can be observed, where now (in order to get a finite number) we fail to distinguish points that are closer together than some positive distance ε. Let us say that x and y are (n, ε)-*separated* if their initial orbit blocks of length n can be distinguished by such a myopic observer, i.e., if

$$d(T^k x, T^k y) > \varepsilon \text{ for some } k = 0, 1, \ldots, n-1.$$

A set $E \subset X$ is called (n, ε)-*separated* if x and y are (n, ε)-separated whenever $x, y \in E$ and $x \neq y$. Then the maximum number of distinguishable orbit n-blocks is

$$s(n, \varepsilon) = \max \{\text{card } E : E \subset X \text{ is } (n, \varepsilon)\text{-separated}\}.$$

We define

$$h(T, \varepsilon) = \limsup_{n\to\infty} \frac{1}{n} \log_2 s(n, \varepsilon)$$

and

$$h_{\text{top}}(T) = \lim_{\varepsilon \to 0^+} h(T, \varepsilon).$$

The latter limit exists because $h(T, \varepsilon)$ clearly increases as ε decreases to 0. It represents the information we gain from watching T if we can make arbitrarily accurate observations.

6.3. Topological entropy

If X is not compact, for each compact $K \subset X$ let

$$s_K(n, \varepsilon) = \max \{\operatorname{card} E : E \subset K \text{ is } (n, \varepsilon)\text{-separated}\},$$

$$h_K(T, \varepsilon) = \limsup_{n \to \infty} \frac{1}{n} \log_2 s_K(n, \varepsilon),$$

$$h_K(T) = \lim_{\varepsilon \to 0^+} h_K(T, \varepsilon),$$

and define

$$h_{\text{top}}(T) = \sup_K h_K(T).$$

In this way the topological entropy is defined for a homeomorphism of any metric space. Similar definitions apply also to uniform spaces, but we will not need to consider such a general situation. This definition of h_{top} is related to Kolmogorov's concept of ε-*entropy*; see Walters (1975), Dinaburg (1970).

One can also take a reverse approach to the definition of $h_{\text{top}}(T)$. For each $\varepsilon > 0$ and $n = 1, 2, \ldots$, call a set $F \subset X$ (n, ε)-*spanning* if for each $x \in X$ there is $y \in F$ with

$$d(T^k x, T^k y) \leq \varepsilon \text{ for all } k = 0, 1, \ldots, n-1.$$

Let E be a *maximal* (n, ε)-separated set. Then E must also be (n, ε)-spanning, since if we try to add any point y to E, $E \cup \{y\}$ is no longer (n, ε)-separated — which is to say that the initial orbit n-blocks of y and some $x \in E$ will no longer be distinguishable by an ε-accurate observer. Therefore, if

$$r(n, \varepsilon) = \min \{\operatorname{card} F : F \subset X \text{ is } (n, \varepsilon)\text{-spanning}\},$$

then

$$r(n, \varepsilon) \leq s(n, \varepsilon).$$

On the other hand, I claim that

$$s(n, \varepsilon) \leq r(n, \varepsilon/2).$$

For fix an $(n, \varepsilon/2)$-spanning set F and let E be an (n, ε)-separated subset of X. We can define a map $E \to F$ by choosing for each $x \in E$ a point $y \in F$ whose initial orbit block of length n '$\varepsilon/2$-shadows' that of x. This map must be one-to-one, since any two points of E have ε-distinguishable initial orbit n-blocks. Thus $\operatorname{card}(E) \leq \operatorname{card}(F)$, and the result follows. Now the equivalence of the following alternative definition of $h_{\text{top}}(T)$ with the one already given is obvious.

Proposition 3.7 $h_{\text{top}}(T) = \lim_{\varepsilon \to 0^+} \limsup_{n \to \infty} 1/n \log_2 r(n, \varepsilon)$.

Next we establish the equivalence of the Adler, Konheim, McAndrew and Bowen definitions of topological entropy.

Proposition 3.8 $\hbar_{top}(T) = h_{top}(T)$.

Proof: Given $\varepsilon > 0$, let $\mathcal{U}$ be an open cover of X with diam $\mathcal{U} = \sup_{U \in \mathcal{U}} \operatorname{diam} U < \varepsilon$. Then

$$s(n, \varepsilon) \leq N(\mathcal{U}_0^{n-1}) \quad \text{for each } n = 1, 2, \ldots,$$

because two points of any (n, ε)-separated set cannot both be in the same element of $\mathcal{U}_0^{n-1}$. It follows that

$$\hbar_{top}(T) \leq h_{top}(T).$$

For the reverse inequality, given any open cover $\mathcal{U}$ of X let ε be the Lebesgue number of $\mathcal{U}$. Fix $n = 1, 2, \ldots$ and let F be a minimal (n, ε)-spanning set. We will use F to find a subcover of $\mathcal{U}_0^{n-1}$ of cardinality less than or equal to card $F = r(n, \varepsilon)$.

For each $y \in F$ and $k = 0, 1, \ldots, n-1$, the ball $B(T^k y, \varepsilon)$ of radius ε centered at $T^k y$ is contained in some single member $U(k, y)$ of $\mathcal{U}$. Given $x \in X$, there is $y \in F$ whose initial orbit-n-block is ε-indistinguishable from that of x; this says

$$x \in U(0, y) \cap T^{-1} U(1, y) \cap \ldots \cap T^{-n+1} U(n - 1, y).$$

Thus

$$\{U(0, y) \cap T^{-1} U(1, y) \cap \ldots \cap T^{-n+1} U(n - 1, y) : y \in F\}$$

is a subcover of $\mathcal{U}_0^{n-1}$, and it clearly has cardinality less than or equal to that of F.

It follows that

$$N(\mathcal{U}_0^{n-1}) \leq r(n, \varepsilon),$$

and hence

$$h_{top}(T) \leq \hbar_{top}(T).$$

We turn now to the proof of the variational principle, which appears as a conjecture in the original paper of Adler, Konheim, and McAndrew (1965). This principle shows clearly the importance and naturalness of the concept of topological entropy. There is an interesting theory of when the supremum is attained, and of when it is attained by a single measure; see Denker, Grillenberger and Sigmund (1976) for an introduction to this subject of current research. First we need one property of h_{top}.

Proposition 3.9 For any $k = 1, 2, \ldots, h_{top}(T^k) = k h_{top}(T)$.

Proof: For an open cover $\mathcal{U}$ of X,

$$h_{top}(T^k) \geq h(\mathcal{U}_0^{k-1}, T^k)$$

$$= k \lim_{n \to \infty} \frac{1}{nk} H(\mathcal{U}_0^{k-1} \vee T^{-k} \mathcal{U}_0^{k-1} \vee \ldots \vee T^{-(n-1)k} \mathcal{U}_0^{k-1})$$

6.3. Topological entropy

$$= k \lim_{n\to\infty} \frac{1}{nk} H(\mathcal{U}_0^{nk-1}) = kh(\mathcal{U}, T).$$

Also, $\mathcal{U} \vee (T^k)^{-1}\mathcal{U} \vee \ldots \vee (T^k)^{-n+1}\mathcal{U} \leq \mathcal{U} \vee T^{-1}\mathcal{U} \vee \ldots \vee T^{-nk+1}\mathcal{U}$ implies that

$$h(\mathcal{U}, T) = \lim_{n\to\infty} \frac{1}{nk} H(\mathcal{U} \vee T^{-1}\mathcal{U} \vee \ldots \vee T^{-nk+1}\mathcal{U})$$

$$\geq \lim_{n\to\infty} \frac{1}{nk} H(\mathcal{U} \vee (T^k)^{-1}\mathcal{U} \vee \ldots \vee (T^k)^{-n+1}\mathcal{U})$$

$$= \frac{1}{k} h(\mathcal{U}, T^k).$$

Combining these inequalities and taking the supremum over all $\mathcal{U}$ yields the result.

Theorem 3.10 Let $T: X \to X$ be a homeomorphism on a compact metric space X, and denote by $\mathcal{M}_T$ the set of all T-invariant Borel probability measures on X. Then

$$h_{\text{top}}(T) = \sup_{\mu \in \mathcal{M}_T} h_\mu(T).$$

Proof: (Misiurewicz, 1976) Fix an invariant measure μ for T; we will show that $h_\mu(T) \leq h_{\text{top}}(T)$.

Let α be a finite measurable partition of X. For each $A_i \in \alpha$ choose a compact set $B_i \subset A_i$ which approximates A_i so closely in measure that if β is the partition $\{B_1, B_2, \ldots, (\bigcup B_j)^c\}$, then

$$H_\mu(\alpha | \beta) < 1.$$

Notice that since $H(\alpha_0^{n-1} | \beta_0^{n-1}) \leq nH(\alpha | \beta)$ for each $n = 1, 2, \ldots$ (see the proof of Proposition 2.13), we have

$$H_\mu(\alpha_0^{n-1}) \leq H_\mu(\alpha_0^{n-1} \vee \beta_0^{n-1})$$

$$= H_\mu(\beta_0^{n-1}) + H_\mu(\alpha_0^{n-1} | \beta_0^{n-1}) \leq H_\mu(\beta_0^{n-1}) + n,$$

and hence

$$h_\mu(\alpha, T) = \lim_{n\to\infty} \frac{1}{n} H_\mu(\alpha_0^{n-1}) \leq h_\mu(\beta, T) + 1.$$

Now form an open cover $\mathcal{U}_1 = \{U_i\}$ of X by letting $U_i = (\bigcup_{j \neq i} B_j)^c$ for each i. Then each set U_i intersects at most two members of β–certainly B_i and possibly also $(\bigcup B_j)^c$. Let $\mathcal{U}$ be a subcover of $\mathcal{U}_1$ of minimal cardinality. Then for all $n = 1, 2, \ldots$,

$$\operatorname{card}(\beta_0^{n-1}) \leq 2^n \operatorname{card}(\mathcal{U}_0^{n-1}).$$

6. More about entropy

For card $(\mathscr{U}_0^{n-1})$ counts the number of different $\mathscr{U}$, n-names of points of X, i.e., the number of different sequences $U_{i_0}, U_{i_1},\ldots,U_{i_{n-1}}$ such that for some $x\in X$, $T^k x \in U_{i_k}$ for $k = 0, 1, \ldots, n-1$; similarly for card (β_0^{n-1}). And once we know the $\mathscr{U}$, n-name of a point $x\in X$, then at each time k we know which element of $\mathscr{U}$ $T^k x$ belongs to, and have a choice of only two elements of β to which $T^k x$ might belong. Therefore the number of different β, n-names is at most 2^n times the number of different $\mathscr{U}$, n-names, and the assertion follows.

Having this, it is easy to compute that

$$H_\mu(\beta_0^{n-1}) \leq \log_2 \text{card}(\beta_0^{n-1}) \leq n + \log_2 N(\mathscr{U}_0^{n-1}),$$

and therefore,

$$h_\mu(\alpha, T) \leq h_\mu(\beta, T) + 1 = \lim_{n\to\infty} \frac{1}{n} H_\mu(\beta_0^{n-1}) + 1 \leq 2 + h(\mathscr{U}, T).$$

Since α was arbitrary, we have

$$h_\mu(T) \leq 2 + h_{\text{top}}(T).$$

This same calculation applies if T is replaced by T^k for any $k = 1, 2, \ldots$. The conclusion then says that

$$kh_\mu(T) \leq 2 + kh_{\text{top}}(T).$$

Since this holds for all k, we must have

$$h_\mu(T) \leq h_{\text{top}}(T).$$

In order to prove the reverse inequality, let $\varepsilon > 0$, for each $n = 1, 2, \ldots$ choose a maximal (n, ε)-separated set E_n, let

$$\mu_{E_n} = \frac{1}{\text{card } E_n} \sum_{x\in E_n} \delta_x$$

(where δ_x denotes the unit point mass at x), and

$$\mu_n = \frac{1}{n} \sum_{k=0}^{n-1} \mu_{E_n} T^{-k}.$$

Choose a subsequence $\{n_k\}$ such that

$$h(T, \varepsilon) = \limsup_{n\to\infty} \frac{1}{n} \log_2 s(n, \varepsilon) = \lim_{k\to\infty} \frac{1}{n_k} \log_2 s(n_k, \varepsilon).$$

By passing to a further subsequence if necessary, we may assume that $\{\mu_{n_k}\}$ converges to some measure μ in the weak * topology:

$$\int_X f\,d\mu_{n_k} \to \int_X f\,d\mu \quad \text{for all } f\in\mathscr{C}(X).$$

6.3. Topological entropy

It is easy to see that $\mu \in \mathcal{M}_T$.

Choose a finite measurable partition α of X such that for all $A \in \alpha$,

   (i) $\operatorname{diam}(A) < \varepsilon$
   (ii) $\mu(\partial A) = 0$.

(Notice that for any $x \in X$, at most countably many balls $B(x, \delta)$ have $\mu(\overline{B(x,\delta)} \setminus B(x,\delta)) > 0$. Cover X by finitely many balls B of diameter less than ε for which $\mu(\partial B) = 0$, and let α be the partition generated by this cover.) The importance of this second condition is the following:

Fact: If $\mu_n \to \mu$ weak $*$ and A is a measurable set with $\mu(\partial A) = 0$, then $\mu_n(A) \to \mu(A)$.

Proof: Choose $0 \leqslant f_k \in \mathscr{C}(X)$ with $f_k \searrow \chi_{\bar{A}}$. Then

$$\limsup_n \mu_n(\bar{A}) \leqslant \limsup_n \int f_k \, d\mu_n = \int f_k \, d\mu \searrow \mu(\bar{A})$$

so that

$$\limsup_n \mu_n(A) \leqslant \limsup_n \mu_n(\bar{A}) \leqslant \mu(\bar{A}) = \mu(A).$$

Similarly

$$\limsup_n \mu_n(X \setminus A) \leqslant \mu(X \setminus A),$$

or

$$1 - \liminf_n \mu_n(A) \leqslant 1 - \mu(A).$$

Together these say that

$$\limsup_n \mu_n(A) \leqslant \mu(A) \leqslant \liminf_n \mu(A).$$

For each $A \in \alpha_0^{n-1}$, $\operatorname{card}(E_n \cap A) \leqslant 1$, because two different points of E_n cannot stay within ε of one another during their initial orbit n-blocks. Therefore

$$H_{\mu_{E_n}}(\alpha_0^{n-1}) = - \sum_{A \in \alpha_0^{n-1}} \mu_{E_n}(A) \log_2 \mu_{E_n}(A)$$

$$= - \sum_{A \in \alpha_0^{n-1}} \left(\frac{1}{\operatorname{card} E_n} \sum_{y \in E_n \cap A} 1 \right) \log_2 \left(\frac{1}{\operatorname{card} E_n} \sum_{y \in E_n \cap A} 1 \right)$$

$$= \sum_{\substack{A \in \alpha_0^{n-1} \\ \text{with } A \cap E_n \neq \emptyset}} \frac{1}{\operatorname{card} E_n} \log_2 \operatorname{card} E_n$$

$$= \sum_{y \in E_n} \frac{1}{\operatorname{card} E_n} \log_2 \operatorname{card} E_n = \log_2 \operatorname{card} E_n.$$

6. More about entropy

Fix $m = 1, 2, \ldots$ and let $n = dm + r$, where $0 \leq r < m$ and $d > 0$ are integers. We may write, for each $k = 0, 1, \ldots, m-1$,

$$\alpha_0^{n-1} = \alpha_0^{dm+r-1} = \bigvee_{j=0}^{d-1} T^{-jm}\alpha_0^{m-1} \vee \alpha_{dm}^{dm+r-1}$$

$$\leq \bigvee_{j=0}^{d-1} T^{-jm-k}\alpha_0^{m-1} \vee (\alpha_{dm}^{dm+r-1} \vee \alpha_0^{k-1}).$$

Because the partition in parentheses has no more than $(\text{card } \alpha)^{2m}$ atoms,

(1)
$$H_{\mu_{E_n}}(\alpha_0^{n-1}) \leq \sum_{j=0}^{d-1} H_{\mu_{E_n}}(T^{-jm-k}\alpha_0^{m-1}) + H_{\mu_{E_n}}(\alpha_{dm}^{dm+r-1} \vee \alpha_0^{k-1})$$

$$\leq \sum_{j=0}^{d-1} H_{\mu_{E_n}}(T^{-jm-k}\alpha_0^{m-1}) + 2m \log_2 \text{card } \alpha.$$

Using the concavity of the function $f(t) = -t \log_2 t$, for each $A \in \alpha_0^{m-1}$

$$f(\mu_n(A)) = f\left(\frac{1}{n}\sum_{k=0}^{n-1} \mu_{E_n}(T^{-k}A)\right) \geq \frac{1}{n}\sum_{k=0}^{n-1} f(\mu_{E_n}(T^{-k}A)),$$

and hence, summing on A and using (1),

$$H_{\mu_n}(\alpha_0^{m-1}) \geq \frac{1}{n}\sum_{k=0}^{n-1} H_{\mu_{E_n}}(T^{-k}\alpha_0^{m-1})$$

$$\geq \frac{1}{n}\sum_{k=0}^{m-1}\sum_{j=0}^{d-1} H_{\mu_{E_n}}(T^{-jm-k}\alpha_0^{m-1})$$

$$\geq \frac{1}{n}\sum_{k=0}^{m-1}\left[H_{\mu_{E_n}}(\alpha_0^{n-1}) - 2m \log_2 \text{card } \alpha\right]$$

$$= \frac{m}{n}H_{\mu_{E_n}}(\alpha_0^{n-1}) - \frac{2m^2}{n}\log_2 \text{card } \alpha$$

$$= \frac{m}{n}\log_2 \text{card } E_n - \frac{2m^2}{n}\log_2 \text{card } \alpha$$

$$= \frac{m}{n}\log_2 s(n, \varepsilon) - \frac{2m^2}{n}\log_2 \text{card } \alpha.$$

Let $n = n_k$ and $k \to \infty$, while m is held fixed. Because $H_{\mu_n}(\alpha_0^{m-1}) \to H_{\mu}(\alpha_0^{m-1})$ by the Fact above (p. 271) and our choice of α, we are able to conclude that

$$H_{\mu}(\alpha_0^{m-1}) \geq mh(T, \varepsilon).$$

When $m \to \infty$, this yields

$$h_{\mu}(\alpha, T) \geq h(T, \varepsilon),$$

6.4. Introduction to Ornstein theory

and letting $\varepsilon \to 0$ (which, incidentally, requires us to take different αs) gives

$$h_\mu(T) \geq h_{\text{top}}(T).$$

Exercises

1. If X is a strictly ergodic (see Section 4.2) subset of the two-shift $(\{0,1\}^{\mathbb{Z}}, \sigma)$, then $h_{\text{top}}(\sigma|_X) < 1$.
2. Compute the topological entropy of the *Morse minimal set*, which is the orbit closure in $(\{0,1\}^{\mathbb{Z}}, \sigma)$ of the sequence $\ldots 0110 0 1.0 1 10 1001\ldots$ (see Exercise 4.4.3).
3. Prove that if (X, T) is an extension of (Y, S), then $h_{\text{top}}(T) \geq h_{\text{top}}(S)$ by using (a) the Adler–Konheim–McAndrew definition (b) the Bowen definition.
4. Let (X_A, σ_A) be the subshift of finite type in $(\{0, 1, \ldots, n-1\}^{\mathbb{Z}}, \sigma)$ determined by an $n \times n$ 0,1-matrix A (see Section 4.2).
   (a) If N_n denotes the number of fixed points in X_A of σ_A^n, then
   $$h_{\text{top}}(\sigma_A) = \lim_{n \to \infty} \frac{1}{n} \log_2 N_n.$$
   (b) $N_n = \text{trace}(A^n)$.
   (c) $h_{\text{top}}(\sigma_A) = \log|\lambda_{\max}|$, where $\lambda_{\max}$ is the (unique) eigenvalue of A which has maximal modulus.
   (d) Describe X_A and find $h(\sigma_A)$ in case $A = \begin{pmatrix} 1 & 1 \\ 1 & 0 \end{pmatrix}$.
5. Prove as easily as possible that
   $$h_{\text{top}}(S \times T) = h_{\text{top}}(S) + h_{\text{top}}(T).$$
6. Show that if $\mathscr{E}_T$ is the set of all ergodic T-invariant Borel probability measures on X, then
   $$h_{\text{top}}(T) = \sup_{\mu \in \mathscr{E}_T} h_\mu(T).$$

6.4. Introduction to Ornstein theory

Ornstein's discovery in 1970 that entropy completely classifies Bernoulli shifts up to isomorphism (Ornstein 1970a) created a revolution in ergodic theory. There are now known several conditions that guarantee that a given m.p.t. is Bernoulli, an extensive list of transformations satisfying these conditions, many interesting examples including K-automorphisms that are not Bernoulli, and parallel theories for finitary maps (see Section 6.5 and Rudolph 1981) and for equivalence when taking induced transformations is also allowed (variously called Kakutani equivalence,

6. More about entropy

monotone equivalence, or loosely Bernoulli (see Feldman 1976). Here we can give only a brief introduction to this vast and quickly developing subject. The central result, that two Bernoulli shifts are isomorphic if and only if they have the same entropy, is completely proved in the next section. For a detailed treatment of the theory, we recommend first Shields (1973) and then Ornstein (1974).

Let $T: X \to X$ be an ergodic m.p.t. on a probability space $(X, \mathscr{B}, \mu)$. Each finite measurable partition $\alpha = \{A_1, A_2, \ldots, A_n\}$ naturally determines a map

$$\phi_\alpha: X \to \{1, 2, \ldots, n\}^{\mathbb{Z}} = \Omega$$

by

$$(\phi_\alpha x)_k = j \quad \text{if and only if} \quad T^k x \in A_j;$$

the kth coordinate of $\phi_\alpha x$ is the number of the atom of α to which $T^k x$ belongs. Obviously

$$\phi_\alpha(Tx) = \sigma(\phi_\alpha x),$$

where as usual σ is the left shift transformation. The map ϕ_α carries μ down to a σ-invariant measure ν on Ω:

$$\nu(E) = \mu(\phi_\alpha^{-1} E) \quad \text{for measurable} \quad E \subset \Omega.$$

In this way α determines a factor map

$$\phi_\alpha: (X, \mathscr{B}, \mu, T) \to (\Omega, \mathscr{C}, \nu, \sigma).$$

This simple concept of using a partition to code a system in terms of sequences on a finite alphabet is extremely important. There is some helpful terminology associated with the technique: the sequence $\phi_\alpha(x)$ is called the α, T-*name of* x (or α-*name* of x), and for each $k = 1, 2, \ldots,$ the initial k-block $(\phi_\alpha x)_0 (\phi_\alpha x)_1 \ldots (\phi_\alpha x)_{k-1}$ is called the α, k-*name* of x (under T). A point $x \in X$ will be called α-*generic for* T (or simply *generic*) if each block $i_1 i_2 \ldots i_k$ in the α, T-name of x appears with the right frequency, namely $\mu(A_{i_1} \cap T^{-1} A_{i_2} \cap \ldots \cap T^{-k+1} A_{i_k})$. By The Ergodic Theorem, almost every point of X is α-generic for T.

Proposition 4.1 The map ϕ_α is an isomorphism if and only if α is a generator with respect to T.

Proof: If ϕ_α is an isomorphism, then $\phi_\alpha^{-1} \mathscr{C} = \mathscr{B}$ up to sets of measure 0. But if $\mathscr{C}_k \subset \mathscr{C}$ is the σ-algebra generated by all cylinder sets

$$\{\omega \in \Omega: \omega_{-k} = i_{-k}, \ldots, \omega_k = i_k\} \quad (i_{-k}, \ldots, i_k \in \{1, 2, \ldots, n\}),$$

then $\phi_\alpha^{-1} \mathscr{C}_k = \mathscr{B}(\alpha_{-k}^k)$. Therefore $\mathscr{B}(\alpha_{-\infty}^\infty) \supset \phi^{-1} \mathscr{B}(\bigcup \mathscr{C}_k) = \mathscr{B}$, and α is a generator.

6.4. Introduction to Ornstein theory

Conversely, suppose that α is a generator. The measure algebra map $\phi_\alpha^{-1}: (\widehat{\mathscr{C}}, \hat{v}) \to (\widehat{\mathscr{B}(\alpha_{-\infty}^\infty)}, \hat{\mu})$ is always an isomorphism, since $\phi_\alpha^{-1} U = \phi_\alpha^{-1} V$ implies $\mu(\phi_\alpha^{-1}(U \Delta V)) = 0$, so that $U = V$ mod 0. If $\mathscr{B}(\alpha_{-\infty}^\infty) = \mathscr{B}$ up to sets of measure 0, then $\phi_\alpha^{-1}: (\widehat{\mathscr{C}}, \hat{v}) \to (\widehat{\mathscr{B}}, \hat{\mu})$ is an isomorphism of measure algebras, which, by von Neumann's theorem (1.4.7), because we are dealing with Lebesgue spaces, arises from a point isomorphism mod 0.

Let us say that a partition α is *independent* if the σ-algebras $\mathscr{B}(T^k \alpha)$, $k \in \mathbb{Z}$, are independent: for any choice of *distinct* powers $i_1, \ldots, i_r \in \mathbb{Z}$ and not necessarily distinct $A_{i_j} \in \alpha, j = 1, \ldots r$,

$$\mu(T^{-i_1} A_{i_1} \cap T^{-i_2} A_{i_2} \cap \ldots \cap T^{-i_r} A_{i_r}) = \mu(A_{i_1}) \mu(A_{i_2}) \ldots \mu(A_{i_r}).$$

(The main example of an independent partition to bear in mind is the time-0 partition of a Bernoulli shift.)

Proposition 4.2 If T has an independent generator, then T is isomorphic to a Bernoulli shift.

Proof: Let $\alpha = \{A_1, A_2, \ldots, A_n\}$ be the independent generator. Then the image v of μ under ϕ_α is product measure. Thus the system $(\Omega, \mathscr{C}, v, \phi_\alpha)$ is the Bernoulli shift on n symbols with weights $\mu(A_1), \mu(A_2), \ldots, \mu(A_n)$. The map is an isomorphism by Proposition 4.1.

Notice that if α is an independent generator for T, then of course $h(T) = h(\alpha, T) = $ the entropy of the Bernoulli shift downstairs.

One form of Ornstein's theorem is a remarkable strengthening of the above Proposition, in which the hypothesis of independence of α is replaced by a much weaker sort of asymptotic independence. We deal henceforth with *ordered* finite measurable partitions. The *distribution* of a partition $\alpha = \{A_1, A_2, \ldots, A_n\}$ is the probability vector

$$\mathrm{dist}(\alpha) = (\mu(A_1), \mu(A_2), \ldots, \mu(A_n));$$

if $B \in \mathscr{B}$, we define

$$\mathrm{dist}(\alpha|B) = (\mu(A_1|B), \mu(A_2|B), \ldots, \mu(A_n|B))$$

$$= \left(\frac{\mu(A_1 \cap B)}{\mu(B)}, \ldots, \frac{\mu(A_n \cap B)}{\mu(B)} \right).$$

Definition 4.3 (1) Let $\varepsilon > 0$ and let α and β be partitions. We say that α is ε-*independent* of β, and write $\alpha \perp^\varepsilon \beta$, in case there is a family $\mathscr{G}$ of atoms of β such that

(i) $\mu(\bigcup \mathscr{G}) > 1 - \varepsilon$.

(ii) If $B \in \mathscr{G}$, then $\| \mathrm{dist}(\alpha|B) - \mathrm{dist}(\alpha) \| = \sum_{A \in \alpha} |\mu(A|B) - \mu(A)| < \varepsilon$.

(2) Let $\varepsilon > 0$. A sequence $\alpha_1, \alpha_2, \ldots$ of partitions is called ε-*independent* in case for each $n = 2, 3, \ldots \alpha_n$ is ε-independent of $\alpha_1 \vee \ldots \vee \alpha_{n-1}$.

276   6. *More about entropy*

Definition 4.4 A partition α is called *weakly Bernoulli* (WB) for an ergodic m.p.t. T if given $\varepsilon > 0$ there is an N such that for all $m = 1, 2, \ldots$,
$$\alpha_0^m \perp^\varepsilon \alpha_{-N-m}^{-N}.$$

Thus for a weakly Bernoulli α, considering the factor space determined by ϕ_α as above, cylinder sets of arbitrary index length are ε-independent provided only that their index sets are separated by distance $N = N(\varepsilon)$.

Theorem 4.5 (*Friedman and Ornstein* 1970) If T has a weakly Bernoulli generator, then T is isomorphic to any Bernoulli shift of the same entropy.

Of course every independent generator is WB and in particular the time 0 partitions of Bernoulli shifts are WB; it follows that any two Bernoulli shifts of the same entropy are isomorphic. The next section contains a proof of this result.

Example 4.6 *The time 0 partition of a mixing Markov shift is WB.* Let $(\Omega, \mathcal{F}, \mu, \sigma)$ be the Markov shift on $\{1, 2, \ldots, r\}^{\mathbb{Z}}$ determined by an $r \times r$ stochastic matrix $A = (a_{ij})$ which fixes a positive (row) probability vector $p: pA = p$ (see Section 1.2.D, 2.4, 2.5). Let α be the partition into the sets $\alpha_i = \{\omega \in \Omega : \omega_0 = i\}$, $i = 1, 2, \ldots, r$. Fix m and N, and choose
$$E = \alpha_{i_0} \cap T^{-1}\alpha_{i_{-1}} \cap \ldots \cap T^{-m}\alpha_{i_{-m}} \in \alpha_0^m,$$
$$F = T^N \alpha_{i_N} \cap \ldots \cap T^{N+m}\alpha_{i_{N+m}} \in \alpha_{-N-m}^{-N}.$$

Then
$$\mu(E \cap F) = \sum_{i_1, i_2, \ldots, i_{N-1}} p_{i_{-m}} a_{i_{-m} i_{-m+1}} \cdots a_{i_{N+m-1} i_{N+m}}$$
$$= p_{i_{-m}} a_{i_{-m} i_{-m+1}} \cdots a_{i_{-1} i_0} a^{(N)}_{i_0 i_N} a_{i_N i_{N+1}} \cdots a_{i_{N+m-1} i_{N+m}},$$

where $a^{(N)}_{i_0 i_N} = (A^N)_{i_0 i_N}$ is the N-step transition probability from state i_0 to state i_N. Recall that strong mixing is characterized by
$$\lim_{N \to \infty} a^{(N)}_{ij} = p_j \quad \text{for all } i \text{ and } j.$$

Thus
$$\sum_E |\mu(E|F) - \mu(E)|$$
$$= \frac{1}{\mu(F)} \sum_E |\mu(E \cap F) - \mu(E)\mu(F)|$$
$$= \frac{1}{\mu(F)} \sum_E \mu(E)\mu(F) \left|1 - \frac{a^{(N)}_{i_0 i_N}}{p_{i_N}}\right| \leq \sum_{i_0} \left|1 - \frac{a^{(N)}_{i_0 i_N}}{p_{i_N}}\right|,$$

which tends to 0 uniformly in m. This proves that α is WB.

6.4. Introduction to Ornstein theory

Corollary 4.7 A mixing Markov shift is isomorphic to every Bernoulli shift of the same entropy.

In order to state weakenings of the WB condition and the associated strengthenings of the above theorem, we need to formulate a definition of a metric, the $\bar{d}$-*distance*, on the set of finite-state stationary processes. The $\bar{d}$-distance is a limit of the Hamming metrics of information theory. Two processes will be thought of as being close if their printouts are similar except possibly on a set of times of low frequency. If $\alpha = \{A_1, A_2, \ldots, A_n\}$ and $\beta = \{B_1, B_2, \ldots, B_n\}$ are ordered partitions of X and Y, respectively, define
$$d(\alpha, \beta) = \sum_{i=1}^{n} \mu(A_i \Delta B_i).$$
Let $T: X \to X$ and $S: Y \to Y$ be ergodic m.p.t.s. To measure the distance between the processes (α, T) and (β, S), we need first to represent them on the same space. To this end, take a Lebesgue space Z and isomorphisms $\phi: X \to Z$ and $\psi: Y \to Z$. We let
$$\bar{d}^{(m)}_{\phi,\psi}((\alpha, T), (\beta, S)) = \frac{1}{m} \sum_{k=0}^{m-1} d(\phi T^{-i}\alpha, \psi S^{-i}\beta) \quad (m = 1, 2, \ldots),$$
$$\bar{d}^{(m)}((\alpha, T), (\beta, S)) = \inf_{\phi,\psi} \bar{d}^{(m)}_{\phi,\psi}((\alpha, T), (\beta, S)),$$
and
$$\bar{d}((\alpha, T), (\beta, S)) = \sup_m \bar{d}^{(m)}((\alpha, T), (\beta, S)) = \lim_{m \to \infty} \bar{d}^{(m)}((\alpha, T), (\beta, S)).$$
(The last equality is the content of Exercise 3.)

Proposition 4.8. The following statements are equivalent:
(1) $\bar{d}((\alpha, T), (\beta, S)) < \delta$.
(2) *The α, T-name of every α-generic point for T can be changed on a set of frequency less than δ to produce the β, S-name of a β-generic point for S.*
(3) There exist an α-generic point x for T and β-generic point y for S such that the α, T-name of x and the β, S-name of y differ on a subsequence of density less than δ.

The $\bar{d}$ distance is involved in another useful criterion for isomorphism.

Proposition 4.9 Two finite-entropy ergodic m.p.t.s, T on X and S on Y, are isomorphic if and only if there exist generators α for T and β for S such that $\bar{d}((\alpha, T), (\beta, S)) = 0$.

Proof: The 'only if' part follows by taking α to be a finite generator for

T and β its image under the given isomorphism. Conversely, given α and β we have isomorphisms $\phi_\alpha:(X,\mathscr{B},\mu,T)\to(\Omega,\mathscr{C},\nu,\sigma)$ and $\phi_\beta:(Y,\mathscr{B}',\mu',S)\to(\Omega,\mathscr{C},\nu',\sigma)$. If $\bar{d}((\alpha,T),(\beta,S))=0$, then by, e.g., Proposition 4.8 (3), ν and ν' agree on cylinder sets in $\mathscr{C}$, so that $\nu=\nu'$ and T is isomorphic to S.

Now we use the $\bar{d}$ distance to formulate a weakening of the WB condition. We will require that given $\varepsilon>0$ there is an N such that for all $m=1,2,\ldots$, most atoms of α_0^m have been moved, on the average during the time interval $[0,N-1]$, in a way that is approximately independent of α. To make this more precise, extend the above definition of the $\bar{d}$ metric to any pair of finite sequences of partitions $\alpha_1,\ldots,\alpha_N$ of X and $\beta_1,\ldots,\beta_N$ of Y by

$$\bar{d}^{(N)}(\{\alpha_i\},\{\beta_i\}) = \inf_{\phi,\psi} \frac{1}{N}\sum_{k=0}^{N-1} d(\phi\alpha_i,\psi\beta_i),$$

where the infimum is again taken over all isomorphisms $\phi:X\to Z$, $\psi:Y\to Z$. Given a partition α of X and a measurable set $E\subset X$, α determines a natural partition $\alpha|E = \{A\cap E:A\in\alpha\}$ of E.

Definition 4.10 A partition α is called *very weakly Bernoulli* (VWB) for an ergodic m.p.t. T if given $\varepsilon>0$ there is an N such that for all $m=1,2,\ldots$, there is a family $\mathscr{G}_m$ of atoms of $\alpha_0^m = \bigvee_{i=0}^m T^{-i}\alpha$ such that
 (i) $\mu(\bigcup \mathscr{G}_m) > 1-\varepsilon$,
 (ii) If $A\in\mathscr{G}_m$, then

$$\bar{d}^{(N)}(\{T^i\alpha|A:i=1,2,\ldots,N\},\{T^i\alpha:i=1,2,\ldots,N\})<\varepsilon.$$

Remark 4.11 Every WB partition is VWB.

Theorem 4.12 (*Ornstein* 1970*b*) If T has a very weakly Bernoulli generator, then T is isomorphic to any Bernoulli shift of the same entropy.

We consider one further weakening of the independence condition and corresponding strengthening of Ornstein's theorem. An independent partition α (also WB and VWB partitions) can be shown to have the curious property that any partition that approximates it well enough in entropy, and in distribution up to a certain time n, must approximate it well, on average (i.e. in the $\bar{d}$ sense), for all time. Ornstein found that this weak property of being 'finitely determined' was still enough to allow him to prove the isomorphism theorem.

6.4. Introduction to Ornstein theory

Definition 4.13 A partition α is *finitely determined* (*FD*) with respect to an ergodic m.p.t. T if given $\varepsilon > 0$ there are δ and N such that if (β, S) is any ergodic process with

(i) card β = card α,
(ii) $|h(\alpha, T) - h(\beta, S)| < \delta$, and
(iii) $\|\text{dist}(\alpha_0^N) - \text{dist}(\beta_0^N)\|$
$$= \sum_{i_0,\ldots,i_{N-1}} |\mu(A_{i_0} \cap T^{-1}A_{i_1} \cap \ldots \cap T^{-N+1}A_{i_{N-1}})$$
$$- \nu(B_{i_0} \cap S^{-1}B_{i_1} \cap \ldots \cap S^{-N+1}B_{i_{N-1}})| < \delta,$$

then

$$\bar{d}((\alpha, T), (\beta, S)) < \varepsilon.$$

Proposition 4.14 WB $\Rightarrow$ VWB $\Rightarrow$ FD.

Theorem 4.15 (*Ornstein 1970b*) The following statements about an ergodic m.p.t. T are equivalent:

1. T is isomorphic to a Bernoulli shift.
2. T has a finitely determined generator.
3. Every partition is finitely determined with respect to T.

Corollary 4.16 A factor of a Bernoulli shift is Bernoulli.

Remark 4.17 Ornstein and Weiss (1974) showed that in fact every finitely determined partition is very weakly Bernoulli. Thus by 4.15 (3), *every* partition of a Bernoulli shift is VWB.

Remark 4.18 *Some systems that have so far been proved to be Bernoulli*. Note: A flow $\{T_t\}$ is Bernoulli if T_1 is, in which case all T_t are.

1. Mixing Markov shifts (Friedman and Ornstein 1970).
2. Ergodic automorphisms of the n-torus (Katznelson 1971).
3. More generally, automorphisms of compact abelian groups (Lind 1977 and Miles and Thomas 1978) and certain ergodic automorphisms of nilmanifolds (Dani 1976).
4. A billiard system (one ball on a square table with finitely many convex obstacles) (Ornstein and Gallavotti 1974).
5. The hard-sphere gas in a rectangular box (Sinai, unpublished – see Ornstein 1978).

6. Geodesic flow on a surface of negative curvature (Ornstein and Weiss 1973).
7. More generally, any mixing Anosov flow with a smooth invariant measure (Ratner 1974 and Bunimovitch 1973, 1974).
8. Brownian motion in a rectangular region with reflecting boundary (Ornstein and Shields 1973).
9. Natural extensions (see 1.3.G) of certain noninvertible maps of the interval, e.g. $x \to \beta x$ mod 1, where $\beta > 1$. (Adler 1973, Smorodinsky 1973).
10. Shifts on certain Ising models (Gallavotti 1973, Ledrappier 1973, Liberto, Gallavotti and Russo 1973).

Thus many classical dynamical systems and algebraically constructed systems not only *contain* random behavior, but from the point of view of measure theory actually *are the same* as repeated random experiments. Whether one views this as a demonstration of the unsuspected complexity of phenomena or of the inappropriateness of the concept of metric isomorphism, the collection of results is revolutionary – and the revolution is continuing.

Exercises

1. Show that there is a function $f(\varepsilon)$, which decreases to 0 as $\varepsilon \to 0$, such that $\alpha \perp^\varepsilon \beta$ implies $\beta \perp^{f(\varepsilon)} \alpha$. (Hint: Consider the relationship between $\alpha \perp^\varepsilon \beta$ and $\sum_{A,B} |\mu(A \cap B) - \mu(A)\mu(B)| < \delta$.)
2. Show that there is a function $g(\delta)$, which decreases to 0 as $\delta \to 0$, such that $H(\alpha) - H(\alpha|\beta) < \delta$ implies $\alpha \perp^{g(\delta)} \beta$. (Hint: If $H(\alpha) - H(\alpha|\beta) < \delta$, then for a set of atoms B of β of total measure at least $1 - \sqrt{\delta}$, $-\sum_{A \in \alpha} \mu(A|B) \log_2 [\mu(A|B)/\mu(A)] < \sqrt{\delta}$. If $\phi(t) = t - \log_2 t - 1$, then the first step implies that $\phi(\mu(A)/\mu(A|B)) < \sqrt[4]{\delta}$ for all 'good' $B \in \beta$ and a set of atoms A of α of total measure at least $1 - \sqrt[4]{\delta}$. Then use continuity of ϕ^{-1} and sum over the 'good' $A \in \alpha$.) cf. Proposition 5.2.9.
3. Prove that $\sup_m \bar{d}^{(m)}((\alpha, T), (\beta, S)) = \lim_{m \to \infty} \bar{d}^m((\alpha, T), (\beta, S))$. (Hint: If $\bar{d}^{(m)} = r$, then $\bar{d}^{(km)} \geq r$ for all k, and so for large n, $\bar{d}^{(n)} > r - \varepsilon$.)
4. Show that the entropy $h(\alpha, T)$ is a continuous function of ergodic finite-state stationary processes with respect to the $\bar{d}$-distance.
5. Show that $\bar{d}((\alpha, T), (\beta, S))$ can be found in the following way. Take an ergodic transformation $R: Z \to Z$ which has (α, T) and (β, S) as marginals, in that there are partitions α' and β' of Z with $(\alpha', R) \approx (\alpha, T)$ and $(\beta', R) \approx (\beta, S)$, and compute the partition distance $d(\alpha', \beta')$; then take the infimum over all such systems. (We may think of the Z process as printing out both an α symbol and a

6.5. Finitary coding between Bernoulli shifts

β symbol at each time. If these symbols agree with high probability then the $\bar{d}$-distance between the two processes is small.)

6.5. Finitary coding between Bernoulli shifts

Let $\mathcal{B}(p) = \mathcal{B}(p_0, \ldots, p_{a-1})$ (with measure μ) and $\mathcal{B}(q) = \mathcal{B}(q_0, \ldots, q_{b-1})$ (with measure ν) be two Bernoulli shifts, on alphabets $A = \{0, 1, \ldots, a-1\}$ and $B = \{0, 1, \ldots, b-1\}$, of equal entropy h:

$$h = H(p) = -\sum_{i=0}^{a-1} p_i \log_2 p_i = -\sum_{j=0}^{b-1} q_j \log_2 q_j = H(q).$$

According to Ornstein's Theorem (Theorem 4.5, Theorem 4.12, Theorem 4.15), the systems $\mathcal{B}(p)$ and $\mathcal{B}(q)$ are metrically isomorphic. Ornstein's proof does not actually produce an isomorphism, although it has the advantage of giving various conditions sufficient for any system to be isomorphic to a Bernoulli shift. Further, the isomorphism, or coding, $\phi: \mathcal{B}(p) \to \mathcal{B}(q)$ may not be effectively realizable in actual practice, i.e., by a real machine, since in order to determine the 0th entry $(\phi x)_0$ of the image of a sequence $x \in \mathcal{B}(p)$, one may in general need to know the *entire* sequence $x = \ldots x_{-1} x_0 x_1 \ldots$. Keane and Smorodinsky (1979a; see also 1977, 1979b, 1979c) gave a new proof of Ornstein's theorem which does construct the isomorphism explicity, and in which the isomorphism is *finitary* – for a.e. $x \in \mathcal{B}(p)$, $(\phi x)_0$ depends only on a finite number (which may depend on x) of entries in x. (Unfortunately, in general the expectation of this coding time may be infinite.) This section is devoted to the major case of the Keane–Smorodinsky proof.

First, let us mention several characterizations of finitary maps. A map $\phi: \mathcal{B}(p) \to \mathcal{B}(q)$ is called *finitary* if it satisfies any one of the following three conditions, all of which can be shown to be equivalent (Exercise 1):

Conditions 5.1

(1) There are sets of measure 0, $N \subset \mathcal{B}(p)$ and $M \subset \mathcal{B}(q)$, such that $\phi: \mathcal{B}(p) \setminus N \to \mathcal{B}(q) \setminus M$ is continuous.

(2) For a.e. $x \in \mathcal{B}(p)$ there is an integer $j(x)$ such that if x and x' agree on their central $j(x)$-blocks, then $(\phi x)_0 = (\phi x')_0$.

(3) The inverse image $\phi^{-1}\{y \in \mathcal{B}(q): y_0 = j\}$ of each time-0 cylinder set in $\mathcal{B}(q)$ is, up to a set of measure 0, a countable union of cylinder sets in $\mathcal{B}(p)$.

The Keane–Smorodinsky proof produces a *metric isomorphism* $\phi: \mathcal{B}(p) \to \mathcal{B}(q)$, when $H(p) = H(q)$, such that *both ϕ and ϕ^{-1} are finitary*. It is not true in general that the inverse of a finitary isomorphism is finitary

(Rudolph), although this may be true of maps between Bernoulli shifts. Meshalkin (1959) and Blum and Hanson (1963) constructed finitary metric isomorphisms between Bernoulli shifts in certain special cases, such as between $\mathscr{B}(\frac{1}{4},\frac{1}{4},\frac{1}{4},\frac{1}{4})$ and $\mathscr{B}(\frac{1}{2},\frac{1}{8},\frac{1}{8},\frac{1}{8},\frac{1}{8})$, each of which has entropy 2. Finitary codes between subshifts of finite type of equal entropy were constructed by Adler and Marcus (1979), and Denker and Keane (1979) and Denker (1977) began a general study of the finitary category. It can be shown (Exercise 2) that $\mathscr{B}(p)$ and $\mathscr{B}(q)$ are isomorphic under a measure-preserving *homeomorphism* if and only if p is a permutation of q. Rudolph (1981) has worked out the theory, parallel to Ornstein's, of sufficient conditions for a given system to be finitarily isomorphic to a Bernoulli shift.

Notice that if both alphabets have two elements ($a = b = 2$), then the isomorphism can be constructed easily. For the function

$$g(t) = -t\log_2 t - (1-t)\log_2(1-t)$$

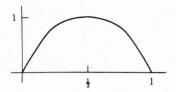

on $[0, 1]$ has the property that $g(t_1) = g(t_2)$ if and only if $t_1 = t_2$ or $t_1 = 1 - t_2$. Thus $H(p_0, 1 - p_0) = H(q_0, 1 - q_0)$ implies that either $p_0 = q_0$ or $p_0 = 1 - q_0$. For the isomorphism, then, we take either the identity map or else the 'dualizing' map which sends 0 to 1 and 1 to 0.

Henceforth we will assume therefore that $a \geq 2$ and $b > 2$. If actually $a = 2$, the same proof that we are about to give for the case when $a > 2$ and $b > 2$ will carry through, provided that a few technical modifications are made here and there (Exercises 5 and 6). For the sake of the exposition, we will limit our attention to the major case when $a \geq 3$ and $b \geq 3$.

A. Sketch of the proof

There is an easy reduction to the case when the probability vectors p and q contain a common weight, say $p_0 = q_0$. Then the symbol 0 appears with the same probability in the sequences of $\mathscr{B}(q)$ as of $\mathscr{B}(p)$, so it is natural (and very helpful) to consider codes $\phi: \mathscr{B}(p) \to \mathscr{B}(q)$ which take 0 to 0; i.e., we will have $(\phi x)_i = 0$ if and only if $x_i = 0$, for all $i \in \mathbb{Z}$. This gives us sort of frame for the code, and we only need to tell which of

6.5. Finitary coding between Bernoulli shifts

the remaining symbols $1,\ldots,b-1$ fill in the blanks in ϕx. The 0s are called *markers*, and the Bernoulli shift $\mathcal{B}(p_0, 1-p_0)$ is called the *marker process*; it is a factor of both $\mathcal{B}(p)$ and $\mathcal{B}(q)$ in the obvious way (code 0 to 0 and any other symbol to 1).

$$
\begin{array}{l}
x..00....0...000...0..\in\mathcal{B}(p) \\
\phi\downarrow \qquad\qquad\qquad\qquad\qquad\quad \searrow^{\pi} \; z = ..0011..11011...11000111..11011 \\
\phi x = y..00....0...000...0...\in\mathcal{B}(q)\;\nearrow_{\tau} \qquad\qquad\qquad\qquad\in\mathcal{B}(p_0, 1-p_0)
\end{array}
$$

The idea now is to look at a very long central block of x, observe the 'filler' in it (i.e., what's in the spaces between 0s), and use this information to fill in many of the corresponding places in y, including at least y_0. This amounts to defining ϕ so that it sends fibers of π to fibers of τ. The long blocks looked at are selected to be of a kind which can inherently never overlap themselves, namely the blocks between two enormously long record (as we look outward from the center) runs of consecutive 0s. Such a block of 0s and spaces is called a *skeleton*. Almost all $x \in \mathcal{B}(p)$ have skeletons of all 'ranks'. We think of a skeleton as appearing in both x and y.

Given a skeleton, its blanks can be filled by finite blocks on the symbols $A_0 = \{1,\ldots,a-1\}$ to get a block in x, or $B_0 = \{1,\ldots,b-1\}$ to get a block in y. These are called the *filler alphabets*. The relative probabilities of the various filler blocks in x are determined by products of the *filler measure*

$$\mu_0(k) = \frac{p_k}{1-p_0} \qquad (k=1,\ldots,a-1),$$

and those for y by products of

$$\nu_0(k) = \frac{q_k}{1-q_0} \qquad (k=1,\ldots,b-1).$$

When there are very many spaces in the skeleton we are looking at, the Shannon–McMillan–Breiman Theorem (Corollary 2.5) implies, because $H(p_1,\ldots,p_{a-1}) = H(q_1,\ldots,q_{b-1})$, that the blocks available to go into the spaces in $\mathcal{B}(p)$ and in $\mathcal{B}(q)$ have roughly the same numbers and sizes; this gives us hope that they can be matched up, and that by looking at the filler in x we will be able to fill in many of the blanks in the corresponding portion of y.

Given a skeleton, the fillers F for x can be thought of as cylinder sets of the Bernoulli shift on the alphabet A_0 with weights given by ν_0, and the fillers G for y as cylinder sets of the Bernoulli shift on the alphabet B_0 with weights given by ν_0; these are the *filler Bernoulli shifts*. Each filler

6. More about entropy

F has associated with it a certain set $J(F)$ of places in the skeleton whose entries it determines; as we pass to longer skeletons, places previously fixed stay fixed, and with the same entries. At any given stage we are interested in the *equivalence classes* of fillers, where two fillers are regarded as equivalent if they agree on their fixed sets, since these form the cylinder sets of the candidates for the x and y that will be matched to one another by the isomorphism.

We look at progressively longer skeletons. At each odd stage, look at the equivalence classes of fillers in points x of $\mathscr{B}(p)$ and associate to each a set of equivalence classes of possible fillers for the corresponding points of $y = \phi x$ in a measure-nondecreasing way; such an association is called a *society*. At each even stage do the same, except with the roles of $\mathscr{B}(p)$ and $\mathscr{B}(q)$ reversed. When we pass from one stage to the next, the association is made in a manner consistent with what went before and in a particularly efficient way. That this is possible follows from an extension of the famous combinatorial *Marriage Lemma* that has found many previous applications throughout mathematics, including in ergodic theory (see Rota and Harper 1971 and Ornstein 1970a). The sequence of skeletons and associations of equivalence classes of fillers produced in this way is called an *assignment*.

The parameters of the procedure can be fixed in such a way that for all $x \in \mathscr{B}(p)$ with the possible exception of a set of measure 0, when we look at a long enough central skeleton of x, the equivalence class of its actual filler in x has associated to it by our assignment a *single* equivalence class of fillers in $\mathscr{B}(q)$. Then all those places in ϕx can be filled in. Further, we will show that the central place of ϕx is among this set of then-determined places. Thus $(\phi x)_0$ is determined by finitely many coordinates of x, and the map ϕ is finitary. The measure-nondecreasing condition in our assignments makes ϕ measure-preserving, and the finitary map ψ constructed from $\mathscr{B}(q)$ to $\mathscr{B}(p)$ simultaneously and consistently with ϕ will be ϕ^{-1} (once sets of atypical points, of measure 0, are discarded). The map ϕ is actually defined on entire orbits, and so is clearly shift-invariant.

Now we go on to look at the details, which involve nothing more than some easy combinatorics and a few estimates arising from the Shannon–McMillan–Breiman Theorem, or, if you prefer to look at its proof rather than its statement, the Weak Law of Large Numbers (i.e., the Ergodic Theorem, with convergence only in probability rather than in measure). A certain amount of notation and bookkeeping is necessary; behind this, though, there is a very simple and beautiful argument.

6.5. Finitary coding between Bernoulli shifts

B. Reduction to the case of a common weight

We are dealing with two Bernoulli shifts $\mathscr{B}(p) = \mathscr{B}(p_0, \ldots, p_{a-1})$ and $\mathscr{B}(q) = \mathscr{B}(q_0, \ldots, q_{b-1})$ on alphabets $A = \{0, 1, \ldots, a-1\}$ and $B = \{0, 1, \ldots, b-1\}$, where $a \geq 3$ and $b \geq 3$, each of entropy h. The following Lemma allows us to assume that p and q have a weight in common, say $p_0 = q_0$. As outlined above, this produces the common factor marker process, makes possible the construction of the frame for the coding, according to which 0 goes to 0, and sets off the search for skeletons in the sequences of the two processes.

Lemma 5.2 Marker Lemma There is a probability vector $r = (r_0, \ldots, r_{c-1})$ with $H(r) = -\sum_{i=0}^{c-1} r_i \log_2 r_i = h$ such that $r_0 = p_i$ for some i and $r_1 = q_j$ for some j.

Proof: Assume that $p_0 \geq p_1 \geq \ldots \geq p_{a-1}$, $q_0 \geq q_1 \geq \ldots \geq q_{b-1}$, and $p_{a-1} \geq q_{b-1}$. Put $r_0 = p_0$ and $r_1 = q_{b-1}$. I claim that

$$H(r_0, r_1, 1 - (r_0 + r_1)) \leq H(p_0, p_{a-1}, 1 - (p_0 + p_{a-1})).$$

Intuitively this is clear, because the vector on the right is more equidistributed than the one on the left (recall that $a \geq 3$):

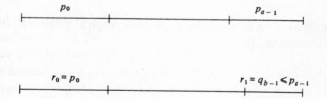

More precisely, it is enough to observe that the function

$$g(t) = -p_0 \log_2 p_0 - t \log_2 t - (1 - (t + p_0)) \log_2 (1 - (t + p_0))$$

increases as t increases to $(1 - p_0)/2$, so that $g(q_{b-1}) \leq g(p_{a-1})$.

Of course $H(p_0, p_{a-1}, 1 - (p_0 + p_{a-1})) \leq H(p) = h$, since $\alpha \leq \beta$ implies $H(\alpha) \leq H(\beta)$. Now we only need to choose $c \geq 3$ and $r_2, \ldots, r_{c-1}$ such that $H(r_0, r_1, \ldots, r_{c-1}) = h$. This can be done as in Exercise 5.3.4: use Lagrange multipliers to see that the continuous function $-\sum_{i=2}^{c-1} r_i \log_2 r_i$, subject to the constraints $r_i \geq 0$, $\Sigma r_i = 1 - r$, where $r = r_0 + r_1$, takes any value between $-(1-r)\log_2(1-r)$ and $-(1-r)\log_2(1-r) + (1-r)\log_2(c-2)$.

If the finitary code can be produced whenever there is a common weight, then given arbitrary p and q of equal entropy we can find an isomorphism from each to a $\mathscr{B}(r)$, where r is as in the Lemma, and composition of one of these maps with the inverse of the other will yield the isomorphism from $\mathscr{B}(p)$ to $\mathscr{B}(q)$. We assume henceforth that $p_0 = q_0$.

C. Framing the code

Let $N_1 < N_2 < \ldots$ be a sequence of integers, to be determined later. For each $r = 1, 2, \ldots$ a *skeleton* S *of rank* $r = r(S)$ is a string of blocks of 0s and spaces

$$S = 0^{n_0} \underbrace{\qquad\qquad}_{l_1} 0^{n_1} \underbrace{\qquad\qquad}_{l_2} \cdots \underbrace{\qquad\qquad}_{l_m} 0^{n_m}$$

(where 0^i means a block of i consecutive 0s) such that

(i) $n_0, n_m \geq N_r$,
(ii) $N_r > n_k$ for $k \neq 0, m$.

Skeletons in a sequence $x \in \mathscr{B}(p)$ or $y \in \mathscr{B}(q)$ are found by beginning at the center and moving out to both the left and right until we first encounter a block of at least N_r consecutive 0s; in such a case we include *all* the 0s in the skeleton: we say a skeleton S *appears* in a sequence x in case the spaces of S can be filled in by nonzero symbols in such a way that the resulting block appears in x neither preceded nor followed by a 0.

The number of spaces $l = l_1 + \ldots + l_m$ is called the *length* $l(S)$ of the skeleton S.

If S is as above, then by a *subskeleton* of S we mean a skeleton S_0 of some rank r_0 that takes the form

$$S_0 = 0^{n_k} \underbrace{\qquad\qquad}_{l_{k+1}} 0^{n_{k+1}} \underbrace{\qquad\qquad}\cdots\underbrace{\qquad\qquad} 0^{n_i}$$

for some choice of k and i.

Two subskeletons of the same skeleton are called *disjoint* if their sets $\{k+1, \ldots, i\}$ (as above) are disjoint (the right endblock of 0s of one could be the left endblock of 0s of the other). We say that a family of subskeletons $S_1, \ldots, S_j$ of S forms a *decomposition* of S, and write $S = S_1 \times \ldots \times S_j$, in case the S_m are pairwise disjoint and the union of their sets of indices $\{k+1, \ldots, i\}$ is all of $\{1, \ldots, m\}$.

Lemma 5.3 Skeleton Lemma

(1) Each skeleton S of rank $r > 1$ admits a unique decomposition, called the *rank decomposition* of S, into subskeletons of rank $r - 1$.

(2) For a.e. $x \in \mathscr{B}(p)$, either $x_0 = 0$ or for each $r \geq 1$ there is a unique skeleton $S_r(x)$ of rank r which includes the central coordinate of x.

(3) Given any sequence $L_1, L_2, \ldots$ of lengths, we can choose $N_1 < N_2 < \ldots$

6.5. Finitary coding between Bernoulli shifts

such that for a.e. $x\in \mathcal{B}(p)$ and a.e. $y\in \mathcal{B}(q)$, $l(S_r(x)) \geq L_r$ and $l(S_r(y)) \geq L_r$ for all sufficiently large r.

Proof:

(1) Starting at the left of S, move to the right until we hit the first block of 0s of length at least N_{r-1}; all the places passed on the way comprise S_1. Continue.

(2) Beginning at the center of x, move out to the left and right until in each direction we hit a string of consecutive 0s of length at least N_r. Almost every point contains infinitely many such strings, so such an $S_r(x)$ can be found with probability 1.

(3) We want to choose the N_r so that if
$$E_r = \{x\in \mathcal{B}(p): l(S_r(x)) \geq L_r\},$$
then
$$\mu\left(\bigcup_{r_0=1}^{\infty} \bigcap_{r\geq r_0} E_r\right) = 1.$$
(Because the search for skeletons involves only the marker process, the same N_rs will do also for $\mathcal{B}(q)$.) This can be done by making the N_r so large that we have to wait a very long time to see repetitions of 0^{N_r}. By the easy half of the Borel–Cantelli Lemma, it is enough to achieve
$$\sum_{r=1}^{\infty} \mu(E_r^c) < \infty,$$
for then
$$\mu\left(\bigcap_{r_0=1}^{\infty} \bigcup_{r\geq r_0} E_r^c\right) = 0.$$

Now $E_r^c = \{x: l(S_r(x)) < L_r\} \subset \{x: 0^{N_r} \text{ appears in } x_0 x_1 \ldots x_{(L_r+1)N_r}\}$, as can be seen from this picture,

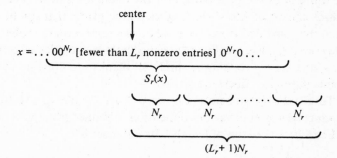

so that, denoting by $[B]$ the cylinder set determined by a block B,

$$\mu(E_r^c) \leq \mu\left(\bigcup_{i=0}^{L_r N_r} \{x : 0^{N_r} \text{ appears at the } i\text{th place in } x\}\right)$$

$$\leq (L_r N_r + 1)\mu[0^{N_r}] = (L_r N_r + 1)p_0^{N_r} < \frac{1}{2^r}$$

for appropriate choice of N_r.

D. What to put in the blanks

We define for $\mathscr{B}(p)$ and $\mathscr{B}(q)$ the

filler alphabets

$A_0 = \{1, \ldots, a-1\}$ and $B_0 = \{1, \ldots, b-1\}$

filler measures

$\mu_0(k) = \dfrac{p_k}{1-p_0} \ (k=1,\ldots,a-1)$ and $\nu_0(k) = \dfrac{q_k}{1-q_0} \ (k=1,\ldots,b-1)$

filler entropies

$$-\sum_{k=1}^{a-1} \mu_0(k)\log_2 \mu_0(k) \quad = g = -\sum_{k=1}^{b-1} \nu_0(k)\log_2 \nu_0(k)$$

filler Bernoulli shifts

$(A_0^{\mathbb{Z}}, \mu_0^{\mathbb{Z}}) = \mathscr{B}(\mu_0(1), \ldots, \mu_0(a-1))$ and $(B_0^{\mathbb{Z}}, \nu_0^{\mathbb{Z}}) = \mathscr{B}(\nu_0(1), \ldots, \nu_0(b-1))$.

(It is easy to check that the two entropies are equal.)

Again a word about what is going on. We will try to pick blocks from these filler Bernoulli shifts to fill in the spaces in skeletons. Look at a single fiber over the marker process in each of $\mathscr{B}(p)$ and $\mathscr{B}(q)$. If r is odd, for each $x \in \mathscr{B}(p)$ in this fiber we associate to the equivalence class of the actual filler $F_r(x)$ in $S_r(x)$ a set of equivalence classes of fillers from $(B_0^{\mathbb{Z}}, \nu_0^{\mathbb{Z}})$, which we think of as the possibilities for the fillers for $S_r(\phi x)$. (The equivalence classes consist of fillers which agree on the places that can be determined from that time on.) If r is even, we do the same with the roles of $\mathscr{B}(p)$ and $\mathscr{B}(q)$ reversed. Eventually the assigned equivalence classes become singletons, and, as $r \to \infty$, almost surely all blanks are filled in and almost all points in the two fibers are paired.

To avoid excessive notation, we will use μ_0 and ν_0 also to denote the product measures of various dimensions obtained from powers of μ_0 or ν_0.

Let S be a skeleton of length l. Its *index set* is

$$I(S) = \{1, 2, \ldots, l\},$$

6.5. Finitary coding between Bernoulli shifts

and its *filler sets* are

$$\mathscr{F}(S) = A_0^l = \{F = f_1 f_2 \ldots f_l : \text{all } f_i \in A_0\} \quad \text{for } \mathscr{B}(p)$$

and

$$\mathscr{G}(S) = B_0^l = \{G = g_1 g_2 \ldots g_l : \text{all } g_i \in B_0\} \quad \text{for } \mathscr{B}(q).$$

Step 5.4 Determining the N_r Fix a sequence $\varepsilon_1 > \varepsilon_2 > \ldots \to 0$. Let α be the time-0 partition of $(A_0^{\mathbb{Z}}, \mu_0^{\mathbb{Z}}, \sigma)$. Fix $r = 1, 2, \ldots$. By the Shannon–McMillan–Breiman Theorem (Corollary 2.5), there is a k_r such that if $k \geq k_r$, then α_0^{k-1} (which is identified with $\mathscr{F}(S)$ in case $l(S) = k$) is the disjoint union

$$\alpha_0^{k-1} = \mathscr{g}_r \cup \mathscr{b}_r,$$

where

(i) $\mu_0(\cup \mathscr{b}_r) < \varepsilon_r$,

(ii) For each $A \in \mathscr{g}_r$, if $\eta = \min_{1 \leq i \leq a-1} \mu_0(i)$, $\frac{1}{\eta} 2^{-k(g+\varepsilon_r)} < 2^{-k(g+\varepsilon_r/2)} < \mu_0(A) < 2^{-k(g-\varepsilon_r/2)} < \frac{1}{\eta} 2^{-k(g-\varepsilon_r/2)} < \frac{1}{\eta} 2^{-k(g-\varepsilon_r)}.$

(We have chosen k_r large enough that $2^{-k\varepsilon_r/2} < \eta$ for $k \geq k_r$.)

This allows us now to *choose the L_r* (depending on the ε_r) in such a way that

(i) $L_r \geq k_r$ for all r,
(ii) $\lim_{r \to \infty} L_r(\varepsilon_{r-1} - \varepsilon_r) = \infty$.

Then *the L_r determine the N_r* by the Skeleton Lemma (5.3(3)).

Step 5.5 Definition of equivalence of fillers Let S be a skeleton of rank r and length l. We will define an equivalence relation $\sim$ on the set $\mathscr{F}(S)$ of fillers for S in $\mathscr{B}(p)$ by attaching to each $F \in \mathscr{F}(S)$ a set $J(F) \subset \{1, 2, \ldots, l\}$ (of (thenceforth) determined places in S) and agreeing that $F \sim F'$ if and only if

(i) $J(F) = J(F')$
(ii) $F = F'$ on $J(F)$.

Actually $J(F)$ is defined by a sort of stopping rule in such a way that (ii) forces (i). A similar definition is made for $\mathscr{G}(S)$ (using ν_0 in place of μ_0), and the resulting sets of equivalence classes are denoted by $\tilde{\mathscr{F}}(S)$ and $\tilde{\mathscr{G}}(S)$. If $F \in \mathscr{F}(S)$, then $\tilde{F}$ denotes the equivalence class to which it belongs. Appropriate powers of μ_0, and ν_0, determine probability measures on $\mathscr{F}(S)$ and $\tilde{\mathscr{F}}(S)$, and $\mathscr{G}(S)$ and $\tilde{\mathscr{G}}(S)$, in the obvious way.

$J(F)$ is defined by induction on the rank r of S. Suppose that $r(S) = 1$, $l(S) = l$, and $F = a_1 \ldots a_l \in \mathscr{F}(S)$. Define $J(F)$ to be the initial segment of $1, 2, \ldots, l$ whose corresponding cylinder set's μ_0-measure first drops below the upper bound found in the preceding paragraph for atoms of α_1^l:

$$J(F) = \{k : 1 \leq k \leq l, \mu_0(a_1) \ldots \mu_0(a_{k-1}) \geq \frac{1}{\eta} 2^{-l(g-\varepsilon_1)}\}.$$

Then usually $J(F) \subsetneq \{1, \ldots, l\}$.

Suppose now that $J(F)$ is already defined for all $F \in \mathscr{F}(S)$ whenever $r(S) < r$. If S is a skeleton of rank r, form the rank decomposition $S = S_1 \times \ldots \times S_j$ of S into disjoint subskeletons of rank $r - 1$. (Throughout, amalgamate, consolidate, and keep track of all index sets correctly.) If $F \in \mathscr{F}(S)$, then in an obvious way F is a concatenation $F = F_1 \times \ldots \times F_j$, with each $F_i \in \mathscr{F}(S_i)$. We define

$$J_0(F) = \bigcup_{i=1}^{j} J(F_i) = \text{all the previously fixed places,}$$

and we also add any new places, up to wherever the measures of atoms fall below the new (smaller) upper bound:

Let $\{1, \ldots, l\} \setminus J_0(F) = \{t_1, \ldots, t_u\}$, where $t_1 < \ldots < t_u$, and define

$$J(F) = J_0(F) \cup \left\{ t_v : 1 \leq v \leq u \text{ and} \right.$$

$$\left. \cdot \left(\prod_{t \in J_0(F)} \mu_0(a_t) \right) \mu_0(a_{t_1}) \ldots \mu_0(a_{t_{v-1}}) \geq \frac{1}{\eta} 2^{-l(g-\varepsilon_r)} \right\}.$$

Thus at each stage r we include all places previously fixed plus throw in as many as possible until the measure of the atom formed starting at the left drops below the r'th bound, *including* also this last place. It is clear that if F' agrees with F on $J(F)$, then $J(F') = J(F)$.

Notice that if S and S' are skeletons of the same length, then $\mathscr{F}(S) = \mathscr{F}(S')$, but the equivalence relations on $\mathscr{F}(S)$ and $\mathscr{F}(S')$ may differ.

The key estimates of the proof are contained in the following Lemma, which gives us some idea of the number and the sizes of the equivalence classes in $\mathscr{F}(S)$. (Bear in mind that eventually a set of possible equivalence classes will be forced to be a singleton.) Here

$$\theta = \max_{1 \leq i \leq a-1} \mu_0(i).$$

Lemma 5.6 Filler Lemma

1. If S is a skeleton of rank r and length l, and if $F \in \mathscr{F}(S)$, then

$$\mu_0(\tilde{F}) \geq 2^{-l(g-\varepsilon_r)}.$$

6.5. Finitary coding between Bernoulli shifts

2. If S is a skeleton of rank r and length $l \geq L_r$, then for all $F \in \mathscr{F}(S)$ except maybe a set of μ_0 (l-fold product) measure less than ε_r,

   (a) $\mu_0(\tilde{F}) \leq \frac{1}{\eta} 2^{-l(g-\varepsilon_r)}$,

   (b) $\frac{1}{l} \operatorname{card} J(F) > 1 - \frac{2}{|\log_2 \theta|} \varepsilon_r$.

Proof: (1) Let $F = a_1 \ldots a_l$ and $J(F) = J_0(F) \cup \{t_1, \ldots, t_w\}$. Then

$$\mu_0(\tilde{F}) = \mu_0 \left(\bigcup_{F' \sim F} F' \right) = \prod_{i \in J(F)} \mu_0(a_i)$$

$$= \left(\prod_{i \in J_0(F)} \mu_0(a_i) \right) \mu_0(a_{t_1}) \ldots \mu_0(a_{t_w}) \geq \mu_0(a_{t_w}) \frac{1}{\eta} 2^{-l(g-\varepsilon_r)} \geq 2^{-l(g-\varepsilon_r)},$$

by the definition of $J(F)$.

(2a) If $J(F) \neq I(S)$, then the product of all the $\mu_0(a_i)$, $i \in J(F)$, is no more than $(1/\eta) 2^{-l(g-\varepsilon_r)}$, by definition of $J(F)$. It remains only to consider the case when $J(F) = I(S) = \{1, 2, \ldots, l\}$. Then $\tilde{F} = F$. But by the Shannon–McMillan–Breiman Theorem (see Step 5.4), the set of fillers $F \in \mathscr{F}(S)$ (i.e., atoms of α_1^l) for which

$$\mu_0(F) > \frac{1}{\eta} 2^{-l(g-\varepsilon_r)}$$

together comprise a set of μ_0 measure less than ε_r.

(2b) It is enough to prove that if $\operatorname{card} J(F)/l \leq 1 - 2\varepsilon_r/|\log_2 \theta|$, then $\mu_0(F) \leq (1/\eta) 2^{-l(g+\varepsilon_r)}$, for then, again by Step 5.4, F will be contained in the same set of measure less than ε_r mentioned in the proof of 2(a). If (b) does not hold for F but (a) does, then

$\operatorname{card}(I(S) \setminus J(F)) \geq 2l\varepsilon_r/|\log_2 \theta|$, so that

$$\mu_0(F) = \left(\prod_{i \in J(F)} \mu_0(i) \right) \left(\prod_{i \in I(S) \setminus J(F)} \mu_0(i) \right) \leq \mu_0(\tilde{F}) \theta^{2l\varepsilon_r/|\log_2 \theta|}$$

$$\leq \frac{1}{\eta} 2^{-l(g-\varepsilon_r)} 2^{-2l\varepsilon_r} = \frac{1}{\eta} 2^{-l(g+\varepsilon_r)},$$

where we have also used the arcane fact that $\theta^{1/|\log_2 \theta|} = 2^{-1}$ if $0 < \theta < 1$.

Of course a similar lemma holds for $\mathscr{G}(S)$ and ν_0. In the following, η and θ will be the minimum and maximum of all the $\mu_0(i)$ and $\nu_0(i)$.

E. Sociology

At an odd stage r, we associate to the equivalence class $\tilde{F}$ of the actual filler of the r-skeleton S of each element x of a fiber in $\mathscr{B}(p)$ a set of 'suitable partners,' namely a set of equivalence classes of possible fillers

of the (same) r-skeleton of ϕx. This is done by means of a map
$$\mathscr{S}: \tilde{\mathscr{F}}(S) \to 2^{\tilde{\mathscr{G}}(S)},$$
which is called a *society* if it also satisfies a measure-nondecreasing property which will eventually imply that ϕ is measure-preserving. At even stages we do the same, except that the roles of $\mathscr{B}(p)$ and $\mathscr{B}(q)$ are reversed. At each step the associations are consistent with ones previously established (this is made possible by forming the *dual society* when we pass from step r to step $r + 1$), and the set of suitable partners tends to decrease as we *refine* the societies by using the Marriage Lemma, until at some stage the set of suitable partners shrinks to a singleton.

In this section we discuss societies on arbitrary finite sets U and V, which for mnemonic purposes, although at the risk of arousing prurient interest, can be thought of as a set of boys and a set of girls, respectively.

Definition 5.7 Let (U, ρ) and (V, σ) be probability spaces of finite cardinality. By a *society* $\mathscr{S}$ *from* U *to* V, denoted $\mathscr{S}: U \rightsquigarrow V$, we mean a map
$$\mathscr{S}: U \to 2^V$$
such that
$$\rho(B) \leq \sigma(\mathscr{S}B) \quad \text{for all } B \subset U.$$

A society $\mathscr{S}$ is thought of as an association to each boy b a set of girls $\mathscr{S}b$ whom he 'knows,' in such a way that the size of any set of boys is no larger than the size of the set of all the girls whom they know. In the situation of the classical Marriage Theorem (see 5.13 below), ρ and σ are just normalized counting measures.

We say that a society $\mathscr{R}: U \rightsquigarrow V$ is a *refinement* of a society $\mathscr{S}: U \rightsquigarrow V$, and write
$$\mathscr{R} < \mathscr{S},$$
in case $\mathscr{R}(b) \subset \mathscr{S}(b)$ for all $b \in U$. The *promiscuity* number $\pi(\mathscr{S})$ of a society $\mathscr{S}$ is defined to be the number of girls who know more than one boy:
$$\pi(\mathscr{S}) = \text{card } \{g \in V : \text{there are } b_1 \neq b_2 \text{ in } U \text{ with } g \in \mathscr{S}(b_1) \cap \mathscr{S}(b_2)\}.$$
The *dual* of a society $\mathscr{S}: U \rightsquigarrow V$ is the map $\mathscr{S}^*: V \to 2^U$ defined by
$$\mathscr{S}^*g = \{b \in U : g \in \mathscr{S}(b)\}.$$

Proposition 5.8 $\mathscr{S}^*$ is a society from V to U.
Proof: Let $G \subset V$. The boys in $(\mathscr{S}^*G)^c$ know no girls in G, so
$$\mathscr{S}[(\mathscr{S}^*G)^c] \subset G^c.$$

6.5. Finitary coding between Bernoulli shifts

Thus

$$\sigma(G) = 1 - \sigma(G^c) \leq 1 - \sigma(\mathscr{S}[(\mathscr{S}*G)^c]) \leq 1 - \rho[(\mathscr{S}*G)^c] = \rho(\mathscr{S}*G).$$

This easy, but mildly surprising, proposition, by establishing the symmetry between the roles of U and V, exculpates the discussion of any suspicion of sexism.

Remark 5.9 One particularly nice way to generate a society is from a *joining* of the probability measure ρ and σ, that is, a probability measure λ on $U \times V$ which projects to ρ and σ:

(1) $\quad \sum_{b \in U} \lambda(b, g) = \sigma(g), \quad \sum_{g \in V} \lambda(b, g) = \rho(b) \quad$ for all b, g.

We then define $\mathscr{S} = \mathscr{S}_\lambda : U \leadsto V$ by
$g \in \mathscr{S}b$ if and only if $\lambda(b, g) > 0$.
Then

$$\sigma(\mathscr{S}B) = \sum_{g \in \mathscr{S}B} \sigma(g) = \sum_{g \in \mathscr{S}B} \sum_{b \in U} \lambda(b, g)$$
$$\geq \sum_{b \in B} \sum_{g \in \mathscr{S}b} \lambda(b, g) = \sum_{b \in B} \sum_{g \in V} \lambda(b, g) = \sum_{b \in B} \rho(b) = \rho(B),$$

so that $\mathscr{S}$ really is a society. Although not all societies arise in this way (even if we allow $\geq$ in one of the equations in (1)), every society has a refinement which does. To prove this we need one preliminary result.

Lemma 5.10 Given a society $\mathscr{S} : U \leadsto V$, there is a society $\mathscr{R} < \mathscr{S}$ such that card $(\mathscr{R}b_1 \cap \mathscr{R}b_2) \leq 1$ whenever $b_1 \neq b_2$.

Proof: Suppose that there are $b_1, b_2 \in U$ and $g_1, g_2 \in V$ such that $\{g_1, g_2\} \subset \mathscr{S}b_1 \cap \mathscr{S}b_2$. Define

$$\mathscr{R}_i b = \begin{cases} \mathscr{S}b & \text{if } b \neq b_i \\ \mathscr{S}b \setminus g_i & \text{if } b = b_i \end{cases} \quad \text{for } i = 1, 2.$$

We claim that at least one of $\mathscr{R}_1, \mathscr{R}_2$ is a society. The required $\mathscr{R} < \mathscr{S}$ can then be produced by repeating this construction as many times as necessary.

Suppose that $\mathscr{R}_1$ is not a society. Then there is $B_0 \subset U$ with $\rho(B_0) > \sigma(\mathscr{R}_1 B_0)$. Notice that if $b_1 \notin B_0$, then clearly $\mathscr{R}_1 B_0 = \mathscr{S}B_0$; and because $\mathscr{S}B_0 \setminus \mathscr{R}_1 B_0 \subset \{g_1\} \subset \mathscr{S}b_2, b_2 \in B_0$ also implies that $\mathscr{R}_1 B_0 = \mathscr{S}B_0$. We must have, therefore, $b_1 \in B_0$ and $b_2 \notin B_0$.

Let $B \subset U$ be given; we will verify that $\rho(B) \leq \sigma(\mathscr{R}_2 B)$, so that $\mathscr{R}_2$ is a society. If $b_2 \notin B$ or $b_1 \in B$, by the preceding argument $\mathscr{R}_2 B = \mathscr{S}B$, and we're done. Assume then that $b_2 \in B$ and $b_1 \notin B$, so that $\mathscr{R}_2 B = \mathscr{S}B \setminus \{g_2\}$. Then $b_1 \notin B$ implies that $\mathscr{R}_2(B \cap B_0) \subset \mathscr{R}_1(B \cap B_0)$, and we may calculate that

$$\rho(B_0) + \rho(B\setminus B_0) \leq \sigma(\mathscr{S}(B \cup B_0)) = \sigma(\mathscr{R}_1 B_0 \cup \mathscr{R}_2 B)$$
$$= \sigma(\mathscr{R}_1 B_0 \cup [\mathscr{R}_2(B\setminus B_0)\setminus \mathscr{R}_1(B \cap B_0)])$$
$$\leq \sigma(\mathscr{R}_1 B_0) + \sigma[\mathscr{R}_2(B\setminus B_0)\setminus \mathscr{R}_1(B \cap B_0)],$$

and hence
$$\rho(B\setminus B_0) \leq \sigma[\mathscr{R}_2(B\setminus B_0)\setminus \mathscr{R}_1(B \cap B_0)].$$

Because $b_1, b_2 \notin B \cap B_0$, we have $\mathscr{R}_1(B \cap B_0) = \mathscr{R}_2(B \cap B_0) = \mathscr{S}(B \cap B_0)$, so that

$$\rho(B \cap B_0) \leq \sigma(\mathscr{R}_1(B \cap B_0)).$$

Adding these inequalities gives
$$\rho(B) = \rho(B\setminus B_0) + \rho(B \cap B_0)$$
$$\leq \sigma(\mathscr{R}_1(B \cap B_0)) + \sigma[\mathscr{R}_2(B\setminus B_0)\setminus \mathscr{R}_1(B \cap B_0)]$$
$$= \sigma(\mathscr{R}_1(B \cap B_0) \cup \mathscr{R}_2(B\setminus B_0))$$
$$= \sigma(\mathscr{R}_2(B \cap B_0) \cup \mathscr{R}_2(B\setminus B_0)) = \sigma(\mathscr{R}_2 B).$$

Proposition 5.11 Every society has a refinement which is generated by a joining.

Proof: Let $\mathscr{S}: U \twoheadrightarrow V$ be a society and fix $n = 1, 2, \ldots$. Split each girl $g_i \in V$ into n equal-weight subgirls g_i^j, $j = 1, 2, \ldots, n$, to form a new set V_n and society $\mathscr{S}_n: U \twoheadrightarrow V_n$. More precisely, we may define $V_n = V \times \{1, 2, \ldots, n\}$, $\sigma_n(g, j) = \sigma(g)/n$ for all j, and $\mathscr{S}_n b = \mathscr{S}b \times \{1, 2, \ldots, n\}$ for all $b \in U$.

By the Lemma, there is a society $\mathscr{R}_n < \mathscr{S}_n$ such that card $(\mathscr{R}_n b_1 \cap \mathscr{R}_n b_2) \leq 1$ whenever $b_1 \neq b_2$. We define

$$\lambda_n(b, g) = \sigma_n(\mathscr{R}_n b \cap (\{g\} \times \{1, 2, \ldots, n\}))$$

(the proportion of $\mathscr{R}_n b$ that falls 'within' the girl g). Then

$$(2) \quad \sum_{g \in V} \lambda_n(b, g) = \sum_{g \in V} \sigma_n(\mathscr{R}_n b \cap (\{g\} \times \{1, 2, \ldots, n\}))$$
$$= \sigma_n(\mathscr{R}_n b) \geq \rho(b) \quad \text{for all } b \in U,$$

and, because the sets $\mathscr{R}_n b \cap (\{g\} \times \{1, 2, \ldots, n\})$, $b \in U$, cover $\{g\} \times \{1, 2, \ldots, n\}$ and can be made disjoint by discarding at most one element of the second, two of the third, etc.,

$$(3) \quad \sum_{b \in U} \lambda_n(b, g) = \sum_{b \in U} \sigma_n(\mathscr{R}_n b \cap (\{g\} \times \{1, 2, \ldots, n\}))$$
$$\leq \sigma(g) + \frac{(\text{card } U - 1)(\text{card } U)\sigma(g)}{2n}.$$

Choose $\lambda'_n(b, g) \leq \lambda_n(b, g)$ for all (b, g) so that equality holds in (2); this does not affect (3). Notice that $\lambda'_n(b, g) > 0$ only if $g \in \mathscr{S}b$. Choose $(\lambda(b, g))$

6.5. Finitary coding between Bernoulli shifts

to be any cluster point of the sequence of matrices $(\lambda'_n(b, g))$. Then clearly λ is a joining of ρ and σ which generates a society that refines $\mathscr{S}$.

Lemma 5.12 Marriage Lemma Given a society $\mathscr{S}: U \rightsquigarrow V$, there is a society $\mathscr{R} < \mathscr{S}$ such that $\pi(\mathscr{R}) <$ card U.

Proof: Choose $\mathscr{R}$ to be a minimal refinement of $\mathscr{S}$. Then $\mathscr{R}$ is given by a joining λ of ρ and σ. By a *cycle* we will mean a sequence $b_1, \ldots, b_k$ of $k \geq 2$ different boys for which there are *different* girls $g_1, \ldots, g_k$ such that

$$g_i \in \mathscr{R}(b_i) \cap \mathscr{R}(b_{i+1}) \quad \text{for all } i,$$

where $b_{k+1} = b_1$. We will show that this minimal $\mathscr{R}$ has no cycles, and then the estimation of $\pi(\mathscr{R})$ will be easy.

If there is a cycle as above, let $m = \min \{\lambda(b_i, g_i) : i = 1, \ldots, k\}$ and define

$$\hat{\lambda}(b, g) = \begin{cases} \lambda(b_i, g_i) - m & \text{if } b = b_i \text{ and } g = g_i \\ \lambda(b_{i+1}, g_i) + m & \text{if } b = b_{i+1} \text{ and } g = g_i \\ \lambda(b, g) & \text{otherwise.} \end{cases}$$

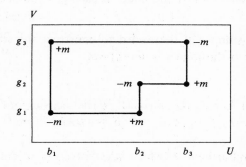

It is clear that $\hat{\lambda}$ is again a probability measure with marginals ρ and σ. Because the support of $\hat{\lambda}$ is a proper subset of the support of λ, the society $\mathscr{S}_{\hat{\lambda}}$ generated by $\hat{\lambda}$ is a strict refinement of $\mathscr{R}$. Since this is impossible, there cannot be any cycles.

Now make a graph whose vertices are all the elements of U. Each girl in V who knows more than one boy determines just one edge of the graph by connecting any one such pair of boys. By the foregoing argument, this graph has no cycles, and hence each of its components is a tree. Every tree has fewer edges than vertices, and hence so does the entire graph. This says exactly that $\pi(\mathscr{R}) <$ card U.

It is curious that the hypothesis of this Lemma involves the measures ρ and σ, but its conclusion refers only to cardinalities. Perhaps one can best understand this Lemma by observing that it is a kind of strengthening of the usual Marriage Lemma.

Lemma 5.13 *Usual Marriage Lemma* Suppose there is a set of n boys, each of whom knows at least one of a set of n girls, and that any set of k boys knows at least k girls, $k = 1, 2, \ldots, n$. Then one can hold a mass wedding in which each boy marries a girl whom he knows.

Proof : When ρ and σ give equal weights $1/n$ to all the boys and girls, then the map $\mathscr{S}$ which assigns to each boy all the girls whom he knows is a society. Find the $\mathscr{R} < \mathscr{S}$ with no cycles as above.

Each girl knows at least one boy; remove from consideration those girls who know exactly one boy. The boys who know the remaining girls are the vertices of a graph; each of the remaining girls is placed down as an edge joining any pair of boys whom she knows. Let B be the set of vertices of any component. Even though we are allowing repeats, still there are no cycles as defined above; this forces $\rho(B) > \sigma(\mathscr{S}B)$, violating the society condition. Thus the graph is empty, and under $\mathscr{R}$ each girl knows exactly one boy.

The following Proposition will be needed to construct the societies involved in the definition of the finitary isomorphism.

Proposition 5.14 Let $\mathscr{S}_i : (U_i, \rho_i) \to (V_i, \sigma_i)$ be societies for $i = 1, 2$. Then the product map $\mathscr{S}(b_1, b_2) = \mathscr{S}_1 b_1 \times \mathscr{S}_2 b_2$ is a society from $U_1 \times U_2$ to $V_1 \times V_2$.

Proof : Each $\mathscr{S}_i$ has a refinement $\mathscr{R}_i$ which is generated by a joining λ_i. Then $\lambda_1 \times \lambda_2$ is a joining of $\rho_1 \times \rho_2$ and $\sigma_1 \times \sigma_2$, and the society it generates is a refinement of $\mathscr{S}$. Therefore $\mathscr{S}$ is also a society.

F. Construction of the isomorphism

A point $z \in \mathscr{B}(p_0, 1 - p_0)$ in the marker process determines the r-skeletons S_r of all points in the fibers over z of both $\mathscr{B}(p)$ and $\mathscr{B}(q)$. By induction on r we will define for all r-skeletons S societies

$$\mathscr{R}_S : \tilde{\mathscr{F}}(S) \to \tilde{\mathscr{G}}(S) \quad \text{if} \quad r \text{ is odd}$$

and

$$\mathscr{R}_S : \tilde{\mathscr{G}}(S) \to \tilde{\mathscr{F}}(S) \quad \text{if} \quad r \text{ is even,}$$

independently of the z in which S appears.

6.5. Finitary coding between Bernoulli shifts

Let $r = 1$. For each 1-skeleton S, $\mathscr{S}_S$ is defined to be the trivial society in which each boy knows all the girls:

$$\mathscr{S}_S \tilde{F} = \tilde{\mathscr{G}}(S) \quad \text{for all } \tilde{F} \in \tilde{\mathscr{F}}(S).$$

(All $l(S)$-blocks in the filler process of $\mathscr{B}(q)$ are candidates for filling in the blanks in ϕx.) Using the Marriage Lemma, choose $\mathscr{R}_S < \mathscr{S}_S$ to be a society with $\pi(\mathscr{R}_S) < \text{card}(\tilde{\mathscr{F}}(S))$.

Suppose now that S is a skeleton of even rank r, and that the $\mathscr{R}$s have been defined for all ranks $> r$. Form the rank decomposition $S = S_1 \times \ldots \times S_j$ of S into skeletons of rank $r - 1$, look at the known societies

$$\mathscr{R}_{S_i} : \tilde{\mathscr{F}}(S_i) \rightsquigarrow \tilde{\mathscr{G}}(S_i),$$

form their duals

$$\mathscr{R}^*_{S_i} : \tilde{\mathscr{G}}(S_i) \rightsquigarrow \tilde{\mathscr{F}}(S_i),$$

and the product society (cf. Proposition 5.14)

$$\mathscr{S}_S : \tilde{\mathscr{G}}(S_1) \times \ldots \times \tilde{\mathscr{G}}(S_j) \to \tilde{\mathscr{F}}(S_1) \times \ldots \times \tilde{\mathscr{F}}(S_j).$$

Recall that each $\tilde{F} \in \tilde{\mathscr{F}}(S)$ is determined by fixing the entries for a certain set of places $J(F)$ among the blanks in S. The elements of

$$\bar{\mathscr{F}}(S) = \tilde{\mathscr{F}}(S_1) \times \ldots \times \tilde{\mathscr{F}}(S_j)$$

are equivalence classes determined by fixing certain places in a (usually) proper subset of these. Because this is a coarser equivalence relation, each element of $\bar{\mathscr{F}}(S)$ is a disjoint union of equivalence classes in $\tilde{\mathscr{F}}(S)$, and so we may regard $\mathscr{S}_S : \tilde{\mathscr{G}}(S) \to \tilde{\mathscr{F}}(S)$ as a society from $\tilde{\mathscr{G}}(S)$ to $\tilde{\mathscr{F}}(S)$. (A subset of $\tilde{\mathscr{F}}(S)$ naturally determines one of $\bar{\mathscr{F}}(S)$.)

Choose $\mathscr{R}_S < \mathscr{S}_S$ by the Marriage Lemma with $\pi(\mathscr{R}_S) < \text{card}\,\tilde{\mathscr{G}}(S)$. Consider $\mathscr{R}_S : \tilde{\mathscr{G}}(S) \rightsquigarrow \tilde{\mathscr{F}}(S)$ as a society from $\tilde{\mathscr{G}}(S)$ to $\tilde{\mathscr{F}}(S)$ by putting

$$\mathscr{R}_S(\tilde{G}) = \mathscr{R}_S(\bar{G}) \quad \text{for each filler } G \in \mathscr{G}(S),$$

clearly a well-defined map. $\mathscr{R}_S$ is a society because

$$\rho(\tilde{G}) \leqslant \rho(\bar{G}) \leqslant \sigma(\mathscr{R}_S \bar{G}) = \sigma(\mathscr{R}_S(\tilde{G}))$$

(ρ and σ are appropriate powers of ν_0 and μ_0).

If S is a skeleton of odd rank r, we perform a similar construction except in the opposite direction, ending up with $\mathscr{R}_S : \tilde{\mathscr{F}}(S) \to \tilde{\mathscr{G}}(S)$.

Now we are prepared to define the isomorphisms $\phi : \mathscr{B}(p) \to \mathscr{B}(q)$ and $\psi : \mathscr{B}(q) \to \mathscr{B}(p)$. For each $x \in \mathscr{B}(p)$, let $i_r(x)$ denote the index in $I(S_r(x)) = \{1, 2, \ldots, l(S_r(x))\}$ of the 0th coordinate of x. Recall that $F_r(x)$ denotes the $l(S_r(x))$-block that occupies in x the places occupied by blanks in $S_r(x)$.

Lemma 5.15 *Assignment Lemma* For almost all $x \in \mathcal{B}(p)$ with $x_0 \neq 0$ there is an even $r = r(x)$ such that

(1) With regard to the society $\mathcal{R}_{S_r(x)} : \tilde{\mathcal{G}}(S_r(x)) \rightsquigarrow \tilde{\mathcal{F}}(S_r(x))$, $\mathcal{R}_{S_r(x)}^{-1}(\tilde{F}_r(x))$ is a singleton, $\bar{G}_r(x)$.

(2) $i_r(x) \in J_0(\bar{G}_r(x))$ (the union of the places fixed in the skeletons of the rank decomposition of S_r).

Before proving this Lemma, let us show how it allows us to *define the isomorphism* ϕ. For almost all $x \in \mathcal{B}(p)$ choose such an even r, and define

$$(\phi x)_0 = \begin{cases} 0 \text{ if } x_0 = 0 \\ \text{the } i_r(x)\text{th entry in } \bar{G}_r(x) \text{ if } x_0 \neq 0. \end{cases}$$

Because subsequent societies arise from duals and refinements of previous ones, $(\phi x)_0$ is independent of r. In words, the map is defined by looking at each stage at the set of girls determined by x, waiting until there is only one suitable boy, and then reading off the 0th coordinate of this boy (which all boys from then on will have in common).

The map ϕ is actually defined on entire orbits by

$$(\phi x)_j = \begin{cases} 0 \text{ if } x_j = 0 \\ i_r(\sigma^j x)\text{th entry in } \bar{G}_r(\sigma^j x) \text{ if } x_j \neq 0, \end{cases}$$

and is clearly shift-invariant. Actually at any stage r we can fill in all the places in $J_0(\bar{G}_r(x))$.

Using odd rs, we obtain a map $\psi : \mathcal{B}(q) \to \mathcal{B}(p)$ which must be the inverse of ϕ (wherever defined), since the societies involved at each stage are duals of one another: ψ has to assign to $\phi(x)$ (equivalence classes of) fillers with the same 0 entry as x.

To show that ϕ is measure-preserving, we need first to disintegrate the measure v on $\mathcal{B}(q)$ with respect to the factor map $\mathcal{B}(q) \to \mathcal{B}(p_0, 1 - p_0)$. In the current situation this is relatively easy, because we are dealing with essentially a product space. If m is the probability measure on $\mathcal{B}(p_0, 1 - p_0)$ and $C \subset \mathcal{B}(q)$ is a cylinder set, then it is easy to check that

$$v(C) = \int_{\mathcal{B}(p_0, 1-p_0)} v_0(C_z) dm(z),$$

where $C_z = \{g \in B_0^{\mathbb{Z}} : (z, g) \in C\}$ is the section of C over z. (C_z may change with z, but $v_0(C_z)$, when not 0, is constant.) A similar analysis applies to $\mathcal{B}(p)$, μ, and μ_0.

Fix a cylinder set $C \subset \mathcal{B}(q)$; we will show that $\mu(\phi^{-1}C) \geq v(C)$. This will imply that $\mu\phi^{-1} \geq v$ and, by considering complements, that $\mu\phi^{-1} = v$.

6.5. Finitary coding between Bernoulli shifts

For a fixed $z \in \mathscr{B}(p_0, 1 - p_0)$, the section C_z is a cylinder set in $B_0^{\mathbb{Z}}$. Fix an r large enough that the r-skeleton $S_r(z)$ of all x and y in the fibers over z includes all the places which determine C. As in the second part of the proof below of the Assignment Lemma (denoting by $G_r(z, g)$ the filler of $S_r(z)$ in $(z, g) \in \mathscr{B}(q)$), unless g is in a subset U_r of $B_0^{\mathbb{Z}}$ having measure $\delta_r \to 0$, then the set of places determining C_z is contained in the fixed set $J_0(G_r(z, g))$. (This will allow us to tell whether or not $(z, g) \in C$ by considering its filler equivalence class in $\bar{\mathscr{G}}(S_r(z))$.) Also, as in the first part of the proof of the Assignment Lemma, if $f \in A_0^{\mathbb{Z}}$ avoids a certain set E_r of measure less than $\delta_r \to 0$, then under the society $\mathscr{R}_{S_r(z)} : \bar{\mathscr{G}}(S_r) \rightsquigarrow \tilde{\mathscr{F}}(S_r)$, $\mathscr{R}_{S_r}^{-1} \tilde{F}_r(z, f)$ is a singleton. (And hence we can identify $\phi(z, f)$ on the fixed set of this singleton equivalence class.)

The set C_z determines a set of fillers G for $S_r(z)$. Let $\bar{G}$ denote the saturation of G (i.e. the set of equivalence classes of members of G). As long as we avoid U_r and E_r, each filler $f \in A_0^{\mathbb{Z}}$ which is in $\mathscr{R}_{S_r} \bar{G}$ has an equivalence class in $\tilde{\mathscr{F}}(S_r)$ which 'knows' a single equivalence class in $\bar{\mathscr{G}}(S_r)$, and the fixed set of this last equivalence class contains all the places which C_z fixes. Since all such f are assigned by ϕ to something in C_z (we are regarding ϕ also as a map of the filler Bernoulli shifts, which is permissible because ϕ respects fibers), we have

$$\mu_0(\phi^{-1} C_z) \geq \mu_0(\mathscr{R}_{S_r} \bar{G}) - 2\delta_r = \sigma(\mathscr{R}_{S_r} \bar{G}) - 2\delta_r \geq \rho(\bar{G}) - 2\delta_r$$
$$\geq v_0(G) - 2\delta_r = v_0(C_z) - 2\delta_r.$$

Applying the disintegration formula gives $\mu(\phi^{-1} C) \geq v(C) - 2\delta_r$, and letting $r \to \infty$ concludes the argument.

Proof of the Assignment Lemma: Fix $z \in \mathscr{B}(p_0, 1 - p_0)$, thereby determining the skeletons S_r for all r. By the Skeleton Lemma (5.3(3)), $l = l_r = l(S_r) \geq L_r$ for sufficiently large r. For even r, let us estimate the measure of the set of complete fillers

$$E_r = \{ f \in A_0^{\mathbb{Z}} : \text{with regard to } \mathscr{R}_{S_r(x)} : \bar{\mathscr{G}}(S_r) \rightsquigarrow \tilde{\mathscr{F}}(S_r),$$
$$\text{card } \mathscr{R}_{S_r(x)}^{-1} \tilde{F}_r(z, f) > 1 \}.$$

(We are still identifying $\mathscr{B}(p_0, 1 - p_0) \times A_0^{\mathbb{Z}}$ with $\mathscr{B}(p)$.) The cardinality of the set of $\tilde{F}$ for which card $\mathscr{R}_{S_r}^{-1} \tilde{F} > 1$ is less than card $\bar{\mathscr{G}}(S_r)$, by the Marriage Lemma (5.12). And by the Filler Lemma (5.6(2a)), most $\tilde{F}$ have measure no more than $(1/\eta) 2^{-l(g - \varepsilon_r)}$. Therefore

$$\mu_0(E_r) \leq \frac{1}{\eta} 2^{-l(g - \varepsilon_r)} \text{ card } \bar{\mathscr{G}}(S_r) + \varepsilon_r.$$

Now look at the rank decomposition of S_r. For each $\tilde{G}_i \in \tilde{\mathscr{G}}(S_i)$, by (1) of the Filler Lemma

$$v_0(\tilde{G}_i) \geq 2^{-l_i(g-\varepsilon_{r-1})},$$

so that

$$\text{card } \tilde{\mathscr{G}}(S_i) \leq 2^{l_i(g-\varepsilon_{r-1})}$$

and

$$\text{card } \tilde{\mathscr{G}}(S_r) = \prod_i \text{card } \tilde{\mathscr{G}}(S_i) \leq 2^{l(g-\varepsilon_{r-1})}.$$

Thus for sufficiently large r (depending on z),

$$\mu_0(E_r) \leq \frac{1}{\eta} 2^{-l(g-\varepsilon_r)} 2^{l(g-\varepsilon_{r-1})} + \varepsilon_r = \frac{1}{\eta} 2^{-l_r(\varepsilon_{r-1}-\varepsilon_r)} + \varepsilon_r$$

$$\leq \frac{1}{\eta} 2^{-L_r(\varepsilon_{r-1}-\varepsilon_r)} + \varepsilon_r \to 0.$$

The sets E_r decrease as r increases; therefore

$$\mu\{x : x \in F_r \text{ for all even } r\} = \mu \bigcap_{k=1}^{\infty} E_k = 0,$$

proving (1).

To prove (2), refer to (2b) of the Filler Lemma (5.6). Again fix a point in the marker process, $z \in \mathscr{B}(p_0, 1 - p_0)$, and consider all the points $x = (z, f)(f \in A_0^{\mathbb{Z}})$ in the fiber of $\mathscr{B}(p)$ over z. This set breaks up into l_r sets of equal measure $1/l_r$ according to where the 0 coordinate is located, since the shift maps any one such set to another. Thus, again using the rank decomposition of S_r,

$$\mu_0\{f \in A_0^{\mathbb{Z}} : i_r(x) \notin J_0(\bar{G}_r(x))\} = 1 - \frac{\text{card } J_0(\bar{G}_r(x))}{l_r}$$

$$= 1 - \frac{1}{l_r} \sum_{i=1}^{j} \text{card } J(G_i)$$

$$\leq 1 - \frac{1}{l_r} \sum_{i=1}^{j} \left(l_i - \frac{2\varepsilon_{r-1}}{|\log_2 \theta|} l_i \right) = \frac{2\varepsilon_{r-1}}{|\log_2 \theta|}.$$

Since this tends to 0 as $r \to \infty$, (2) follows by the same argument used for (1).

The sequence ε_r tending to 0 has remained completely arbitrary.

6.5. Finitary coding between Bernoulli shifts

Exercises

1. Prove the equivalence of Conditions 5.1.
2. Show that $\mathscr{B}(p)$ and $\mathscr{B}(q)$ are metrically isomorphic under a homeomorphism if and only if p is a permutation of q.
3. Show that not every society arises from a joining, even if $\geqslant$ is allowed in one of the equations (1) on p. 293.
4. Extend Proposition 5.11 to societies on arbitrary probability spaces.
5. Show that if $a = 2, b \geqslant 3$, there is a probability vector r with at least 3 entries and $H(r) = h$ and $k \geqslant 1$ such that $p_0^k p_1 = r_0^k r_1$. (Hint: choose $r_0 > \max\{p_0, p_1\}$, let $r_1 = (p_0/r_0)^k p_1$, with k large enough that $H(r_0, r_1, 1 - (r_0 + r_1)) < h$.)
6. Prove that $\mathscr{B}(p) \approx \mathscr{B}(q)$ if $a = 2, b \geqslant 3$. (Hint: It is enough to consider the case $p_0^k p_1 = q_0^k q_1$. Replace 0 in the definition of skeleton by $0^k 1$. What other modifications will allow the proof to carry through?
7. Let (U, ρ) and (V, σ) be probability spaces and q a measure on $U \times V$. According to a theorem of Strassen (1965) (see Jacobs 1978) there is a probability measure $\lambda \leqslant q$ on $U \times V$ if and only if

$$q(E \times F) + 1 \geqslant \rho(E) + \sigma(F)$$

for all measurable $E \subset U, F \subset V$. Use this to prove Proposition 5.11. (Hint: let $q(b, g) = \operatorname{card}(\mathscr{S}b \cap \{g\})$.)

References

The following list, while not a complete set of references for ergodic theory, does include all the articles and books mentioned in the text. Asterisks indicate books and survey articles, several of which contain extensive bibliographies.

ABRAMOV, L. M.
   (1959) Entropy of induced automorphisms, *Dokl. Akad. Nauk SSSR* **128**, 647–50.
   (1962) Metric automorphisms with quasi-discrete spectrum, *Izv. Akad. Nauk SSSR Ser. Mat.* **26**, 513–30. *Amer. Math. Soc. Transl.* **39** (1964), 37–56.

ABRAMOV, L. M. and ROKHLIN, V. A.
   (1962) The entropy of a skew product of measure-preserving transformations, *Vestnik Leningrad Univ.* **17**, 5–13. *Amer. Math. Soc. Transl. (Ser. 2)* **48** (1965), 225–65.

ADLER, ROY L.
   (1963) A note on the entropy of skew product transformations, *Proc. Amer. Math. Soc.* **14**, 665–669.
   (1973) F-expansions revisited, *Recent Advances in Topological Dynamics*, Lecture Notes in Mathematics 318, Springer-Verlag, New York, 1–5.

ADLER, ROY L., KONHEIM, A. G., and McANDREW, M. H.
   (1965) Topological entropy, *Trans. Amer. Math. Soc.* **114**, 309–19.

ADLER, ROY L. and MARCUS, BRIAN
   (1979) *Topological Entropy and Equivalence of Dynamical Systems*, Memoirs Amer. Math. Soc. **219**.

ADLER, ROY L. and WEISS, BENJAMIN
   (1967) Entropy, a complete metric invariant for automorphisms of the torus, *Proc. Nat. Acad. Sci. USA* **57**, 1573–6.
   (1970) *Similarity of Automorphisms of the Torus*, Memoirs Amer. Math. Soc. **98**.

AKCOGLU, M. A.
   (1975) A pointwise ergodic theorem in L_p-spaces, *Canad. J. Math.* **27**, 1075–82.

ALPERN, STEVE
   (1978) Approximation to and by measure preserving homeomorphisms. *J. London Math. Soc.* **18**, 305–15.

ANOSOV, D. V. and KATOK, A. B.
  (1970) New examples in smooth ergodic theory: Ergodic diffeomorphisms, *Trudy Moskov. Mat. Obsc.* **23**. *Trans. Moscow Math. Soc.* **23**, 1–35.
ANZAI, HIROTADA
  (1951) Ergodic skew product transformations on the torus, *Osaka Math. J.* **3**, 83–99.
ARNOLD, V. I. and AVEZ, A.
  *(1968) *Ergodic Problems of Classical Mechanics*, W. A. Benjamin, Inc., New York.
AUSLANDER, L., GREEN, L. and HAHN, F.
  *(1963) *Flows on Homogeneous Spaces*, Ann. of Math. Studies **53**, Princeton University Press, Princeton, New Jersey.
AVEZ, A. (see Arnold)
BANACH, M. STEFAN
  (1926) Sur la convergence presque partout de fonctionelles linéaires, *Bull. Sci. Math.* (2) **50**, 27–32, 36–43.
BELLOW, A. and FURSTENBERG, H.
  (1979) An application of number theory to ergodic theory and the construction of uniquely ergodic models, *Israel J. Math.* **33**, 231–40.
BERG, KENNETH R.
  (1968) Entropy of torus automorphisms, *Topological Dynamics, An International Symposium*, Joseph Auslander and Walter Gottschalk, eds., W. A. Benjamin, New York.
BESICOVITCH, A. S.
  *(1932) *Almost Periodic Functions*, Cambridge University Press, Dover Publications, New York, 1954.
BILLINGSLEY, PATRICK
  *(1965) *Ergodic Theory and Information*, John Wiley & Sons, New York.
BIRKHOFF, GEORGE D.
  *(1927) *Dynamical Systems*, Colloquium Publications IX, American Mathematical Society, Providence, Rhode Island; also 1966.
  (1931) Proof of the ergodic theorem, *Proc. Nat. Acad. Sci. USA* **17**, 656–60.
BLUM, J. R. and HANSON, D. L.
  (1963) On the isomorphism problem for Bernoulli schemes, *Bull. Amer. Math. Soc.* **69**, 221–3.
BOCHNER, S.
  (1927) Beiträge zur Theorie der fastperiodischen Funktionen I, II, *Math. Ann.* **96**, 119–47, 383–409.
BOCHNER, S. and von NEUMANN, J.
  (1935) Almost periodic functions in groups, II, *Trans. Amer. Math. Soc.* **37**, 21–50.
BOGOLIOUBOFF, NICOLAS (see Krylov)

BOHL, P.
(1909) Über ein in der Theorie der säkularen Störungen vorkommendes Problem, *Journ. Reine u. Angew. Math.* **135**, 189–283.

BOHR, HARALD
(1925a) Zur Theorie der fastperiodischen Funktionen, *Acta Math.* **45**, 29–127.
(1925b) Zur Theorie der fastperiodischen Funktionen, II, *Acta Math.* **46**, 101–214.
(1926) Zur Theorie der fastperiodischen Funktionen, III, *Acta Math.* **47**, 237–81.
*(1932) *Fastperiodiche Funktionen*, Ergebnisse der Math. und ihrer Grenzgebiete I 5, J. Springer, Berlin, Eng. transl. Chelsea Pub. Co., New York, 1947.

BOOLE, G.
(1857) On the comparison of transcendents with certain applications to the theory of definite integrals, *Philos. Trans. R. Soc. London* **147**, Part III, 745–803.

BOWEN, RUFUS
(1970a) Markov partitions and minimal sets for Axiom A diffeomorphisms, *Amer. J. Math.* **92**, 907–18.
(1970b) Markov partitions for Axiom A diffeomorphisms, *Amer. J. Math.* **92**, 725–47.
(1971) Entropy for group endomorphisms and homogeneous spaces, *Trans. Amer. Math. Soc.* **153**, 401–14.
*(1978) *On Axiom A diffeomorphisms*, CBMS No. 35, American Mathematical Society, Providence, Rhode Island.

BREIMAN, LEO
(1957) The individual ergodic theorem of information theory, *Ann. Math. Stat.* **28**, 809–11. Correction, *ibid.* **31** (1960), 809–10.

BREZIN, JONATHAN and MOORE, CALVIN C.
(1981) Flows on homogeneous spaces: a new look, *Amer. J. Math.* **103**, 571–613.

BRONSTEIN, I. U.
*(1979) *Extensions of Minimal Transformation Groups*, Sijthoff & Noordhoff, Alphen an den Rijn, The Netherlands.

BROWN, JAMES R.
*(1976) *Ergodic Theory and Topological Dynamics*, Academic Press, New York.

BUNIMOVITCH, L. A.
(1973) Inclusion of Bernoulli shifts in some special flows, *Uspehi Mat. Nauk* **28** (3), 171–2.
(1974) On a class of special flows, *Izv. Akad. Nauk SSSR Ser. Mat.* **38** (1), 213–27.

BURKHOLDER, D. L.
(1962) Successive conditional expectations of an integrable function, *Ann. Math. Stat.* **33**, 887–93.

BURKHOLDER, D. L.; GUNDY, R. F. and SILVERSTEIN, M. L.
(1971) A maximal function characterization of the class H^p, *Trans. Amer. Math. Soc.* **157**, 137–53.

CALDERÓN, A. P.
(1950) On the behavior of harmonic functions at the boundary, *Trans. Amer. Math. Soc.* **68**, 47–54.

(1968) Ergodic theory and translation-invariant operators, *Proc. Nat. Acad. Sci. U.S.A.* **59**, 349–53.

(1977) Cauchy integrals on Lipschitz curves and related operators, *Proc. Nat. Acad. Sci. U.S.A.* **74**, 1324–1327.

CARATHÉODORY, CONSTANTIN

(1939) Die Homomorphieen von Somen und die Multiplikation von Inhaltsfunktionen, *Annali della R. Scuola Normale Superiore di Pisa* (Ser. 2) **8**, 105–30.

CARLESON, LENNART

(1958) Two remarks on the basic theorems of information theory, *Math. Scand.* **6**, 175–80.

CAVÄLLIN

Contributions to the theory of the secular perturbations of the planets, *Meddelanden från Lunds astronomiska observatorium* **19**.

CHACON, R. V.

(1964) A class of linear transformations, *Proc. Amer. Math. Soc.* **15**, 560–4.

(1969) Weakly mixing transformations which are not strongly mixing, *Proc. Amer. Math. Soc.* **22**, 559–62.

CHACON, R. V. and ORNSTEIN, D. S.

(1960) A general ergodic theorem, *Illinois J. Math.* **4**, 153–60.

CHUNG, K. L.

(1961) A note on the ergodic theorem of information theory, *Ann. Math. Stat.* **32**, 612–14.

CORNFELD, I. P., FOMIN, S. V., and SINAI, YA. G.

*(1981) *Ergodic Theory*, Springer-Verlag, New York.

COTLAR, M.

(1955) A unified theory of Hilbert transforms and ergodic theorems, *Rev. Mat. Cuyana* **1**, 105–67.

DANI, S. G.

(1976) Bernoullian translations and minimal horospheres on homogeneous spaces, *J. Indian Math. Soc.*, **40**, 245–84.

DEKKING, F. M.

*(1980) *Combinatorial and Statistical Properties of Sequences Generated by Substitutions*, Mathematisch Instituut Katholieke Universiteit van Nijmegen.

DEKKING, F. M. and KEANE, M.

(1978) Mixing properties of substitutions, *Z. Wahrsch. verw. Gebiete* **42**, 23–33.

del JUNCO, ANDRES

(1978) A simple measure-preserving transformation with trivial centralizer, *Pacific J. Math.* **79**, 357–62.

(1981*a*) Disjointness of measure-preserving transformations, minimal self-joinings and category, *Ergodic Theory and Dynamical Systems* **1**, Proc. special year Maryland, 1979–80, A. Katok, ed., *Progress in Math.* **10**, Birkhäuser, Boston, 81–9.

(1981*b*) A family of counter examples in ergodic theory, to appear.

del JUNCO, A., RAHE, M. and SWANSON, L.
  (1980) Chacon's automorphism has minimal self-joinings, *J. d'Analyse Math.* **37**, 276–84.
del JUNCO, A. and ROSENBLATT, J.
  (1979) Counterexamples in ergodic theory and number theory, *Math. Ann.* **245**, 185–97.
DENKER, MANFRED
  (1973) On strict ergodicity, *Math. Z.* **134**, 231–53.
  (1977) Generators and almost topological isomorphisms, Société Mathématique de France, Astérisque **49**, 23–5.
DENKER, MANFRED and EBERLEIN, ERNST
  (1974) Ergodic flows are strictly ergodic, *Adv. in Math.* **13**, 437–73.
DENKER, MANFRED; GRILLENBERGER, CHRISTIAN and SIGMUND, KARL
  *(1976) *Ergodic Theory on Compact Spaces*, Lecture Notes in Mathematics 527, Springer-Verlag, New York.
DENKER, MANFRED and KEANE, MICHAEL
  (1979) Almost topological dynamical systems, *Israel J. Math.* **34**, 139–60.
DERRIENNIC, Y.
  (1973) On the integrability of the supremum of ergodic ratios, *Ann. Prob.* **1**, 338–40.
  (1980) Quelques applications du theorème ergodique sous-additif, Société Mathématique de France, Astérisque **74**, 183–201.
DINABURG, E. I.
  (1970) The relation between topological entropy and metric entropy, *Dokl. Akad. Nauk SSSR* **190**, 19–22. *Soviet Math. Dokl.* **11**, 13–16.
DOOB, J. L.
  (1940) Regularity properties of certain families of chance variables, *Trans. Amer. Math. Soc.* **47**, 455–86.
  *(1953) *Stochastic Processes*, John Wiley & Sons, New York.
EBERLEIN, ERNST (see Denker)
EHRENFEST, PAUL and TATIANA
  *(1957) *The Conceptual Foundations of the Statistical Approach in Mechanics*, Cornell University Press, Ithaca, New York.
ELLIS, ROBERT
  (1958) Distal transformation groups, *Pacific J. Math.* **8**, 401–5.
  *(1969) *Lectures on Topological Dynamics*, W. A. Benjamin, Inc., New York.
  (1973) The Veech structure theorem, *Trans. Amer. Math. Soc.* **186**, 203–18.
ENGEL, DAVID and KAKUTANI, SHIZUO
  (1981) Maximal ergodic equalities, to appear.
ENGLAND, JAMES W. (see Martin)
ENGLAND, JAMES W. and MARTIN, N. F. B.
  (1968) On weak mixing metric automorphisms, *Bull. Amer. Math. Soc.* **74**, 505–7.

ERDÖS, P.
  (1964) Problems and results on Diophantine approximations, *Comp. Math.* **16**, 52–65.
ERDÖS, PAUL and TURÁN, PAUL
  (1936) On some sequences of integers, *J. London Math. Soc.* **11**, 261–4.
FATHI, ALBERT and HERMAN, MICHAEL R.
  (1977) Existence de difféomorphismes minimaux, *Société Mathématique de France, Astérique* **49**, 37–59.
FEFFERMAN, C. and STEIN, E. M.
  (1972) H^p spaces of several variables, Acta Math. **129**, 137–193.
FELDMAN, J.
  (1976) New K-automorphisms and a problem of Kakutani, Israel J. Math. **24**, 16–38.
FELLER, WILLIAM
  *(1950) *An Introduction to Probability Theory and its Applications, Vol. I.*, John Wiley & Sons, New York.
FIELDSTEEL, A.
  (1979) An uncountable family of prime transformations not isomorphic to their inverses, unpublished.
FOGUEL, SHAUL R.
  *(1980) *Selected Topics in the Study of Markov Operators*, Carolina Lecture Series, Department of Mathematics, University of North Carolina.
FOMIN, S. V. (see Cornfeld)
FRIEDMAN, NATHANIEL A.
  *(1970) *Introduction to Ergodic Theory*, Van Nostrand Reinhold Company, New York.
FRIEDMAN, N. A. and ORNSTEIN, D. S.
  (1970) On isomorphism of weak Bernoulli transformations, *Adv. in Math.* **5**, 365–94.
FURSTENBERG, HARRY (see also Bellow)
  (1960) *Stationary Processes and Prediction Theory*, Annals of Math. Studies **44**, Princeton University Press, Princeton, N.J.
  (1963) The structure of distal flows, *Amer. J. Math.* **85**, 477–515.
  (1967) Disjointness in ergodic theory, minimal sets, and a problem in Diophantine approximation, *Math. Systems Theory* **1**, 1–49.
  (1977) Ergodic behavior of diagonal measures and a theorem of Szemerédi on arithmetic progressions, *J. d'Analyse Math.* **31**, 204–56.
  *(1981) *Recurrence in Ergodic Theory and Combinatorial Number Theory*, Princeton University Pres , Princeton, New Jersey.
FURSTENBERG, H. and KATZNELSON, Y.
  (1978) An ergodic Szemerédi theorem for commuting transformations, *J. d'Analyse Math.* **34**, 275–91.
FURSTENBERG, HARRY; KEYNES, HARVEY and SHAPIRO, LEONARD
  (1973) Prime flows in topological dynamics, *Israel J. Math.* **14**, 26–38.

FURSTENBERG, H. and WEISS, B.
  (1978) Topological dynamics and combinatorial number theory, *J. d'Analyse Math.* **34**, 61–85.
GALLAVOTTI, GIOVANNI (see also Liberto, Ornstein)
  (1973) Ising model and Bernoulli schemes in one dimension, *Comm. Math. Phys.* **32**, 183–90.
GARSIA, ADRIANO M.
  (1965) A simple proof of E. Hopf's maximal ergodic theorem, *J. Math. Mech.* **14**, 381–2.
  *(1970) *Topics in Almost Everywhere Convergence*, Markham Pub. Co., Chicago.
  *(1973) *Martingale Inequalities. Seminar Notes on Recent Progress*, W. A. Benjamin, Inc., Reading, Massachusetts.
GIRSANOV, I. V.
  (1958) Spectra of dynamical systems generated by stationary Gaussian processes, *Dokl. Akad. Nauk SSSR* **119**, 851–3.
GLASNER, SHMUEL
  (1975) Compressibility properties in topological dynamics, *Amer. J. Math.* **97**, 148–71.
  *(1976) *Proximal Flows*, Lecture Notes in Mathematics 517, Springer-Verlag, New York.
GOODMAN, T. N. T.
  (1971) Relating topological entropy and measure entropy, *Bull. London Math. Soc.* **3**, 176–80.
GOODWYN, L. WAYNE
  (1969) Topological entropy bounds measure-theoretic entropy, *Proc. Amer. Math. Soc.* **23**, 679–88.
  (1972) Comparing topological entropy with measure-theoretic entropy, *Amer. J. Math.* **94**, 366–88.
GOTTSCHALK, W. H.
  (1944) Orbit-closure decompositions and almost periodic properties, *Bull. Amer. Math. Soc.* **50**, 915–19.
GOTTSCHALK, WALTER HELBIG and HEDLUND, GUSTAV ARNOLD
  *(1955) *Topological Dynamics*, Colloquium Publications XXXVI, American Mathematical Society, Providence, Rhode Island.
GRAHAM, R. L. and ROTHSCHILD, B. L.
  (1971) Ramsey's theorem for n-parameter sets, *Trans. Amer. Math. Soc.* **159**, 257–92.
GRAHAM, RONALD L.; ROTHSCHILD, BRUCE L. and SPENCER, JOEL H.
  *(1981) *Ramsey Theory*, John Wiley & Sons, New York.
GRANT, EDWARD (see also Oresme)
  *(1971) *Nicole Oresme and the Kinematics of Circular Motion*, University of Wisconsin Press, Madison, Wisconsin.
GREEN, L. (see Auslander)
GRILLENBERGER, CHRISTIAN (see also Denker)
  (1970) Zwei kombinatorische Konstruktionen für strikt ergodische Folgen, Thesis, Universität Erlangen-Nürnberg.

(1973a) Constructions of strictly ergodic systems I: Given entropy, *Z. Wahrsch. verw. Geb.* **25**, 323–34.

(1973b) Constructions of strictly ergodic systems II. K-systems, *Z. Wahrsch. verw. Geb.* **25**, 335–42.

GRILLENBERGER, CHRISTIAN and SHIELDS, PAUL

(1975) Construction of strictly ergodic systems III. A Bernoulli system, *Z. Warsch. verw. Geb.* **33**, 215–17.

GUNDY, R. F. (see also Burkholder)

(1969) On the class $L \log L$, martingales, and singular integrals, *Studia Math.* **33**, 109–18.

GUREVICH, B. M.

(1961) The entropy of horocycle flows, Dokl. Akad. Nauk SSSR **136**, 768–70. *Soviet Math. Dokl.* **2**, 124–30.

GYLDÉN, J. A. H.

(1895) Sur la transformation des agrégata périodiques, *Astronomiska iakttagelser och undersökningar på Stockholms Observatorium* **5**, no. 4.

HAHN, FRANK J. (see also Auslander)

(1965) Skew product transformations and the algebras generated by exp (p (n)), *Illinois J. Math.* **9**, 178–89.

(1967) A fixed point theorem, *Math. Systems Theory* **1**, 55–57.

HAHN, FRANK and KATZNELSON, YITZHAK

(1967) On the entropy of uniquely ergodic transformations, *Trans. Amer. Math. Soc.* **126**, 335–60.

HAJIAN, ARSHAG B. and KAKUTANI, SHIZUO

(1964) Weakly wandering sets and invariant measures, *Trans. Amer. Math. Soc.* **110**, 136–51.

HALMOS, PAUL, R.

(1943) On automorphisms of compact groups, *Bull. Amer. Math. Soc.* **49**, 619–24.

(1944) In general a measure preserving transformation is mixing, *Ann. of Math.* **45**, 786–92.

(1947) Invariant measures, *Ann. of Math.* **48**, 735–54.

(1949a) A non-homogeneous ergodic theorem, *Trans. Amer. Math. Soc.* **66**, 284–88.

*(1949b) Measurable transformations, *Bull. Amer. Math. Soc.* **55**, 1015–34.

*(1950) *Measure Theory*, D. Van Nostrand Co., Inc., Princeton, New Jersey.

*(1956) *Lectures on Ergodic Theory*, Chelsea Publishing Co., New York.

HALMOS, PAUL R. and von NEUMANN, JOHN

(1942) Operator methods in classical mechanics, II, *Ann. of Math.* **43**, 332–50.

HANSEL, GEORGES

(1974) Strict uniformity in ergodic theory, *Math. Z.* **135**, 221–48.

HANSEL, G. and RAOULT, J. P.

(1973) Ergodicity, uniformity, and unique ergodicity, *Indiana Univ. Math. J.* **23**, 221–37.

HANSON, D. L. (see Blum)

HARDY, G. H. and LITTLEWOOD, J. E.
  (1930) A maximal theorem with function-theoretic applications, *Acta Math.* **54**, 81–116.
HARPER, L. H. (see Rota)
HARTMAN, PHILIP
  (1947) On the ergodic theorems, *Amer. J. Math.* **69**, 193–99.
HECKE, E.
  (1922) Über analytische Funktionen und die Verteilung von Zahlen mod. eins, *Abh. Math. Sem. Hamburg Univ.* **1**, 54–76.
HEDLUND, G. A. (see also Gottschalk)
  (1969) Endomorphisms and automorphisms of the shift dynamical system, *Math. Systems Theory* **3**, 320–75.
HERMAN, MICHAEL R. (see Fathi)
HILBERT, D.
  (1892) Ueber die Irreducilität ganzer rationaler Functionen mit ganzzahligen Coefficienten, *J. Math.* **110**, 104–29.
HINDMAN, N.
  (1974) Finite sums from sequences within cells of a partition of N, *J. Combinatorial Theory* **A17**, 1–11.
HOCKING, JOHN G. and YOUNG, GAIL S.
  *(1961) *Topology*, Addison-Wesley, Reading, Massachusetts.
HOPF, EBERHARD
  *(1937) *Ergodentheorie*, J. Springer, Berlin. Also Chelsea, New York, 1948.
  (1944) Über eine Ungleichung der Ergodentheorie, *S.-B. Math.-Nat. Abt. Bayer. Akad. Wiss.*, 171–76.
  (1954) The general temporally discrete Markoff process, *J. Rat. Mech. Anal.* **3**, 13–45.
HUREWICZ, WITHOLD
  (1944) Ergodic theorem without invariant measure, *Ann. of Math.* **45**, 192–206.
IONESCU TULCEA, ALEXANDRA (see also Bellow)
  (1960) Contributions to information theory for abstract alphabets, *Arkiv. Mat.* **4**, 235–47.
JACOBS, KONRAD
  (1957) Fastperiodizitätseigenschaften allgemeiner Halbgruppen in Banach-Räumen, *Math. Z.* **67**, 83–92.
  *(1960) *Neuere Methoden und Ergebnisse der Ergodentheorie*, Springer-Verlag, Berlin.
  *(1962/ *Lecture Notes on Ergodic Theory, Parts I and II*, Matematisk Institut, 1963) Aarhus Universitet.
  (1970a) Lipschitz functions and the prevalence of strict ergodicity for continuous time flows, *Contributions to Ergodic Theory and Probability*, Lecture Notes in Mathematics 160, Springer-Verlag, New York, 87–124.

(1970b) Systèmes dynamiques Riemanniens, *Czechos. Math. J.* **20(90)**, 628–31.
(1978) *Measure and Integral*, Academic Press, New York.

JEWETT, ROBERT I.
(1970) The prevalence of uniquely ergodic systems, *J. Math. Mech.* **19**, 717–29.

JONES, LEE KENNETH
(1971) A mean ergodic theorem for weakly mixing operators, *Adv. in Math.* **7**, 211–16.

JONES, R. L.
(1977) Inequalities for the ergodic maximal function, *Studia Math.* **60**, 111–29.

KAC, M.
(1947) On the notion of recurrence in discrete stochastic processes, *Bull. Amer. Math. Soc.* **53**, 1002–10.

KAKUTANI, SHIZUO (see also Engel, Hajian, Yosida)
(1938a) Two fixed-point theorems concerning bicompact convex sets, *Proc. Japan Acad.* **14**, 242–5.
(1938b) Iteration of linear operations in Banach spaces, *Proc. Japan Acad.* **14**, 295–300.
(1943) Induced measure preserving transformations, *Proc. Japan Acad.* **19**, 635–41.
*(1952) Ergodic theory, *Proc. Int. Cong. Math. Cambridge, Mass. 1950*(2), 128–42.
(1973) Examples of ergodic measure preserving transformations which are weakly mixing but not strongly mixing, *Recent Advances in Topological Dynamics*, Lecture Notes in Mathematics 318, Springer-Verlag, New York, 143–9.

KAKUTANI, SHIZUO and PETERSEN, KARL
(1981) The speed of convergence in the Ergodic Theorem, *Monatsh. Math.* **91**, 11–18.

KATOK, A. B. (see Anosov)
KATOK, A. B.; SINAI, YA. G. and STEPIN, A. M.
*(1975) Theory of dynamical systems and general transformation groups with invariant measure, Progress in Science and Technology, *Math. Analysis* **13**. *J. Soviet Math.* **7** (1977), 974–1065.

KATOK, A. B. and STEPIN, A. M.
(1967) Approximations in ergodic theory, *Uspehi Mat. Nauk* **22**, 81–106. *Russ. Math. Surveys* **22**, 77–102.
(1970) Metric properties of measure preserving homeomorphisms, *Uspehi Mat. Nauk* **25**, 193–220. *Russ. Math. Surveys* **25**, 191–220.

KATZNELSON, YITZHAK (see also Furstenberg, Hahn)
(1971) Ergodic automorphisms of T^n are Bernoulli shifts, *Israel J. Math.* **10**, 186–95.

KEANE, MICHAEL (see Dekking, Denker)
KEANE, MICHAEL and SMORODINSKY, MEIR
(1977) A class of finitary codes, *Israel J. Math.* **26**, 352–71.

(1979a) Bernoulli schemes of the same entropy are finitarily isomorphic, *Ann. of Math.* **109**, 397–406.

(1979b) Finitary isomorphism of irreducible Markov shifts, *Israel J. Math.* **34**, 281–6.

(1979c) The finitary isomorphism theorem for Markov shifts, *Bull. Amer. Math. Soc.* **1**, 436–8.

KEYNES, HARVEY B. (see also Furstenberg)

*(1972) *Lectures on Ergodic Theory*, School of Mathematics, University of Minnesota.

KEYNES, HARVEY B. and ROBERTSON, JAMES B.

(1968) On ergodicity and mixing in topological transformation groups, *Duke Math. J.* **35**, 809–19.

(1969) Eigenvalue theorems in topological transformation groups, *Trans. Amer. Math. Soc.* **139**, 359–69.

KEYNES, H. B. and NEWTON, D.

(1976) Real prime flows, *Trans. Amer. Math. Soc.* **218**, 237–55.

KHINTCHINE, A. I.

(1923) Über dyadische Brüche, *Math. Z.* **18**, 109–16.

(1933) Zu Birkhoffs Lösung des Ergodenproblems, *Math. Ann.* **107**, 485–8.

(1934) Eine Verschärfung des Poincaréschen "Wiederkehrsatzes", *Comp. Math.* **1**, 177–9.

*(1949) *Mathematical Foundations of Statistical Mechanics*, Dover Publications, New York.

*(1948) *Three Pearls of Number Theory*, Graylock Press, New York.

*(1957) *Mathematical Foundations of Information Theory*, Dover Publications, New York.

KOLMOGOROV, A. N.

(1925) Sur les fonctions harmoniques conjuguées et les séries de Fourier, *Fund. Math.* **7**, 24–9.

(1928) Über die Summen durch den Zufall bestimmter unabhängiger Größen, *Math. Ann.* **99**, 309–19.

(1929) Über das Gesetz des iterierten Logarithmus, *Math. Ann.* **101**, 126–35.

(1930) Bemerkungen zu meiner Arbeit 'Über die Summen zufälliger Größen', *Math. Ann.* **102**, 484–8.

(1937) Ein vereinfachter Beweis des Birkhoff-Khintchineschen Ergodensatzes, *Recueil Math. (Mat. Sb.)* **44**, 367–8.

(1953) On dynamical systems with an integral invariant on the torus, *Dokl. Akad. Nauk SSSR* **93**, 763–6.

(1958) New metric invariants of transitive dynamical systems and automorphisms of Lebesgue spaces, *Dokl. Akad. Nauk SSSR* **119**, 861–4.

KONHEIM, A. G. (see Adler)

KOOPMAN, B. O.

(1931) Hamiltonian systems and transformations in Hilbert space, *Proc. Nat. Acad. Sci. U.S.A.* **17**, 315–18.

KOOPMAN, B. O. and von NEUMANN, J.
(1932) Dynamical systems of continuous spectra, *Proc. Nat. Acad. Sci. U.S.A.* **18**, 255–63.

KRENGEL, ULRICH
(1970) On certain analogous difficulties in the investigation of flows in a probability space and of transformations in an infinite measure space, *Functional Analysis*, Proceedings of a symposium held at Monterey, California, October 1969, Carroll O. Wilde, ed., Academic Press, New York, 75–91.
(1978) On the speed of convergence in the ergodic theorem, *Monatsh. Math.* **86**, 3–6.

KRIEGER, WOLFGANG
(1970) On entropy and generators of measure-preserving transformations, *Trans. Amer. Math. Soc.* **149**, 453–64.
(1972) On unique ergodicity, *Proc. Sixth Berkeley Symp. (1970)* **I**, University of California Press, Berkeley and Los Angeles, 327–46.

KRONECKER, LEOPOLD
(1884) Näherungsweise ganzzahlige Auflösung linearer Gleichungen, *S.-B. Preuss. Akad. Wiss.*, 1179–93, 1271–99. *Werke* III (1), 47–109.

KRYLOV, NICOLAS and BOGOLIOUBOFF, NICOLAS
(1937) La théorie générale de la mesure dans son application à l'étude des systèmes dynamiques de la mécanique non linéaire, *Ann. of Math.* **38**, 65–113.

KUIPERS, L. and NIEDERREITER, H.
*(1974) *Uniform Distribution of Sequences*, John Wiley & Sons, New York.

LAGRANGE, J. L.
(1870) *Oeuvres de Lagrange*, Tome V, Paris.

LEBESGUE, H.
*(1904) *Leçons sur l'intégration et la recherche des fonctions primitives*, Gauthiers-Villars, Paris.

LEDRAPPIER, F.
(1973) Mesures d'équilibre sur un reseau, *Comm. Math. Phys.* **33**, 119–28.

LÉVY, PAUL
*(1937) *Théorie de l'Addition des Variables Aléatoires*, Gauthier-Villars, Paris.

LIBERTO, FRANCESCO di; GALLAVOTTI, GIOVANNI and RUSSO, LUCIO
(1973) Markov processes, Bernoulli schemes, and Ising model, *Comm. Math. Phys.* **33**, 259–82.

LIND, D. A.
(1977) The structure of skew products with ergodic group automorphisms, *Israel J. Math.* **28**, 205–48.

LITTLEWOOD, J. E. (see Hardy)

LOOMIS, LYNN H.
(1946) A note on the Hilbert transform, *Bull. Amer. Math. Soc.* **52**, 1082–6.

MACKEY, GEORGE W.
  *(1974) Ergodic theory and its significance for statistical mechanics and probability theory, *Adv. in Math.* **12**, 178–268.
MARCUS, BRIAN (see also Adler)
  (1975) Unique ergodicity of the horocycle flow: variable negative curvature case, *Israel J. Math.* **21**, 133–44.
  (1976) Reparametrization of uniquely ergodic flows, *J. Differential Equations* **22**, 227–35.
MARCUS, BRIAN and PETERSEN, KARL
  (1979) Balancing ergodic averages, *Ergodic Theory*, Lecture Notes in Mathematics 729, Springer-Verlag, New York, 126–43.
MARTIN, NATHANIEL F. G. (see England)
MARTIN, NATHANIEL F. G. and ENGLAND, JAMES W.
  *(1981) *Mathematical Theory of Entropy*, Addison-Wesley Pub. Co., Reading, Massachusetts.
McANDREW, M. H. (see Adler)
McMILLAN, BROCKWAY
  (1953) The basic theorems of information theory, *Ann. Math. Stat.* **24**, 196–219.
MESHALKIN, L. D.
  (1959) A case of isomorphism of Bernoulli schemes, *Dokl. Akad. Nauk SSSR* **128**, 41–4.
MILES, G. and THOMAS, R. K.
  (1978) Generalized torus automorphisms are Bernoullian, *Studies in Probability and Ergodic Theory*, *Adv. in Math.*, Supplementary Studies 2, Academic Press, New York.
MISIUREWICZ, M.
  (1976) A short proof of the variational principle for a Z_+^n action on a compact space, *Int. Conf. Dyn. Systems in Math. Physics, Société Mathématique de France, Astérique* **40**, 147–58.
MOORE, CALVIN C. (see Brezin)
NEMYTSKII, V. V. and STEPANOV, V. V.
  *(1960) *Qualitative Theory of Differential Equations*, Princeton University Press, Princeton, New Jersey.
NEVEU, J.
  (1969) Une démonstration simplifiée et une extension de la formule d'Abramov sur l'entropie des transformations induites, *Z. Wahrsch. verw. Geb.* **13**, 135–140.
  *(1975) *Discrete-Parameter Martingales*, North-Holland, Amsterdam, American Elsevier, New York.
  (1979) The filling scheme and the Chacon-Ornstein theorem, *Israel J. Math.* **33**, 368–77.
NEWTON, D. and PARRY, W.
  (1966) On a factor automorphism of a normal dynamical system, *Ann. Math. Stat.* **37**, 1528–33.

NIEDERREITER, H. (see Kuipers)

ORESME, NICOLE (see also Grant)

*(1351?) *De proportionibus proportionum* and *Ad pauca respicientes*, Edward Grant, ed., University of Wisconsin Press, Madison, 1966.

ORNSTEIN, DONALD S. (see also Chacon, Friedman)

(1970a) Bernoulli shifts with the same entropy are isomorphic, *Adv. in Math.* **4**, 337–52.

(1970b) Imbedding Bernoulli shifts in flows, *Contributions to Ergodic Theory and Probability*, Lecture Notes in Mathematics 160, Springer-Verlag, New York, 178–218.

(1970c) Two Bernoulli shifts with infinite entropy are isomorphic, *Adv. in Math.* **5**, 339–48.

(1971) A remark on the Birkhoff ergodic theorem, *Illinois J. Math.* **15**, 77–9.

(1972) On the root problem in ergodic theory, *Proc. Sixth Berkeley Symposium (1970)*, University of California Press, Berkeley and Los Angeles, 345–56.

*(1974) *Ergodic Theory, Randomness, and Dynamical Systems*, Yale Mathematical Monographs 5, Yale University Press, New Haven, Connecticut.

*(1978) A survey of some recent results in ergodic theory, *Studies in Probability Theory*, Murray Rosenblatt, ed., Mathematical Association of America, 229–62.

ORNSTEIN, DONALD S. and GALLAVOTTI, GIOVANNI

(1974) The billiard flow with a convex scatterer is Bernoulli, *Comm. Math. Phys.* **38**, 83–101.

ORNSTEIN, D. S. and SHIELDS, P. C.

(1973) Mixing Markov shifts of kernel type are Bernoulli, *Adv. in Math.* **10**, 143–6.

ORNSTEIN, D. S. and SMORODINSKY, M.

(1978) Continuous speed changes for flows, *Israel J. Math.* **31**, 161–8.

ORNSTEIN, DONALD and WEISS, BENJAMIN

(1973) Geodesic flows are Bernoullian, *Israel J. Math.* **14**, 184–98.

(1974) Finitely determined implies very weak Bernoulli, *Israel J. Math.* **17**, 94–104.

OXTOBY, JOHN C.

*(1952) Ergodic sets, *Bull. Amer. Math. Soc.* **58**, 116–36.

(1952) Approximation by measure-preserving homeomorphisms, *Recent Advances in Topological Dynamics*, Lecture Notes in Mathematics 318, Springer-Verlag, New York, 206–217.

OXTOBY, J. C. and ULAM, S. M.

(1941) Measure-preserving homeomorphisms and metrical transitivity, *Ann. of Math.* **42**, 874–920.

PARASJUK, O. S.

(1953) Horocycle flows on surfaces of constant negative curvature, *Uspehi Mat. Nauk* **8**, 125–6.

PARRY, WILLIAM (see also Newton)
  (1969a) Compact abelian group extensions of discrete dynamical systems, *Z. Wahrsch. Verw. Geb.* **13**, 95–113.
  *(1969b) *Entropy and Generators in Ergodic Theory*, W. A. Benjamin, Inc., New York.
  (1969c) Ergodic properties of affine transformations and flows on nilmanifolds, *Amer. J. Math.* **91**, 757–71.
  (1971) Metric classification of ergodic nilflows and unipotent affines, *Amer. J. Math.* **93**, 819–28.
  *(1981) *Topics in Ergodic Theory*, Cambridge University Press, Cambridge.
PARTHASARATHY, K. L.
  *(1967) *Probability Measures on Metric Spaces*, Academic Press, New York.
PETERSEN, KARL E. (see also Kakutani, Marcus)
  (1969) *Prime Flows*, Ph.D. Dissertation, Yale University.
  (1970a) A topologically strongly mixing symbolic minimal set, *Trans. Amer. Math. Soc.* **148**, 603–12.
  (1970b) Disjointness and weak mixing of minimal sets, *Proc. Amer. Math. Soc.* **24**, 278–80.
  (1971) Extension of minimal transformation groups, *Math. Systems Theory* **5**, 365–75.
  (1973a) On a series of cosecants related to a problem in ergodic theory, *Comp. Math.* **26**, 313–17.
  (1973b) Spectra of induced transformations, *Recent Advances in Topological Dynamics*, Lecture Notes in Math. 318, Springer-Verlag, New York, 226–30.
  *(1977) *Brownian Motion, Hardy Spaces and Bounded Mean Oscillation*, London Math. Soc. Lecture Notes Series 28, Cambridge University Press, Cambridge.
  (1979) The converse of the dominated ergodic theorem, *J. Math. Anal. Appl.* **67**, 431–6.
PETERSEN, KARL and SHAPIRO, LEONARD
  (1973) Induced flows, *Trans. Amer. Math. Soc.* **117**, 375–90.
PLANTE, J.
  *(1976) *Introduction to Qualitative Theory of Differential Equations*, Carolina Lecture Series, Department of Mathematics, University of North Carolina.
POINCARÉ, HENRI
  Les méthodes nouvelles de la mécanique céleste, I (1892), II (1893), and III (1899), Gauthiers-Villars, Paris. Also Dover, New York, 1957 and NASA TTF 450-2, Washington, D.C., 1967.
POLIT, STEVE
  (1974) Weakly isomorphic maps need not be isomorphic, Ph.D. Dissertation, Stanford University.
POSTNIKOV, A. G.
  *(1966) Ergodic problems in the theory of congruences and of Diophantine

approximations, *Trudy Mat. Instituta Proc. im. V. A. Steklova* **82**, 3–112. *Proc. Steklov Institute of Math* **82**, American Mathematical Society, Providence, Rhode Island, 1967.

PRASAD, V. S.

(1979) Ergodic measure preserving homeomorphisms of $\mathbb{R}^n$ *Indiana Univ. Math. J.* **28**, 859–67.

PROHOROV, JU. V.

(1956) Convergence of random processes and limit theorems in probability theory, *Teor. Verojatnost. i Primenen* **1**, 177–238.

RADO, R.

(1943) Note on combinatorial analysis, *Proc. London Math. Soc.* **48**, 122–60.

RAHE, M. (see del Junco)

RAOULT, J. P. (see Hansel)

RATNER, MARINA

(1974) Anosov flows with Gibbs measures are also Bernoullian, *Israel J. Math.* **17**, 380–91.

(1978) Horocycle flows are loosely Bernoulli, *Israel J. Math.* **31** 122–32.

RIESZ, FRÉDERIC

(1931) Sur un théorème de maximum de MM. Hardy et Littlewood, *J. London Math. Soc.* **7**, 10–13.

(1932) Sur l'existence de la dérivée des fonctions monotones et sur quelques problèmes qui s'y rattachent, *Acta Sci. Mat. (Szeged)* **5**, 208–21.

RÉNYI, A.

(1958) On mixing sequences of sets, *Acta Math. Acad. Sci Hungar.* **9**, 215–28.

ROBERTSON, JAMES B. (see Keynes)

ROKHLIN, V. A. (see also Abramov)

(1948) A 'general' measure-preserving transformation is not mixing, *Dokl. Akad. Nauk SSSR* **60**, 349–51.

*(1949a) On the fundamental ideas of measure theory, *Mat. Sb.* **25**, 107–50. *Amer. Math. Soc. Transl.* **71** (1952).

*(1949b) Selected topics from the metric theory of dynamical systems, *Uspehi Mat. Nauk* **4**, 57–128. *Amer. Math. Soc. Transl. Series* 2 **49**, (1966), 171–240.

*(1960) New progress in the theory of transformations with invariant measure, *Uspehi Mat. Nauk* **94**, 3–26. *Russ. Math. Surveys* **15** (1960), 1–22.

*(1967) Lectures on the theory of entropy of transformations with invariant measure, *Uspehi Mat. Nauk* **22**, 3–56. *Russ. Math. Surveys* **22**, 1–52.

ROSENBLATT, J (see del Junco)

ROTA, GIAN-CARLO and HARPER, L. H.

(1971) Matching theory, an introduction, *Advances in Probability I*, Peter Ney, ed., Marcel Dekker, New York, 171–215.

ROTH, KLAUS

(1952) Sur quelques ensembles d'entiers, *C. R. Acad. Sci. Paris Sér. A-B* **234**, 388–90.

ROTHSCHILD, BRUCE L. (see Graham)
ROYDEN, H. L.
*(1968) *Real Analysis*, Macmillan Pub. Company, New York.
RUDIN, WALTER
*(1967) *Fourier Analysis on Groups*, Interscience Publishers, New York.
*(1976) *Principles of Mathematical Analysis*, McGraw–Hill, New York.
RUDOLPH, DANIEL J.
(1979) An example of a measure preserving map with minimal self-joinings, and applications, *J. d'Analyse Math.* **35**, 97–122.
(1981) A characterization of those processes finitarily isomorphic to a Bernoulli shift, *Ergodic Theory and Dynamical Systems 1, Proceedings Special Year, Maryland, 1979-80*, A. Katok, ed., *Progress in Math.* **10**, Birkhäuser, Boston, 1–64.
RUSSO, LUCIO (see Liberto)
RYLL-NARDZEWSKI, CZESLAW
(1962) Generalized random ergodic theorems and weakly almost periodic functions, *Bull. Acad. Polon. Sci. Sér. Sci. Math. Astronom. Phys.* **10**, 271–5.
(1967) On fixed points of semigroups of endomorphisms of linear spaces, *Proc. Fifth Berkeley Symp. Math. Stat. Prob. (1965–66), II (1)*, University of California Press, Berkeley, 55–61.
SAWYER, S.
(1966) Maximal inequalities of weak type, *Ann. of Math.* **84**, 157–74.
SCHUR, ISSAI
*(1973) *Gesammelte Abhandlungen*, Springer-Verlag, New York.
SHANNON, C. E.
(1948) A mathematical theory of communication, *Bell System Tech. J.* **27** 379–423, 623–56.
SHAPIRO, LEONARD (see Petersen)
SHIELDS, PAUL C. (see also Ornstein)
*(1973) *The Theory of Bernoulli Shifts*, University of Chicago Press, Chicago.
SIGMUND, KARL (see Denker)
SILVERSTEIN, M. L. (see Burkholder)
SINAI, YA. G. (see also Katok, Cornfeld)
(1959a) The notion of entropy of a dynamical system, *Dokl. Akad. Nauk SSSR* **125**, 768–71.
(1959b) Flows with finite entropy, *Dokl. Akad. Nauk SSSR* **125**, 1200–2.
(1968) Construction of Markov partitons, *Funktsional'ngi Analiz i Ego Prilozheniya 2 (no. 3)*, 70–80. *Functional Anal. Appl.* **2**, 245–53.
(1970) Dynamical systems with elastic reflections, *Uspehi Mat. Nauk* **25**. *Russian Math. Surveys* **25**, 137–89.
*(1976) *Introduction to Ergodic Theory*, Princeton University Press, Princeton, New Jersey.
SMORODINSKY, MEIR (see also Keane, Ornstein)
*(1971) *Ergodic Theory, Entropy*, Lecture Notes in Mathematics 214, Springer-Verlag, New York.

(1973) β-automorphisms are Bernoulli shifts, *Acta Math. Acad. Sci. Hungar.* **24**, 273-8.

SPENCER, JOEL H. (see Graham)

STEIN, ELIAS M. (see also Fefferman)
(1961) On limits of sequences of operators, *Ann. of Math.* **74**, 140-70.
(1969) Note on the class $L \log L$, *Studia Math.* **32**, 305-10.
*(1970) *Singular Integrals and Differentiability Properties of Functions*, Princeton University Press, Princeton, N.J.

STEIN, ELIAS and WEISS, GUIDO
*(1971) *Introduction to Fourier Analysis on Euclidean Spaces*, Princeton University Press, Princeton, New Jersey.

STEPANOV, V. V. (see Nemytskii)

STEPIN, A. M. (see Katok)

STRASSEN, V.
(1965) The existence of probability measures with given marginals, *Ann. Math. Stat.* **36**, 423-39.

SWANSON, L. (see del Junco)

SZEMERÉDI, E.
(1969) On sets of integers containing no four elements in arithmetic progression, *Acta Math. Acad. Sci Hungar.* **20**, 89-104.
(1975) On sets of integers containing no k elements in arithmetic progression, *Acta Arith.* **27**, 199-245.

TCHEBYCHEF, P. L.
(1866) Sur une question arithmétique, *Denkschr. Akad. Wiss. St. Petersburg Nr.* **4**, *Oeuvres I*, 637-84.

THOMAS, R. K. (see Miles)

TOTOKI, HARUO
*(1970) *Ergodic Theory*, Lecture Note Series 14, Matematisk Institut, Aarhus Universitet.

TURÁN, PAUL (see Erdös)

ULAM, S. M. (see Oxtoby)

ULAM, S. M. and von NEUMANN, JOHN
(1945) Random ergodic theorems, *Bull. Amer. Math. Soc.* **51**, 660.

van der WAERDEN, BARTEL L.
(1927) Beweis einer Baudet'schen Vermutung, *Nieuw. Arch. Wisk.* **15**, 212-16.
(1971) How the proof of Baudet's conjecture was found, *Studies in Pure Mathematics presented to Richard Rado*, L. Mirsky, ed., Academic Press, London, 251-60.

VARADARAJAN, V. S.
(1958) Weak convergence of measures on separable metric spaces, *Sankhya* **19**, 15-22.
*(1962) *Special Topics in Probability Theory*, Lecture Notes, Courant Inst. Math. Sci.

VARGA, RICHARD S.
*(1962) *Matrix Iterative Analysis*, Prentice-Hall, Englewood Cliffs, New Jersey.

VEECH, WILLIAM A.
(1970) Point-distal flows, *Amer. J. Math.* **92**, 205-42.
*(1977) Topological dynamics, *Bull. Amer. Math. Soc.* **83**, 775-830.

VERSHIK, A. M. and YUZVINSKII, S. A.
*(1970) Dynamical systems with invariant measure, *Progress in Mathematics* **8**, Plenum, New York.

VILLE, JEAN
*(1939) *Étude Critique de la Notion de Collectif*, Gauthier-Villars, Paris.

von NEUMANN, JOHN (see also Bochner, Halmos, Ulam, Koopman)
(1932a) Einige Sätze über messbare Abbildungen, *Ann. of Math.* **33**, 574-86.
(1932b) Proof of the quasi-ergodic hypothesis, *Proc. Nat. Acad. Sci. USA* **18**, 70-82.
(1932c) Über einen Satz von Herrn M. H. Stone, *Ann. of Math.* **33**, 567-73.
(1932d) Zur Operatorenmethode in der klassischen Mechanik, *Ann. of Math.* **33**, 587-642.
(1934) Almost periodic functions in a group, I, *Trans. Amer. Math. Soc.* **36**, 445-92.

WALTERS, PETER
*(1975) *Ergodic Theory: Introductory Lectures*, Lecture Notes in Mathematics 458, Springer-Verlag, New York.
*(1982) *An Introduction to Ergodic Theory*, Springer-Verlag, New York.

WEISS, BENJAMIN (see also Adler, Furstenberg, Ornstein)
*(1972) The isomorphism problem in ergodic theory, *Bull. Amer. Math. Soc.* **78**, 668-84.

WEISS, GUIDO (see Stein)

WEYL, HERMANN
(1916) Über die Gleichverteilung von Zahlen mod. Eins, *Math. Ann.* **77**, 313-52.
(1926) Integralgleichungen und fastperiodische Funktionen, *Math. Ann.* **97**, 338-56.

WHITE, H. E., JR.
(1974) The approximation of one-one measurable transformations by measure preserving homeomorphisms, *Proc. Amer. Math. Soc.* **44**, 391-4.

WIENER, NORBERT
(1939) The ergodic theorem, *Duke Math. J.* **5**, 1-18.

WRIGHT, FRED B.
(1961a) Mean least recurrence time, *J. London Math. Soc.* **36**, 382-4.
(1961b) The recurrence theorem, *Amer. Math. Monthly* **68**, 247-8.
*(1963) *Ergodic Theory* (Proc. Int. Symp. Tulane Univ. 1961), Academic Press, New York.

YOSIDA, KÔSAKU
(1938) Mean ergodic theorem in Banach spaces, *Proc. Japan Acad.* **14**, 292-4.

YOSIDA, KÔSAKU and KAKUTANI, SHIZUO
(1939) Birkhoff's ergodic theorem and the maximal ergodic theorem, *Proc. Japan Acad.* **15**, 165-8.

(1941) Operator-theoretical treatment of Markoff process and mean ergodic theorem, *Ann. Math.* **42**, 188–228.

YOUNG, GAIL S. (see Hocking)

YUZVINSKII, S. A. (see Vershik)

ZIMMER, ROBERT J.

(1975) Extensions of ergodic actions and generalized discrete spectrum, *Bull. Amer. Math. Soc.* **81**, 633–6.

(1976a) Extensions of ergodic group actions, *Illinois J. Math.* **20**, 373–409.

(1976b) Ergodic actions with generalized discrete spectrum, *Illinois J. Math.* **20**, 555–88.

ZYGMUND, A.

*(1959) *Trigonometric Series I*, Cambridge University Press, Cambridge.

Index

Bold page numbers indicate formal definitions. Greek letters are entered according to the initial letter of their name: α (alpha) under 'a'; β (beta) under 'b' etc.

α-generic 274
α-name 274
α, k-name 274
α, T-name 274
Abramov, L. M. 55, 254, 258
Adler, Roy L. 253, 264, 267, 268, 273, 280, 282
affine map 10, **154**
Akcoglu, M. A. 127
Alpern, Steve 72
almost periodic
  function **139**, 133
  point **136**, 150
  pointwise 159
  sequence **136**, 139, 140, 148
  system 182
  uniformly 136, 137, 154
Ambrose, W. 12, 64
Anosov, D. V. 72
Anosov flow 280
Anzai, Hirotada 53, 55
aperiodic
  matrix 59
  transformation 48
approximate eigenvalue 56
approximation proof of the Ergodic Theorem 92
Artin, E. 162
assignment 284
Assignment Lemma 298, 299
atom 16
Auslander, L. 10
automorphism
  of a compact group (*see also* of a torus) 8, 21, 60, 153
    higher mixing of 64
    Bernoulli property of 279
  of a nilmanifold 61, 279
  of a torus
    entropy of 249, 259
    ergodicity of 60, 61
    strong mixing and Bernoulli properties of 61, 254, 279
autocorrelation 184

β-transformation 280
Baker's transformation 21
ballot problem 85
Banach, M. Stefan 91
Banach's Principle **91**, 99
Baudet, Pierre Joseph Henry 162
Bellow, A. (*see also* Ionescu Tulcea) 187, 194
Bellow-Furstenberg Theorem **196**, 200, 210
Berg, Kenneth R. 249
Besicovitch, A. S. 142
Bessel's Inequality 148
Bernoulli flow 279
Bernoulli shift (*or* scheme) **7**, 21, 60, 61, 85, 86, 94, 232, 248, 278
  classification up to isomorphism 273
  closure under factors 279
  conditions for isomorphism to 275, 276, 278, 279
  construction of strictly ergodic models for 187
  countable Lebesgue spectrum of 62
  ergodicity of 49
  higher mixing of 64
  isomorphism with 254, 277, 279, 280
  relation to K-automorphisms 62, 63
  strong mixing of 58
  unilateral 21
Billingsley, Patrick 17
Birkhoff, George D. 3, 30, 90, 119, 151, 161, 168, 186

Index

Birkhoff Recurrence Theorem **151**, 168
Blum, J. R. 282
billiards 279
binary bits 230
Bochner, S. 139
Bogoliouboff, Nicolas 45
Bohl, P. 157
Bohr, Harald 139, 142, 148
Bohr–Fourier series 149
Boltzmann, L. 3, 42, 186, 228
Boltzmann's constant 228
Boltzmann's ergodic hypothesis 42
Boltzmann's H-Theorem 35
Boole, G. 107, 109
Borel, E. 3
bounded gaps (*see also* relatively dense) 37, 136
Bowen, Rufus 168, 252, 264, 267, 273
Bowen definition of topological entropy 266, 267
Breiman, Leo 260
Brownian motion 280
Bunimovitch, L. A. 280
Burkholder, D. L. 88, 89

Calderón, A. P. 91, 107, 108, 114
Calderón–Zygmund decomposition 107
Carathéodory, Constantin 16
Carleson, Lennart 260
cascade 150
Cavällin 157
center 151
Chacon, R. V. 119, 127, 211, 216
Chacon–Ornstein Theorem 76, 119, 121, **126**, 132
character group 20
Clausius, Rudolf 227
coboundary 92
code 274
conditional
 distribution 275
 entropy **235**, 238
 expectation **17**, 235
 information function **235**, 237, 242, 243, 263
 probability 18, 235
conjugate function 90, 107
conservative **38**, 39, 41, 121, 123, 125, 126
conservative part (*see also* Hopf decomposition) 41, 125
contraction 24, 75, **120**
converse dominated theorems 88
Cotlar, M. 108
countable partition 234, 242
Covering Lemma 107, 108
crossings 82
crossing set 80

cutting and stacking 219, 223
cycle 295

$\bar{d}$-distance 221, 277, 281
Dani, S. G. 61, 279
decomposition (of a skeleton) 287
del Junco, Andrés 94, 211, 221, 224
Dekking, F. M. 211
Denker, Manfred 161, 187, 244, 258, 268, 282
density zero **65**, 70, 73, 163
derivative transformation (*see also* induced transformation) 47, 200
Derriennic, Y. 87, 89
diagonal measure **176**, 178
diffeomorphism (realization of m.p.t.s by) 186
Dinaburg, E. I. 264, 267
Dirichlet, P. G. Lejeune 157
discrepancy 94
discrete spectrum **44**, 55, 64, 154
disintegration of a measure 183, 298
disjointness of skeletons 286
dissipative 126
dissipative part (*see also* Hopf decomposition) 41, 123, **125**
distal **154**, 158, 159, 160
distribution (of a partition) 275
Dominated Ergodic Theorem 75, **87**
Doob, J. L. 7, 91, 103
dual (of a society) 292, **293**

ε-entropy 267
ε-independent 275
ε-period 136
Eberlein, Ernst 187
Ehrenfest, Paul and Tatiana 35, 41
eigenfunction 65, 133, 148
 and weak mixing 64, 72
 continuous 153, 160
 generalized 55
 of a group action 182
eigenvalue 43, 65
 approximate 56
Ellis, Robert 158, 159, 160, 161
Ellis semigroup 158, 173
endomorphisms of shift dynamical system 221
Engel, David 76, 81, 82
England, James W. 73
entropy (*see also* topological entropy, variational principle) 4, 186, 187, 227, 244
 as a function of the measure 243
 as an isomorphism invariant 233
 computations
  automorphism of the torus 249, 259

Bernoulli shift 245, 248
discrete spectrum system 245
induced transformation 257, 259
infinite product 248
inverse limit 248
Markov shift 246, 248
product transformation 247
rotation of the circle 245
skew product 254, 259
strictly ergodic system 273
subshift of finite type 273
convergence theorem for 241
infinite 259
of a K-automorphism 63
of a Lebesgue spectrum system 63
of a partition 232
of a source 231, 242, 259
of a transformation
  definition 233
  convergence theorem for 247
of a transformation with respect to a partition 241, 243
  definition 233
  existence 240
  inequality for 242
  properties of 238
  topological 264, **265**, 266, 267
  zero 244
entropy metric 244, 248
enveloping semigroup (*see* Ellis semigroup)
equicontinuous **153**, 154, 156
equidistribution 156
equilibrium 34
equimeasurable functions 114
equivalence of fillers 284, 289
Erdos, P. 162, 167
ergodic decomposition 81
ergodic Hilbert transform (*see* Hilbert transform, ergodic)
ergodic hypothesis 3, 42, 186
ergodicity 3, 41, **42**, 56, 58, 60, 61
  conditions for 57
  examples of 49–54
  relation to topological ergodicity 152
  topological **151**, 152
Ergodic Szemeredi Theorem (*see also* Furstenberg–Katznelson Theorem) 163
Ergodic Theorem (individual *or* pointwise; *see also* Hurewicz, Chacon–Ornstein, Dominated, Hilbert transform, Local, Maximal, Mean, Random, and Shannon–McMillan–Breiman Theorems) 3, 23, 27, **30**, 44, 52, 91, 92, 93, 103, 119, 198, 222, 262
  approximation proof of 92
  for Markov operators 33

expected return time 35, 46
extension 11
  of a cascade 159
  with respect to a group action 182

factor (*see also* homomorphism) **10**, 21
  of a cascade 159, 162
  with respect to a group action 182
  of a Bernoulli shift 279
Fathi, Albert 72
Fefferman, C. 89, 91
Feller, William 85
Feldman, J. 274
fiber entropy 254
fiber measures 183
fiber product 183
Fieldsteel, A. 211
filler 283
  alphabet 283, **288**
  Bernoulli shift 284, **288**
  entropy 288
  equivalence 289
  Lemma **291**, 299, 300
  measure 283, **288**
  set (of a skeleton) 289
filling scheme 76, 119, 120, 122, 123, 126, 127, 129, 130
finitary 249, 273, **282**
finite code 221
finitely determined 279
flow **1**, 76, 150
flow built under a function **11**, 21
Foguel, Shaul R. 33, 123, 127
Fourier conjugate 107
Fourier transform 20
Friedman, N. A. 59, 276, 279
Fundamental Theorem of Calculus 90, 91
Furstenberg, H. 55, 133, 151, 157, 159, 161, 163, 164, 165, 173, 186, 187, 194, 211
Furstenberg–Katznelson Theorem **165**, 166, 185
Furstenberg Structure Theorem for distal cascades **160**, 162
Furstenberg–Weiss Theorem **164**, 170, 171

Gallavotti, Giovanni 279, 280
Garsia, Adriano M. 75, 89, 91
gauge (of a partition) 196
Gaussian system 8, 63
Gelfand, T. M. 189
Gelfand's problem 50
generalized discrete spectrum 55
generalized eigenfunction 55
generic 4, 72, **175**, 274
generating semialgebra 14
generator 234, **243**, 244, 248, 274
geodesic flow 9, 280

Index

geodesic cascade 153
Girsanov, I. V. 63
Glasner, Shmuel 154, 155
Goodman, T. N. T. 264
Goodwyn, L. Wayne 264
Gottschalk, W. H. 9, 10, 136, 150, 161
Gottschalk's Theorem 168
Graham, Ronald L. 172
Grant, Edward 156, 157
Green, L. 10
Grillenberger, Christian 161, 187, 244, 258
group of automorphisms of [0, 1] 71
group rotation (*see also* rotations, translations, discrete spectrum) 64, 181
Grünwald's Theorem **172**, 186
Gundy, R. F. 88, 89
Gurevich, B. M. 63
Gyldén 157

H-Theorem 35
Haar measure 20
Hahn, Frank 10, 154, 155, 186
Hajian, Arshag B. 41
Halmos, Paul R. 17, 34, 39, 48, 60, 64, 71, 72
Hamilton's equations 5
Hamiltonian function 5
Hamming metric 277
Hansel, G. 187
Hanson, D. L. 282
hard-sphere gas 42, 279
harmonic function 108
Harper, L. H. 284
Hartman, Philip 77
Hardy, G. H. 89, 90
Hardy–Littlewood Maximal Function 91
Hardy–Littlewood Maximal Theorem 90
Hardy spaces 89, 90
Hecke, E. 162
Hedlund, G. A. 9, 10, 161, 221
Herman, Michael R. 72
higher-degree mixing 63, 64, 73
higher-dimensional van der Waerden Theorem 172
Hilbert, D. 173
Hilbert transform
 ergodic 90, 108, 113, **116, 119**
 maximal inequality for 110, 113, 115
 real-variable 90, 107, 108, 115
Hindman's Theorem **172**, 173, 187, 195, 196
hole 120, 121, 128
homogeneous
 space 10
 pair 168
 subset 168
homomorphism (*see also* factor) **11**, 21
 of cascades 159
 of measure algebras 15

Hopf, Eberhard 74, 75, 123
Hopf decomposition (*see also* conservative, dissipative parts) 123, **125**
Hopf Maximal Ergodic Theorem **75**, 123, **130**, 131
horocycle cascade 153
horocycle flow **9**, 63, 211
Horowitz, J. 103
Hurewicz' Ergodic Theorem 131

idempotent 158
incompressible transformation **38**, 39, 126
independence 239, 242, 275
index set (of a skeleton) 289
induced transformation (*see also* derivative *and* primitive transformations) 12, 34, **39**, **40**, 41, 45, 56, 210, 257, 259
infinite product transformation (entropy of) 248
infinitely recurrent transformation **38**, 39
information function (*see also* conditional information function) **235**, 238, 243
invariant set 42, 125, 126
inverse limit **12**, 21, 160
 entropy 248
 of cascades **160**, 162
inversion formula 20
Ionescu Tulcea, Alexandra (*see also* Bellow) 260
IP-set **173**, 195, 196
irreducible stochastic matrix 52, 59
Ising model 280
isometric extension 160
isomorphism **4**, 16, 58, 60, 61, 63, 234, 246, 254, 273, 274, 275, 276, 277, 278, 279, 280, 301
Isomorphism Theorem for Lebesgue spaces **16**, 71, 187

Jacobs, Konrad 12, 187, 301
Jensen's Inequality **18**, 22
Jewett, Robert I. 186
Jewett–Krieger Theorem 175, 187, **188**
join
 of open covers 264
 of partitions 233, 237
joining 281, 293, 301
Jones, Lee Kenneth 70, 163
Jones, R. L. 89, 100

K-automorphism **62**, 63, 187, 273
Kac, M. 46
Kac' Theorem 37, **46**
Kakutani, Shizuo (*see also* von Neumann–Kakutani adding machine) 12, 26, 27, 39, 41, 64, 70, 76, 82, 91, 94, 154, 163, 187, 210, 216
Kakutani equivalence 273

Kakutani Fixed Point Theorem 154
Kakutani–Hahn Theorem 155
Kakutani–Rokhlin Lemma (*see also* skyscraper, tower) **48**, 94, 95
Kakutani skyscraper decomposition 45, 48
Katok, A. B. 72, 211
Katznelson, Yitzhak (*see also* Furstenberg–Katznelson Theorem) 27, 61, 133, 165, 186, 253, 279
Keane, Michael 211, 281, 282
Keplerian element 157
Keynes, Harvey B. 152, 153, 157, 211
Khintchine, A. I. 6, 34, 37, 98, 162, 231
Khintchine Recurrence Theorem **37**, 195
Kolmogorov, A. N. 4, 27, 62, 90, 94, 98, 152, 211, 227, 234, 246
Kolmogorov Consistency Theorem **22**, 36
Kolmogorov–Sinai Theorem 234, **244**, 248
Konheim, A. G. 264, 267, 268, 273
Koopman, B. O. 64, 65
Krengel, Ulrich 99, 258
Krickeberg Decomposition 104
Krieger, Wolfgang (*see also* Jewett–Krieger Theorem) 187
Krieger Generator Theorem **244**
Kronecker, Leopold 158
Kronecker system 182
Kronecker–Weyl Theorem 250
Krylov, Nicolas 45
Kuipers, L. 161

Lagrange, J. L. 157
Law of the Iterated Logarithm 98
Law of Random Signs 94
least common refinement 233, 264
Lebesgue, H. 101
Lebesgue Differentiation Theorem **101**, 102
Lebesgue set 102
Lebesgue space 16
Lebesgue spectrum 61, 62, 63
Ledrappier, F. 280
Lehrer, E. 211
Liberto, Francesco di 280
Lind, D. A. 22, 61, 279
Liouville number 94
Liouville's Theorem **5**, 22, 186
Littlewood, J. E. 89, 90, 91
Local Ergodic Theorem 79, 100, **102**, 114
Loomis, Lynn H. 107
Loosely Bernoulli 211, 274

Mackey, George W. 55
Marcus, Brian 76, 211, 282
markers 283
Marker Lemma 285
marker process 283
Markov, A. A. 42
Markov operator 33, **120**, 121

Markov partition 252, 253
Markov process 119
Markov shift **7**, 35, 52, 56, 60
  entropy of 246, 248
  ergodicity of 51
  higher mixing of 64
  isomorphism to Bernoulli shifts 276, 279
  strong mixing of 59
  weak Bernoulli property of 276
  weak mixing of 64
Martin, Nathaniel F. G. 73
martingale 103
Martingale Convergence Theorem (*see also* Submartingale Covergence Theorem) **103**, 262
marriage lemmas 284, 292, **295**, **296**, 297, 299
maximal almost periodic factor 182
maximal equality 74, **76**, 80, 81, **85**, **87**
maximal function 89, 91, 119
  eventual 119
  ergodic 74, 75
  ergodic Hilbert 113
  Hardy–Littlewood **90**, 100
  Hilbert 107
  nontangential 90
Maximal Ergodic Theorem **27**, 31, 91, 92, 93
  for operators 75, 130
  for flows 76
maximal inequality 74, 76, 90, 91, 100, 103, 107, 110, 113, 115, 116, 260
maximum spectral type **19**, 61
Maxwell–Boltzmann velocity distribution 228
McAndrew, M. H. 264, 267, 268, 273
McMillan, Brockway 260
Mean Ergodic Theorem 3, **23**, 37, 176, 177
  in Banach space 26
  in Hilbert space 24
mean motion, problem of 157
mean sojourn time **44**, 45
mean value (of almost-periodic sequence) 137, 139
measure algebra 15
measure algebra of a measure space 15
measure-preserving homeomorphism 72
measure-preserving transformation 2
Meshalkin, L. D. 282
metric indecomposability 42
metric space of a measure algebra 16
metrically prime 221
metrically transitive 42
Miles, G. 61, 279
minimal 42, **136**, **150**
mixing (*see also* higher mixing, strong mixing, topological mixing, uniform mixing, weak mixing) 73, 210

Misiurewicz, M. 269
monothetic group 154
monotone equivalence 274
Morse sequence (or set) 210, 273
Multidimensional Szemerédi Theorem 165
multiple recurrence (*see also* Furstenberg, Furstenberg–Katznelson, Furstenberg–Weiss, Ergodic Szemerédi Theorems) 46, 163, 164, 186

(n, ε)-separated 266
(n, ε)-spanning 267
natural extension 13, 21, 280
Nemytskii, V. V. 161
Neveu, J. 104, 258
Newton, D. 211
Niederreiter, H. 161
nonatomic 16
nonsingular 2, **41**, 131
nonwandering set 151

one-parameter group (*see also* flow) 132
optional sampling 104, 105
orbit 2, **150**
Oren, Ishai 211
Oresme, Nicole 156, 157
Ornstein, Donald S. 4, 27, 59, 63, 73, 88, 119, 211, 273, 274, 275, 276, 278, 279, 280, 284
oscillations 80, 82
oscillation set 80
Oxtoby, J. C. 45, 72

Parry, William 10, 55, 211
Parseval's Formula 149
Parthasarathy, K. L. 161
partition 232
 countable 234, 242
 distribution of 275
Petersen, Karl E. 55, 76, 87, 89, 94, 152, 153, 157, 187, 210, 211
phase space 5, 228
physical system 34
Pigeonhole Principle 157
Pinsker's formula 243
Plancherel Theorem 20
Plante, J. 6
Poincaré, Henri 161, 186
Poincaré Recurrence Theorem **34**, 35, 37, 38, 151, 163, 229
point-distal 161
point homomorphism mod 0 17
pointwise almost periodic 154, 159
Poisson integral 108
Polit, Steve 233
Pontryagin Duality Theorem 20
positive contraction 89, **120**
positive upper density 163

Prasad, V. S. 72
Prasjuk, O. S. 63
prime transformation **211**, 216, 217
proximal **154**, 158, 161, **173**
product 11
Prohorov, Ju. V. 161
promiscuity number 293

quasi-discrete spectrum 55
quasi-eigenfunction 55
quasi-ergodic hypothesis 42

r-recurrent point 186
Rado, R. 162
Rahe, M. 211, 224
Random Ergodic Theorem 99
random walk 85, 86
rank decomposition (of a skeleton) 287
Raoult, J. P. 187
rationally independent 51
Ratner, Marina 211, 280
realization problem 4, 186
recurrence 4, 33, 41
 topological 150
 regional 151
recurrent
 point 34, 151
 transformation **38**, 39
refinement
 of an open cover 264
 of a partition 233
 of a society 293
refining sequence of open covers 265
regional transitivity 152
relatively compact extension 184
relatively dense 37, **136**, 150
relatively ergodic extension 183
relatively weakly mixing extension **183**, 185
Rényi, A. 58
return time 56
reverse maximal inequality 88
Riesz, Fréderic 77, 90, 91, 100
Rising Sun Lemma **77**, 78, 85 90
Robertson, James B. 152, 153
Rokhlin, V. A. (*see also* Kakutani–Rokhlin Lemma) 17, 63, 72, 95, 254
Rosenblatt, J. 94
Rost, H. 131
Rota, Gian-Carlo 284
rotations of compact abelian groups (*see also* group rotation, translation) 8, 133
rotation of the circle (*see also* group rotation, translation) 8, 49, 156, 157
Roth, Klaus 162
Rothschild, Bruce L. 172
Rudin, Walter 21

Rudolph, Daniel J. 211, 233, 273, 282
Russo, Lucio 280
Ryll-Nardzewski, Czeslaw 155
Ryll-Nardzewski Fixed Point Theorem 155

σ-ideal 15
Sand 120, 121, 128
Sawyer, S. 91
Scheller, H. 258
Schreier, O. 162
Schur, Issai 162
Second Law of Thermodynamics 35, 229
semialgebra **14**, 58
semilocal operator 114
separable measure algebra 16
Shannon, C. E. 227, 230, 232, 260
Shannon-McMillan-Breiman Theorem 245, **261**, 263, 284, 285, 289, 291
Shapiro, Leonard 157, 210, 211
shift transformation 7
Shields, Paul C. 274, 280
Sierpinski, Waclaw 108
Sigmund, Karl 161, 244, 258
Silverstein, M. L. 89
Sinai, Ya. G. (*see also* Kolmogorov-Sinai Theorem) 4, 42, 234, 248, 249, 252, 279
singular integral 107
skeleton 283, **286**
  decomposition of 287
Skeleton Lemma 287
skew product **11**, 21, **53**, 55, 100, 254
skyscraper (*see also* Kakutani-Rokhlin Lemma, Tower) 40, 45, 48, 200
Smorodinsky, Meir 211, 280, 281, 282
society 284, **292**, 301
  dual of 293
source 230, 231, 242, 259
space mean 42, 44
spectral measure 19, 69
Spectral Theorem **19**, 65, 68, 133, 148
spectral type 55
speed of approximation 72
speed of convergence in Ergodic Theorem 90, 93
stationary stochastic process 6, 85, 87, 91, 98, 248
Stein, Elias 88, 89, 91, 107, 119
Stepanov, V. V. 161
Stepin, A. M. 72, 221
stochastic matrix 36, 51, 52, 59
stopping time 104, **105**, 106
Strassen, V. 301
strictly ergodic (*see also* uniquely ergodic) **152**, 156, 187, 194, 217
Strong Law of Large Numbers 3
stong mixing (*see also* higher mixing,

mixing) 4, **57**, 58, 59, 60, 61, 62, 210, 211
failure of, with weak mixing 217, 219, 220
Ornstein's condition for 73
of a sequence of sets 73
topological 151, 152, 210, 212, 214
strong topology (on group of m.p.t.s) 71
subadditive sequence 240
sub-Markov operator 120
submartingale 91, **103**
Submartingale Convergence Theorem 104, 106
subshift of finite type 153, 162, 273
subskeleton 286
supermartingale **103**, 105, 106
supremum (of a family of factors) 185
Swanson, L. 221, 224
Symbolic cascades 153
SZ 185
Szemerédi, E. 162
Szemerédi Theorem 159, **163**
Szemerédi-Furstenberg Theorem (*see also* Ergodic Szemerédi Theorem) 150

Tchebychef, P. L. 158
Thomas, R. K. 61, 279
three-series theorem 94
time mean 42, 44
toeplitz sequence 209
topological dynamics 133, 150
topological entropy 264, **265**
  Bowen definition of 266, 267
  of a symbolic cascade 266
topological ergodicity **151**, 152
topological weak mixing 151
topological weak mixing without strong mixing 210, 212, 214
topological strong mixing 151
total function **189**, 191, 202, 208
Totoki, Haruo 9
tower (*see also* skyscraper, Kakutani-Rokhlin Lemma) 197, 200
transition matrix 36, 51, 52, 59
transitivity
  metric 42
  regional 152
translation on a compact group (*see also* group rotation, rotation) 153
translation of the torus (ergodicity of) 51
trigonometric polynomial 148
Turán, Paul 162

$\mathcal{U}$, n-name 270
Ulam, S. M. 72, 99
uniform distribution mod 1, 156
uniform function **189**, 190, 202, 208, 210

uniform mixing 64
uniformly almost periodic 136, 137, 154
uniquely ergodic (see also strictly ergodic) 42, 138, **152**, 186, 187, 217
unitary operator determined by a m.p.t. **24**, 43
upcrossing lemma 104

van der Waerden, Bartel L. 162
van der Waerden's Theorem **162**, 167, **171**, 172
Varadarajan, V. S. 161
Varadhan, S. 133
Varga, Richard . 7
Variational Principle 264, 268, **269**, 273
Veech, William A. 160, 161
very weakly Bernoulli **278**, 279
Ville, Jean 91
virtual group 55
von Neumann, John 3, 17, 23, 55, 64, 65, 80, 99, 139, 188
von Neumann–Kakutani adding machine 209, 211
von Neumann's Theorems on Lebesgue spaces **17**, 71

Walters, Peter 267
wandering point 151
wandering set **38**, 41, 126
Weak Law of Large Numbers 285
weak mixing (see also mixing) **57**, 58, 64, 65, 152
and density 0 73
and eigenfunctions 64, 72
characterizations of 65, 180
of higher degree **151**, 163, 179
of roots and powers 72
topological **151**, 152, 180, 210, 212, 214
without strong mixing 71, 210, 212, 214, 217, 219, 220
weak topology (on group of m.p.t.s) 71, 73
weak-type (1, 1) inequality (see also maximal inequality, Maximal Ergodic Theorem) 113, 115, 119
weak-type (p, q) inequality 91
weakly Bernoulli **276**, 278, 279
weakly wandering set 41
Weiss, Benjamin (see also Furstenberg–Weiss Theorem) 151, 164, 173, 187, 253, 279, 280
Weiss, Guido 107
Weyl, Hermann (see also Kronecker–Weyl Theorem) 50, 149, 156, 158
White, H. E., Jr. 72
Wiener, Norbert 27, 75, 76, 87, 88, 91, 102, 108, 114
Wright, Fred B. 39, 46

Yosida, Kôsaku 25, 27, 76, 91

Zermelo, E. 35
zero entropy 244
Zimmer, Robert J. 55
Zygmund, A. 107